Ilse Zolle
Editor

Technetium-99m Pharmaceuticals

Gedruckt mit Unterstützung des Bundesministeriums
für Bildung, Wissenschaft und Kultur in Wien
und der Kulturabteilung der Stadt Wien, Wissenschafts-
und Forschungsförderung

Ilse Zolle
Editor

Technetium-99m Pharmaceuticals

Preparation and Quality Control in Nuclear Medicine

With 66 Figures and 29 Tables

Springer

Ilse Zolle
Department of Medicinal/Pharmaceutical Chemistry
University of Vienna
Althanstraße 14
1090 Vienna
Austria

Library of Congress Control Number 2006925440

ISBN-10 3-540-33989-2 Springer Berlin Heidelberg New York
ISBN-13 978-3-540-33989-2 Springer Berlin Heidelberg New York

Springer is a part of Springer Science+Business Media

springer.com

Editor: Dr. Ute Heilmann
Desk Editor: Wilma McHugh
Production: LE-TeX Jelonek, Schmidt & Vöckler GbR, Leipzig
Typesetting: K + V Fotosatz, Beerfelden, Germany
Cover: Frido Steinen-Broo, eStudio Calamar, Spain

Printed on acid-free paper 21/3100/YL 5 4 3 2 1 0

Foreword

Some persons have thought that the increasing emphasis and development of positron-emitting radiotracers in nuclear medicine would result in a decrease in the development and use of single photon-emitting radiotracers. That this is not the case is illustrated by the fact that there were 302 presentations involving technetium-99m at the June 2006 annual meeting of the Society of Nuclear Medicine in the United States. Iodine-123 accounted for 88 presentations, and indium-111 for 81.

Among the most recent advances in the fusion of nuclear medicine images with computed tomography (SPECT/CT) and computed tomography angiography (SPECT/CTA) in basic science studies in small animals, SPECT/CT was the topic of 11 instrumentation presentations, while PET/CT in small animals accounted for 5 presentations.

The success of molecular imaging in medicine and biomedical research is the result of the diversity of imaging technologies, the integrating and collaboration of imaging and pharmaceutical development experts, and friendly competitition. The advantage of single photon tracers is that many of them, such as technetium-99m, emit only photons, rather than positrons or negative beta particles that increase the radiation exposure of the patient and limit the doses of the tracers that can be administered with acceptable radiation exposure. Also, the range of positrons in tissue before they encounter an electron and emit 511 keV photons limits the spatial resolution that can be obtained in living animals and patients. Theoretically, the spatial resolution of single photon tracer studies is limitless, especially with the use of special pinhole collimation.

Radioactive tracers used in medicine today provide information, and need to be judged by how reliably they provide this information. Some of the safeguards built into the drug review process by the FDA for regulating pharmaceuticals are not needed in the case of the mass of injected material used in radiopharmaceuticals. With radiopharmaceuticals, the criterion should be whether the information provided by the diagnostic procedure is valid and valuable.

Today, we need to promote a "fast track" regulatory approval process to make diagnostic procedures more readily available. A major difference between therapeutic and diagnostic drugs is that the efficacy of the diagnostic procedure in providing the required information can be assessed shortly after the performance of the imaging procedure. The patients do not need to be followed for longer periods of time to identify any untoward side effects. Multi-institutional randomized control studies can determine whether the procedure provides the diagnostic information provided by the study, as well as identify any untoward side-effects.

Radiotracers, by definition, have no effect on the patients' biochemistry or body functions, which should be a major simplifying factor along the road to their regulatory approval. There is ever-increasing evidence of their great value in answering the questions: What is wrong? What is going to happen? What can be done about it? How did it happen?

This book covers the past, present and future of single photon tracers in medical practice and biomedical research. It is likely to become a standard textbook for those persons entering the exciting career of a radiopharmacist or researcher in biomedical research using radioactive tracers.

The pathway to assuring the safe and effective use of short-lived radiotracers is to place the responsibility for quality assurance in the hands of nuclear pharmacists who fill physician's prescriptions for radiotracers as they do for other drugs under state pharmacy laws. Radiochemists or others working in institutions with radiopharmacies must have the expertise for the preparation of the radiotracers as well as for quality control under good manufacturing practices. This book provides guidance and safety standards applicable to Tc-99m pharmaceuticals.

Baltimore, June 2006

Henry N. Wagner, Jr. M.D.

Foreword

Viewpoint of the Clinician

Radiopharmaceuticals labeled with ^{99m}Tc are commercially available and are employed in more than 80% of all nuclear medicine investigations. Among the radionuclides, technetium-99m is most attractive to the nuclear medicine physician because of its optimal gamma energy for SPECT, its availability, its relatively low cost, and its easy-to-label kit preparations for in-house use. Another advantage is the low radiation burden to patients, due primarily to its short half-life. The decay within hours also facilitates the handling of waste.

Professor Dr. Ilse Zolle – together with other leading international radiochemists and radiopharmacists – made an effort to collect the available data on ^{99m}Tc-labeled compounds with respect to their chemistry, labeling methods, quality control procedures and clinical applications. The comprehensive text is presented in two parts. The first part comprises chapters on technetium compounds in medicine, including advances in labeling biomolecules with technetium, the advantages of sterile kit formulations, and analytical methods to verify pharmaceutical quality. Emphasis is given to the rules governing the manufacture of radiopharmaceuticals and the importance of specifications given by the pharmacopoeia, which are obligatory. A special chapter is devoted to the performance of the ^{99}Mo/^{99m}Tc generator and to the characteristics of the ^{99m}Tc eluate.

For the clinician, Part II offers 25 monographs relating to ^{99m}Tc-pharmaceuticals, which describe the pharmaceutical particulars of each radiotracer as well as relevant information on its clinical application, concerning the pharmaceutical dosage, contraindications and interference with other pharmaceuticals, quality control, pharmacokinetic data, radiation dose and valuable references. In addition, recommendations for storage and criteria of stability are also given.

The multidisciplinary properties of ^{99m}Tc-pharmaceuticals are presented in the form of a highly structured text with informative tables, which enables the clinician to find clinically relevant data very easily.

The book should therefore not only be recommended for radiochemists and radiopharmacists, but also for nuclear medicine physicians using ^{99m}Tc-labeled pharmaceuticals in daily practice.

Würzburg, July 2006 Christoph Reiners

Preface

^{99m}Tc pharmaceuticals mark the beginning of diagnostic nuclear medicine and have contributed to patient care worldwide. Since the short-lived radionuclide was introduced for thyroid imaging in 1964, it has attracted much attention and stimulated clinical research. Little was known about element 43, except that it was an artificial radionuclide obtained by the radioactive decay of molybdenum-99. No wonder specialists from all fields joined medical institutions in the United States to participate in the exploration of its chemistry.

This textbook gives an account of the accomplishments related to the development of ^{99m}Tc pharmaceuticals and their application in diagnostic nuclear medicine. Since radioactive drug development is a multidisciplinary task, experts working in nuclear medicine and research institutions have contributed valuable information concerning the preparation with sterile kits, methods of quality control, and the use of ^{99m}Tc pharmaceuticals in patients. In addition, the legal aspects governing production and clinical application are also considered.

^{99m}Tc pharmaceuticals in nuclear medicine are presented in two parts. Part 1 includes basic principles and methods used for preparation and analysis, in particular the chemistry of technetium-99m and methods described for the synthesis of complexes and conjugates of technetium-99m, the characteristics and performance of the ^{99}Mo/^{99m}Tc generator system, the importance of kits and formulations for one-step labeling, and safety aspects for labeling blood cells. Special emphasis is given to analytical methods verifying pharmaceutical quality. The quality standards of good manufacturing practice (GMP) for ^{99m}Tc pharmaceuticals and purity standards of the pharmacopeia (European Pharmacopeia and United States Pharmacopeia) have been considered as part of the concept of quality assurance.

Part 2 presents 26 monographs of ^{99m}Tc pharmaceuticals, concerning the preparation and safe clinical application. Each monograph provides information on the characteristics of the radiotracer based on chemistry, factors affecting the preparation and in vivo stability, pharmacokinetics and elimination properties, as well as details concerning the clinical application. For each clinical procedure, the effective radiation dose of the patient has been calculated. Methods recommended for quality control and actual results are included.

Each radiopharmaceutical is listed under the name used in daily practice. The chemical name, the abbreviated name, and the officinal name in the pharmacopeia are also stated. Listing the trade names may facilitate understanding, especially when relating to products in the literature, which over the years have changed manufacturers.

The presentation of ^{99m}Tc pharmaceuticals as monographs serves a practical purpose: it offers relevant information on kit preparation and clinical application at a glance. These monographs provide a wide spectrum of information on ^{99m}Tc pharmaceuticals for daily practice in nuclear medicine, serving as a reference source as well as a teaching tool.

Acknowledgments

Dr. Ferenc Rakiás, Deputy Chief of Drug Quality, The National Institute of Pharmacy in Budapest, Hungary and Co-ordinator of a European Working Group on the Quality Control of ^{99m}Tc-Radiopharmaceuticals (COST Action B3) has gained high recognition for major contributions, which resulted from the collective effort of the Working Group. Based on studies of quality control methods, and a collection of manuscripts, Dr. Rakiás and his dedicated staff have produced a first version of this book, documenting the multidisciplinary nature of the cooperation. The National Institute of Pharmacy in Budapest has also hosted most Working Group meetings, their hospitality is appreciated.

The participation of Austrian scientists in COST Action B3 was supported by:
The Federal Ministry of Education, Science, and Culture, actively promoted by:

Dr. Norbert Rozsenich, Head, Section V, Commerce and Technology
Hofrat Dr. Raoul Kneucker, Head, Section VI, Scientific Research and International Affairs
Ministerialrat Dr. Helga Mieling, National COST Co-ordinator
Dipl. Ing. Otto Zellhofer, Science, Energy and Space Program 'AUSTROMIR'
Ministerialrat Alois Söhn, Presidial Division 4
Hofrat Mag. Anna Kolde, and Regierungsrat Franz Gerersdorfer

The Department of Nuclear Medicine and the Ludwig-Boltzmann-Institute of Nuclear Medicine, Professor Emeritus Dr. Rudolf Höfer, and
the Department of Biomedical Engineering and Physics, Professor Dr. Helmar Bergmann, all at the University Hospital AKH, Vienna.

Editorial work was kindly supported by Hofrat Dipl.-Ing. Gerhard Wolf and his team, Mrs. Elisabeth Abromeit and Mrs. Sandra Jama

The assistance of Dr. Yu Jie (Julia) and Mag. pharm. Georg Kropf is gratefully acknowledged.

Contents

Part I

List of Contributors

Andreas G. Bratouss
Nuklearmedizinische Klinik
Klinikum Buch
Helios Kliniken Berlin
Wiltbergstraße 50
13125 Berlin, Germany

Per Oscar Bremer
Radioformulation Development
GE Healthcare
Instituttveien 18
PO Box 65
2027 Kjeller, Norway

Trygve Bringhammar
Medical Products Agency
Radiopharmaceuticals
75103 Uppsala, Sweden

Efstratios Chiotellis
Dept. of Pharmaceutical Chemistry
School of Pharmacy
Aristotelian University of Thessaloniki
G-54124 Thessaloniki, Greece

Clemens Decristoforo
Medical Department of Nuclear Medicine
Medical University Innsbruck
Anichstraße 35
6020 Innsbruck, Austria

Stuart R. Hesslewood
Department of Physics
and Nuclear Medicine
City Hospital NHS Trust
Birmingham B18 7QH
United Kingdom

János Imre
National Institute of Pharmacy
Radiochemistry Section
1051 Budapest, Hungary

Győző A. Jánoki
Department of Applied Radioisotopes
Frederic Joliot-Curie National Research
Institute of Radiobiology
and Radiohygiene
1221 Budapest, Hungary

Silvester Kladnik
Department of Nuclear Medicine
University Medical Centre Ljubljana
1525 Ljubljana, Slovenia

Zoltán Kovács
Radiochemistry Group
Institute of Nuclear Research
Hungarian Academy of Sciences
4001 Debrecen, DOB 51, Hungary

József Környei
Research and Development
Radiopharmaceutical Business Line
Institute of Isotopes Co., Ltd.
Konkoly Thege M. Street, 29–33
1121 Budapest, Hungary

Jens-Uwe Künstler
Research Centre Rossendorf
Institute of Radiopharmacy
PF 510 119
01314 Dresden, Germany

Jesus Mallol Escobar
Universidad de La Laguna, Tenerife
Schering España, SA
28045 Madrid, Spain

Ulderico Mazzi
Professor of Pharmaceutical Sciences
University of Padova
Via F. Marzolo, 5
35131 Padova, Italy

Hans-Juergen Pietzsch
Research Centre Rossendorf
Institute of Radiopharmacy
PF 510 119
01314 Dresden, Germany

Ferenc Rakiás
Drug Quality Department
National Institute of Pharmacy
1051 Budapest, Hungary

Margardia Rodrigues
Department of Nuclear Medicine
Medical University Vienna
1090 Vienna, Austria

Roger Schibli
Center for Radiopharmaceutical Science,
OIPA
Paul Scherrer Institute
ETH Zürich
5232 Villigen PSI, Switzerland

Helmut Sinzinger
Department of Nuclear Medicine
Medical University Vienna
1090 Vienna, Austria

Hartmut Spies
Research Centre Rossendorf
Institute of Radiopharmacy
PF 510119
01314 Dresden, Germany

Alfons Verbruggen
Professor of Radiopharmaceutical
Chemistry
Onderwijs en Navorsing 2, Box 821
Herestraat 49
3000 Leuven, Belgium

Ilse Zolle
Department of Medicinal/
Pharmaceutical Chemistry
University of Vienna
Althanstraße 14
1090 Vienna, Austria

Part I

Chapter 1

Drug Safety

1

I. Zolle

The unification of European states has made it necessary to harmonize legislation governing medicinal products in the member states. Guidelines have been published on the detailed requirements for verification of quality, safety, and efficacy, and on good manufacturing practice (GMP) for the manufacture of medicinal products for human use, as laid down in the Community Directives and authorized by the European Agency for the Evaluation of Medicinal Products in London (EMEA). Directive 89/343/EEC is extending the scope of pharmaceutical directives to radiopharmaceuticals. Account must be taken of the relevant CPMP guidelines (Committee of Proprietary Medicinal Products) and of guidelines developed by the International Conference on Harmonization (ICH).

The harmonization process affected all levels of drug handling, and thus has put strict requirements on the preparation and use of radioactive drugs in hospital facilities. The transition period may be characterized as a major communicative effort between European institutions in order to exchange scientific knowledge and gain practical experience for the implementation of GMP standards for the preparation of ^{99m}Tc pharmaceuticals (Directive 91/356/EEC).

1.1 Quality Control and COST

The Austrian Ministry of Science and Research recognized the importance of quality standards for radiopharmaceuticals, and gave full support for cooperation between the Department of Nuclear Medicine at the Medical University in Vienna (AKH-Wien) and the National Institute of Pharmacy in Budapest. This bilateral cooperation between Austria and Hungary during 1991–1994 formed the basis for initiating an European cooperation (COST).

COST Action B3 (1992–1997) was devoted to the development of new radiotracers for nuclear medicine application and methods of quality assurance. National institutions of sixteen European states participated in five Working Groups (WG).

Working Group 1 was concerned with the standardization of methods for labeling and quality control:

- Working Group 1(a) Quality control of ^{99m}Tc pharmaceuticals, Coordinator F. Rakias (Hungary)
- Working Group 1(b) Standardization of radioiodination reactions, Coordinator J. Mertens (Belgium)

The scientific goal of Working Group 1(a) has been the development and updating of quality control methods to assure safety of ^{99m}Tc pharmaceuticals for parenteral application in nuclear medicine. Scientific institutions in 12 European countries have contributed their experience and results for comparison of the available methods. The European cooperation COST B3 offered the framework for the exchange of young scientists in the field and mobilized national activities; without the spirit of a common goal, this task would not have been accomplished. The best analytical methods to demonstrate radiochemical purity and factors affecting product stability and safe application in patients have been described for each ^{99m}Tc pharmaceutical in monographs presented in Part 2.

1.2 Quality Standards

Identity and purity, stability, and sterility and apyrogenicity. The identity and purity of radiopharmaceuticals is verified by determining the radionuclidic and radiochemical purity. Stability concerns the radioactive label, which is related to radiochemical purity at a certain time after preparation. Since ^{99m}Tc pharmaceuticals are formulated as sterile, pyrogen-free solutions, the safety requirements of drugs for parenteral use do apply. Safe handling of the radionuclide is equally important and must comply with Euratom Directives, regulated by national law for radiation protection, which also concerns the application of radionuclides in adults and in children for diagnostic procedures.

1.3 Quality Assurance

An inherent difference between radiopharmaceuticals and nonradioactive medicinal products is their radioactive decay. In fact, ^{99m}Tc pharmaceuticals have an expiration of hours. After expiration, the radiopharmaceutical should no longer be used because its quality has changed. Radioactive decay puts stringent requirements on the production and use of radiopharmaceuticals.

Short-lived radiopharmaceuticals have to be manufactured, quality-tested, and dispensed within a short time, adding constraint on safety procedures. In order to comply with the strict requirements of GMP, special methods for synthesis and "in process" quality control have been developed, assuring high quality of radiopharmaceuticals, without actually testing sterility and apyrogenicity before dispensing the labeled product.

1.4 Quality Control of Kit Products

In the case of ^{99m}Tc pharmaceuticals, chemistry and safety have been compounded into kits, which have overcome the limitations set by radioactive decay and the risk of bacterial contamination. Kits are manufactured in advance in accordance with GMP requirements for the manufacture of sterile medicinal products, have a long shelf life, and facilitate ad hoc labeling whenever there is a demand in nuclear medicine. Kits provide safety and ease of preparation of highly complex molecules by using aseptic techniques for labeling. Consequently, quality control requirements for kit preparations rely merely on testing the radiochemical purity of a ^{99m}Tc pharmaceutical to demonstrate stability in compliance with the purity requirements stated in the pharmacopeia.

Sterile labeling units (kits) and closed systems offer flexibility for new synthetic concepts with radionuclides and play a major role in the production of short-lived radiopharmaceuticals. High labeling efficiency, short synthesis time, and high safety standards are excellent characteristics for routine production on site, where radiopharmaceuticals are administered.

1.5 European Economic Community Directives and Regulations

Council Directive 65/65/EEC	(26 January 1965)
Council Directive 75/318/EEC	(20 May 1975)
Council Directive 75/319/EEC	(20 May 1975)
Council Directive 89/342/EEC	(3 May 1989)
Council Directive 89/343/EEC	(3 May 1989, radiopharmaceuticals)
Council Directive 91/356/EEC	(13 June 1991, GMP)
Council Regulation (EEC) No 2309/93	(22 July 1993, authorization of EMEA)
Euratom Directives 84/466 and 84/467	(3 September 1984)

Chapter 2

Technetium in Medicine

2

2.1 ^{99m}Technetium Chemistry

U. Mazzi

Technetium is an artificial element obtained by the radioactive decay of molybdenum. Element 43, named technetium in 1947, had been discovered in 1937 by Carlo Perrier and Emilio Segrè in a sample obtained from the Berkely Radiation Laboratory (now Lawrence Berkeley National Laboratory) in California (Perrier and Segrè 1937, 1947). By bombarding a molybdenum strip with 8-MeV deuterons in a 37-in. cyclotron, a radioactive molybdenum species (half-life, 65 h) had been obtained which decayed by β-emission to a short-lived isotope (half-life, 6 h) with novel properties, identified as technetium-99m (Segrè and Seaborg 1938).

In 1965, Richards and his collaborators at Brookhaven National Laboratories (N.Y.) have introduced the ^{99}Mo/^{99m}Tc generator for clinical application (Richards 1966). This radionuclide system made technetium-99m available for clinical research and has stimulated the development of the first labeled compounds, which had a considerable impact on radiochemistry and nuclear medicine (Andros et al. 1965; Harper et al. 1966; McAfee et al. 1964a, b; Stern et al. 1965, 1966). In the years to follow, diagnostic nuclear medicine procedures based on ^{99m}Tc pharmaceuticals increased to approximately 85%. The reasons for this rapid growth were the ideal nuclear properties of technetium-99m, its availability worldwide as a radionuclide generator system, and the development of new labeling techniques.

Labeling procedures have been greatly facilitated by kit preparations (Eckelman et al. 1971). Sterile kits for labeling contain the chemical ingredients in lyophilized form are commercially available and used to prepare ^{99m}Tc pharmaceuticals shortly before application to the patient. Manipulation is minimal, since all that needs to be done is adding the ^{99m}Tc activity to the kit. In some cases, heating of the reaction mixture is performed to increase the labeling yield.

^{99m}Tc pharmaceuticals are organ specific and available to delineate blood flow in organs such as the lung (embolism), heart (ischemia/infarction), and brain (perfusion defects); to evaluate the functional state of the thyroid, liver (phagocytic function), kidney, or the hepatobiliary system (acute cholecystitis); and to detect tumor and metastatic growth in bone structures and more specifically, somatostatin-expressing tumors. Accordingly, the demands on chemical structure and biological performance vary considerably and need a sophisticated approach to radiopharmaceutical design.

Research on new molecules has been growing steadily, stimulated by the demand for new medical applications. However, the low concentration of carrier-free ^{99m}Tc (1 Ci $\sim 10^{-9}$ *M*) in most ^{99m}Tc pharmaceuticals poses difficulties when determining their chemistry. Therefore, structural characterization of new ^{99m}Tc complexes is preferably studied with isotope ^{99}Tc, a long-lived β-emitter ($T_{1/2} = 2.12 \times 10^5$ years), which is commercially available in macroscopic amounts. Analogous ^{99}Tc complexes may be identified using standard analytical techniques such as mass spectrometry, nuclear magnetic resonance (NMR), x-ray crystallography, UV, and elemental analysis.

2.1.1 Technetium Compounds and Their Structures

The knowledge of the chemical properties of technetium has grown over the years, as indicated by review articles and books (Dewanjee 1990; Lever 1995; Nowotnik 1994; Peacock 1966; Schwochau 1983; Steigman and Eckelman 1992). Of particular interest are the Proceedings of the International Symposium on Technetium in Chemistry and Nuclear Medicine, presenting new developments in complex chemistry of technetium and rhenium, with state-of-the-art lectures, listed at the end of this chapter under "Further Reading".

The element technetium belongs to group VIIB of the periodic table, between manganese and rhenium. The atomic radius of technetium is similar to rhenium; thus, many similarities are found in the chemistry between the two elements. The electronic configuration of the neutral atom 43 is described by $[Kr]4d^6 5s^1$, indicating the 4d and 5s orbitals that contribute to several oxidation states. Technetium can exist in eight oxidation states, varying from (VII) to (-I). Considering carrier-free chemistry, the most stable states are (VII), (V), (IV), (III), (I), and 0. Most difficult to stabilize are states (VI), (II), and (-I) (Mazzi 1989).

The highest oxidation state (VII) is occupied by a pertechnetate anion (TcO_4^-) (Fig. 2.1.1), which is eluted from the $^{99}Mo/^{99m}Tc$ generator. The chemical reactivity of the pertechnetate anion is negligible; it does not bind directly to any ligand. Thus, for the production of ^{99m}Tc pharmaceuticals, reduction to lower oxidation states in the presence of a suitable ligand is a prerequisite for the synthesis of ^{99m}Tc-labeled molecules. During reduction, the ligand stabilizes the lower oxidation state, otherwise, colloidal TcO_2 is formed in aqueous media (Lever 1995; Nowotnik 1994).

An exception is technetium sulfide (Tc_2S_7), known as ^{99m}Tc-sulfur colloid (Stern et al. 1966). Scavenging molecules like phosphinimine ($R_3P=N\text{-}SiMe_3$) have been reported to incorporate TcO_4^-, producing organic molecules containing Tc(VII) (Katti et al. 1993; Singh et al. 1995).

With the exception of ^{99m}Tc colloids, ^{99m}Tc pharmaceuticals used in nuclear medicine are metal complexes, prepared by reducing ^{99m}Tc-pertechnetate to a lower oxidation state. The so-called coordination complexes of technetium (central metal) are formed by means of bonds between technetium acting as Lewis acid, and atoms or functional groups, which act as Lewis bases (they donate electron pairs). Typical ligands for technetium complex formation may have one donor group (monodentate) such as amine, amide, thiol, phosphine, oxime, or isonitrile. With two donor groups, the complex is bidentate; when more than two donor groups from a single molecule bind to one Tc core, is it a chelate (Nowotnik 1994).

The redox potential of TcO_4^-/TcO_2 was found to be +0.738 V, and that of TcO_4^-/Tc, 0.477 V (Mazzi 1989). In the presence of suitable ligands, the redox potentials of the TcO_4^-/Tc complex are dependent upon the stability of the complex itself. It depends on the ligand, in which oxidation state a complex will be stabilized. In presence of oxygen atoms Tc(VI) is not stable, it rather disproportionates to (IV) and (VII). However, if a Tc(VI) complex is very stable, no further reduction is possible. In any case, pertechne-

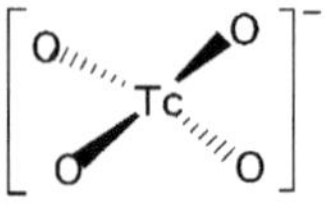

Fig. 2.1.1. Pertechnetate anion

tate is a weak oxidant, certainly weaker than permanganate; in acid medium, it is reduced by weak reductants. In kits, $SnCl_2$ is commonly used as reductant. In certain cases, excess ligand may also act as reductant.

The pertechnetate anion, when reduced in the presence of ligands, usually does not release all the oxygen atoms, leading to complexes in which a TcO^{3+} or a TcO_2^+ core is identified.

Complexes containing a TcO^{3+} core show an octahedral six-coordinated or a square pyramidal five-coordinated spatial configuration; complexes containing a TcO_2^+ core form octahedral six-coordinated complexes. In the presence of suitable ligands, other cores and complexes of lower oxidation states (IV, III, I) may be achieved (Jones and Davison 1982).

2.1.2 Technetium(V) complexes

The majority of ^{99m}Tc pharmaceuticals contain technetium as Tc(V) (Table 2.1.1).

2.1.2.1 Tc-Gluconate

Tc(gluconate) and Tc(glucoheptonate) have been the earliest products of Tc(V) used as radiotracers for renal imaging (De Kieviet 1981; Johannsen and Spies 1988). The compounds were shown to contain a $Tc{=}O^{3+}$ core, but their structures are not completely defined, probably because of more than one stable species at carrier-added level (De Kieviet 1981). However, the x-ray structure of similar Tc complexes with similar ligands (Davison et al. 1987; DePamphilis et al. 1974; Fig. 2.1.2), indirectly supports the formulation as $TcO(Glu)_2^-$, even though the structure of $TcO(Ox)(OxH)^-$ demonstrates the possibility of another species with a $Tc{=}O^{3+}$ core (Abrams et al. 1991).

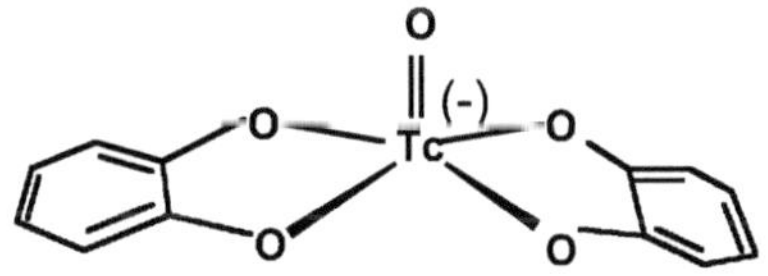

Fig. 2.1.2. $[TcO(cathecol)_2]^-$

These compounds are easily obtained in high yields at no carrier-added level, and they are suitable precursors in the synthesis of new ^{99m}Tc complexes by ligand exchange (or transchelation) (Spies et al. 1980). Other polyhydroxy or hydroxyl acids have been under investigation, such as glycolate, glucarate, tartrate, or citrate, which have been used in transchelation procedures.

Table 2.1.1. Chemical state of ^{99m}Tc-pharmaceuticals in clinical or preclinical use

Compound	Oxidation state core	Geometry	Coordinated number	Charge	Reference
Gluconate	$Tc(V)O_3^+$	Square pyramid	5	−1	Johannsen and Spies 1988
Glucoheptonate	$Tc(V)O_3^+$	Square pyramid	5	−1	De Kieviet 1981
DMSA	TcO_3^+ or Tc(III)	Octahedral	5	0 or −1	Bandoli et al. 1984; Ikeda et al. 1977
Penicillamine	$Tc(V)O_3^+$	Octahedral	6	0	Franklin et al. 1982
EDTA	$Tc(V)O_3^+$	Heptahedral	6	0	Davison and Jones 1982
HMPAO (Ceretec)	$Tc(V)O3^+$	Heptahedral	6	0	Fair et al. 1984
MRP20 (Neuroscint)	$Tc(V)O_3^+$	Heptahedral	6	0	Morgan et al. 1990
DADS	$Tc(V)O_3^+$	Square pyramid	5	0	Davison et al. 1980
DADT	$Tc(V)O_3^+$	Square pyramid	5	0	Watson et al. 1987
ECD (Neurolite)	$Tc(V)O_3^+$	Square pyramid	5	0	Edwards et al. 1990
MAG (MAG_3)	$Tc(V)O_3^+$	Square pyramid	5	0	Nosco et al. 1989
Tetrofosmin (Myoview)	$Tc(V)O_2^+$	Octahedral	5	+1	Kelly et al. 1993
NOEt	$Tc(V)N_2^+$	Octahedral	5	0	Pasqualini et al. 1994
EDTA	Tc(IV) or Tc(III)	Dimeric	7 or 6	0 or −1	Davison and Jones 1982; Burgi et al. 1981
DTPA	Tc(IV) or Tc(III)	Monomeric	?	−1(?)	Gorski and Koch 1970
MDP	Tc(IV)	Monomeric	?	0	Lisbon et al. 1980
HIDA (Choletec)	Tc(III)	Octahedral	6	−1	Loberg and Fields 1978
DMPE	Tc(III)	Octahedral	6	+1	Deutsch et al. 1981
Q12 (Technecard)	Tc(III)	Octahedral	6	+1	Deutsch et al. 1987
BATO (Cardiotec)	Tc(III)	Octahedral	6	0	Bandoli et al. 1982
MIBI (Cardiolite)	Tc(I)	Octahedral	6	+1	Abrams et al. 1983

DMSA dimercaptosuccinic acid, *EDTA* ethylenediaminetetraacetic acid, *HMPAO* hexamethyl propyleneamine oxime, *DADS* *N,N*-bis(mercaptoacetyl)ethylenediamine, *DADT* diaminodithiol, *ECD* ethylcysteinate dimer, *MAG_3* mercaptoacetyltriglycine, *NOEt* $Et(OEt)NCS_2$, *DTPA* diethylene triamine pentaacetate, *MDP* methylenediphosphonate, *HIDA* *N*-(2,6-dimethylphenylcarbamoylmethyl) iminodiacetic acid, *DMPE* 1,2-bis(dimethylphosphino) ethane, *BATO* boronic acid technetium oxime, *MIBI* methoxyisobutyl isocyanide

2.1.2.2 Tc-Dimercaptosuccinic Acid

Two different ^{99m}Tc-dimercaptosuccinic acid (DMSA) complexes are in clinical use, ^{99m}Tc(III)-DMSA with high binding affinity for renal tubuli and ^{99m}Tc(V)-DMSA with tumor affinity.

At acidic pH, at least four ^{99m}Tc-DMSA complexes have been identified in dependence of pH and stannous ion concentration (Ikeda et al. 1977b). Formation of a ^{99m}Tc(III)-DMSA complex is favored at pH 2.5, using an excess amount of stannous ion (Ikeda et al. 1976). This formulation is used for renal scintigraphy (Ikeda et al. 1977a). The coordination characteristics of the ^{99m}Tc(III)-DMSA complex have not yet been established.

At an elevated pH (pH 7.5–8.0), a ^{99}Tc-DMSA complex was produced, which accumulated in the skeleton (Johannsen et al. 1979). Further studies performing ligand exchange with Tc(V)-gluconate (Spies et al. 1980) led to the identification of a pentavalent ^{99m}Tc-DMSA complex with two DMSA molecules coordinated to a Tc(V)oxo core (Bandoli et al. 1984) (Fig. 2.1.3).

Pentavalent ^{99m}Tc-DMSA has been evaluated as a soft tumor-imaging agent (Yokoyama et al. 1985).

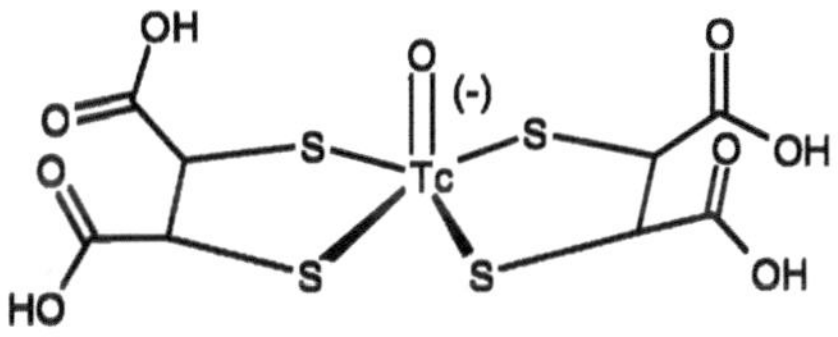

Fig. 2.1.3. Meso-[TcO(dimercaptosuccinic acid)$_2$]$^-$

2.1.2.3 Tc-Penicillamine

Figure 2.1.4 shows the structure of a ^{99}Tc complex with two molecules of penicillamine, confirmed by x-ray crystallography (Franklin et al. 1982). Yet when reduction of ^{99m}Tc-pertechnetate had been performed with $SnCl_2$, a complex with a Tc(IV) oxidation state was reported (Yokoyama et al. 1979).

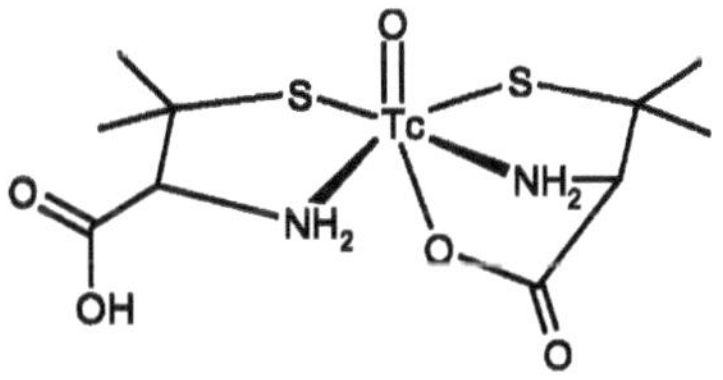

Fig. 2.1.4. [TcO(penicil)$_2$]

2.1.2.4 Tc-Ethylenediaminetetraacetic Acid and Tc-Diethylene Triamine Pentaacetate

Ethylenediamine tetraacetic acid (EDTA) and diethylene triamine pentaacetate (DTPA) are strong coordinating ligands that are administered to reduce in vivo toxicity of heavy metals. Nevertheless, the coordination behavior of EDTA and DTPA ligands with respect to technetium is rather complicated (Steigman et al. 1975).

^{99}Tc-EDTA chemistry studies demonstrated at least two types of stable complexes, one containing a Tc(V)O_3^+ core in a hepta-coordinated environment (Davison and Jones 1982), and the other is a complicated dimer in which technetium can be present as Tc(IV) or as Tc(III) (Linder 1986; Noll et al. 1980; Seifert et al. 1982). The crystal structure of a Tc(IV) complex has been reported (Burgi et al. 1981).

The exact structure of the ^{99m}Tc species was not yet found, mainly because at very low concentrations dimerization is very improbable, and in these solutions Tc(IV) or Tc(III) was detected. The participation of tin in the dimer formation cannot be excluded. In any case, the products pass rapidly through the kidneys.

These facts underline the difficulty of defining the chemical species present in the injection solution. To date, no complex with DTPA has been characterized at the ^{99}Tc level.

Undoubtedly, the production of a monomeric species is expected when EDTA and DTPA are used as chelating moiety for monoclonal antibody labeling.

2.1.2.5 Tc-Hexamethyl Propyleneamine Oxime (Ceretec™)

The structure of D,L-TcO (hexamethyl propyleneamine oxime [HMPAO]) is shown in Fig. 2.1.5.

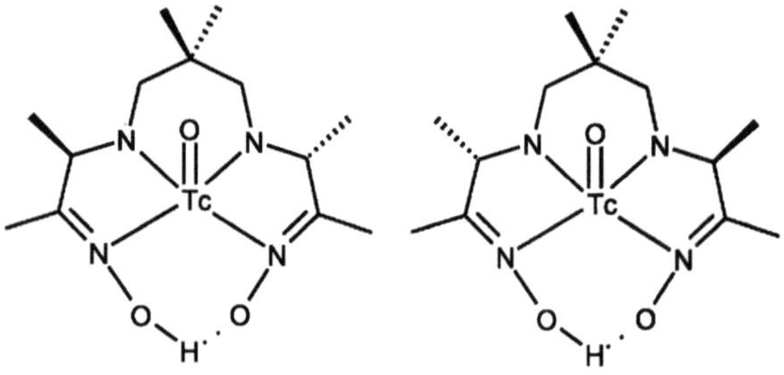

Fig. 2.1.5. D,L-TcO-hexamethyl propyleneamine oxime (HMPAO)

HMPAO is coordinated to a TcO^{3+} core with four nitrogen atoms. Ring closure of the oxime functionalities by hydrogen bonding increases the stability of the lipophilic complex.

^{99m}Tc-D,L-HMPAO was characterized at the ^{99}Tc level (Fair et al. 1984; Jurisson et al. 1987) and is the first neutral ^{99m}Tc complex for brain perfusion imaging (Troutner et al. 1984). The structural configuration has considerable effect on cerebral extraction, the D,L isomers pass the blood–brain barrier (BBB) while the mesoform is excluded (Sharp et al. 1986).

However, lipophilic D,L-HMPAO is easily transformed into a charged complex, which cannot pass the BBB. Once inside the brain, this "secondary" complex is trapped and is released very slowly (Neirinckx et al. 1987). The ^{99m}Tc-HMPAO complex is also used for labeling leukocytes with technetium.

2.1.2.6 Tc-MRP20

MRP20 is one of a series of tetradentate ligands, which incorporate donor sets containing pyrrole, amine, imine, and ketone moieties. The complex is neutral and lipophilic, similar to HMPAO. The chemical structure (Fig 2.1.6.) shows a five coordinated Tc-oxo complex in which the TcO^{3+} core is surrounded in the horizontal plane by the triple deprotonated ligand (Morgan et al. 1990, 1991). Tc-MRP20 was under investigation as a brain perfusion agent.

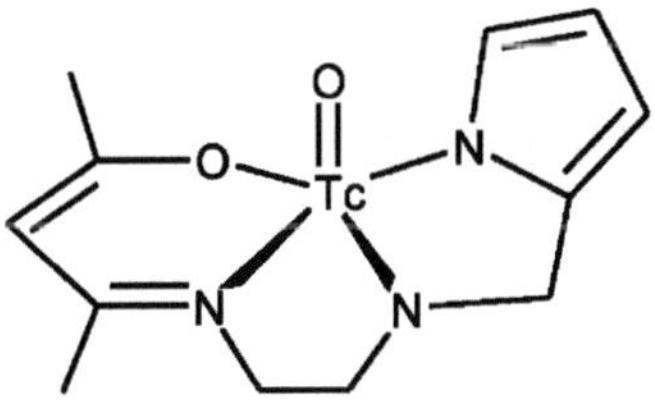

Fig. 2.1.6. Tc-MRP20

2.1.2.7 Tc-*N,N'*-bis(mercaptoacetyl)ethylenediamine and Tc-Diaminodithiol

Tc-*N,N'* bis(mercaptoacetyl)ethylenediamine (DADS) was introduced as a chelate, based on amide nitrogen and thiolate donor groups (Davison et al. 1979 and 1981). Tetradentate diaminodithiol (DADT) ligands form very stable complexes with oxo-technetium; the introduction of two carbonyl oxygen groups resulted in an overall negative charge. Several DADS-derived ^{99m}Tc complexes have been evaluated as renal agents (Brenner et al. 1984; Fritzberg 1986) (Fig. 2.1.7).

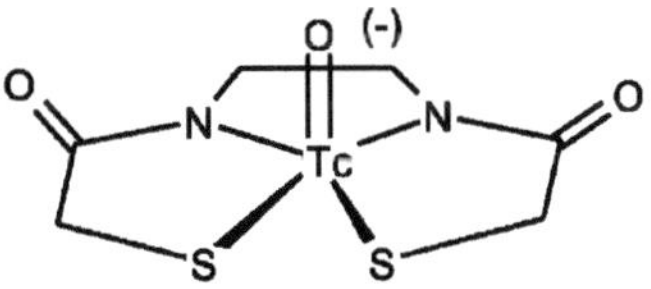

Fig. 2.1.7. Tc-*N,N'*-bis(mercaptoacetyl)ethylenediamine (DADS)

Substitution of the ethylene bridge (center chelate ring) with a carboxylate group produced ^{99m}Tc-CO_2DADS as two stereoisomers (Costello et al. 1983), with one isomer resembling the tubular agent iodohippurate. ^{99m}Tc-CO_2DADS is an important link in the development of the tubular agent ^{99m}Tc-mercaptoacetyltriglycine (MAG_3).

Structural modification of the N_2S_2 ligand has produced several ^{99m}Tc-DADT complexes. An example of substitution of the amine nitrogen (aminoalkyl-DADT) is ^{99m}Tc-

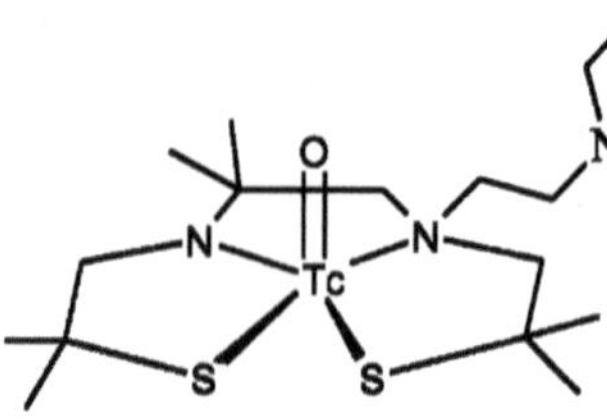

Fig. 2.1.8. TcO-*N*-ethylpiperidinyl-tetradentate diaminodithiol (NEP-DADT)

N-ethylpiperidinyl (NEP)-DADT (Epps et al. 1978). Functionalization of hexamethyl-DADT with a NEP side chain was shown to enhance brain accumulation of neutral, lipophilic ^{99m}Tc-*syn*-NEP-DADT (Lever et al. 1985) (Fig. 2.1.8).

^{99m}Tc-NEP-DADT is an example of systematic derivatization to optimize the structure–biodistribution relationship.

2.1.2.8 Tc-Ethylcysteinate Dimer

The ethylcysteinate dimer (ECD) belongs to the family of neutral, lipophilic tetradentate diaminedithiol ligands. The x-ray structure (Fig. 2.1.9) shows the functionalized ester dimer with a Tc(V)oxo core in a square pyramidal configuration (Watson et al. 1987). In fact, its high-performance liquid chromatography (HPLC) behavior is the same for the ^{99}Tc and ^{99m}Tc-ECD complex (Edwards et al. 1990).

The L,L stereoisomer can cross the BBB and is retained in the brain, presumably due to hydrolysis of the ester function (Leveille et al. 1989). If hydrolysis happens in blood before the molecule has crossed the BBB, the resulting dicarboxylate anion is rapidly excreted by the kidneys. No difference was observed with the monoester monoacid derivatives (Verbruggen et al. 1989 a).

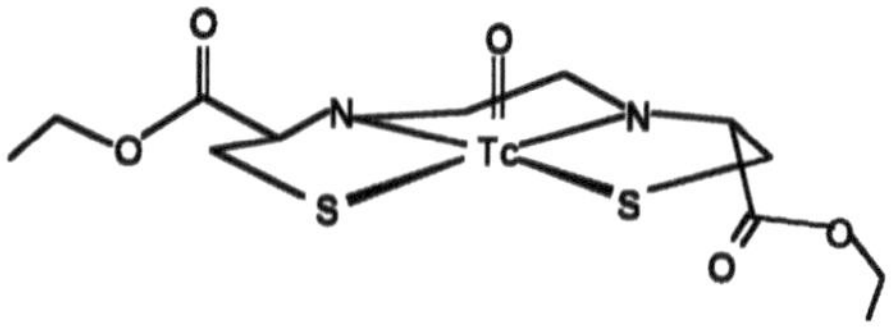

Fig. 2.1.9. Tc(V)O-ECD (ethylcysteinate dimer)

2.1.2.9 Tc-Mercaptoacetyltriglycine

Replacement of one thiolate donor group by a planar amide carrying a carboxylate anion avoids formation of stereoisomers, as observed with ^{99m}Tc-CO_2DADS. MAG_3 is a suitable ligand for producing ^{99m}TcO-MAG_3, a negatively charged complex (Fig. 2.1.10), structurally defined at carrier-added (CA) (Davison et al.1981) and no-carrier added (NCA) (Fritzberg et al. 1986; Verbruggen et al. 1989 b) levels.

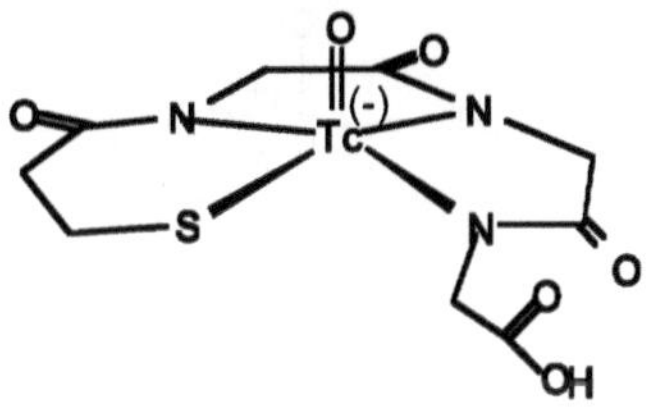

Fig. 2.1.10. Tc(V)O-mercaptoacetyltriglycine$^-$, or Tc(V)O-MAG_3^-

Steric arrangement of the carboxylate group in *syn* position with respect to Tc=O is responsible for active tubular secretion (Coveney and Robbins 1987; Fritzberg et al. 1986). A series of positional isomers of ^{99m}Tc-CO_2DADS were synthesized in order to produce MAG_3 as a ligand with suitable biological properties in man (Fritzberg 1986).

2.1.2.10 Tc-Tetrofosmin (P53)

Myoview is a TcO_2^+ complex, obtained by functional derivatization of 1,2-bis(dimethylphosphino)ethane (DMPE) (Kelly et al. 1993). The chemical structure of ^{99m}Tc-tetrofosmin shows four phosphorus atoms of the bidentate diphosphine ligands, arranged in a plane (Fig. 2.1.11). However, tetrofosmin contains four ethoxyethyl groups, which ensure a rapid clearance of activity from the liver. The cationic charge facilitates myocardial uptake.

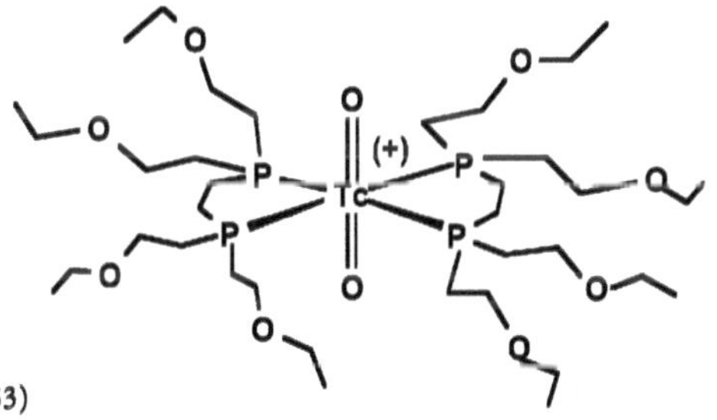

Fig. 2.1.11. $TcO_2P_4^+$ (P53)

Trans-octahedral configuration of the donor atoms has been confirmed by x-ray single crystal analysis of the ^{99}Tc analog. The HPLC behavior is the same for the ^{99}Tc and ^{99m}Tc tetrofosmin complex (Kelly et al. 1993).

Phosphine ligands are interesting coordinating groups because they stabilize complexes with technetium at several oxidation states (from V to I) (Deutsch et al. 1983). They may act as Lewis bases (soft groups stabilizing *trans*-TcO_2^+ core) in the highest technetium oxidation states, and as π-acceptor ligands in the lowest oxidation states in which technetium possesses high electron density. Other π-acceptor ligands are isonitriles, nitrosyl, and carbon monoxide.

Typical DMPE complexes have been reported (Deutsch et al. 1981, 1983). $[Tc^{V}O_2(DMPE)_2]^+$, $[Tc^{III}Cl_2(DMPE)_2]^+$ (Fig. 2.1.12), and $[Tc^{I}DMPE)_3]^+$ were found to be present both at CA and NCA levels, depending on the reaction conditions (Bandoli et al. 1982). As a cationic species, $[Tc^{III}Cl_2(DMPE)_2]^+$ showed myocardial uptake; however, it is species dependent, and there is considerable liver uptake (Deutsch et al. 1989).

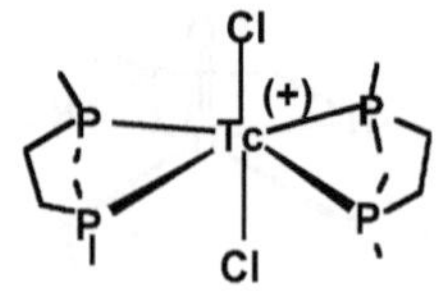

Fig. 2.1.12. $Tc(III)Cl_2$(1,2-bis(dimethylphosphino)ethane)$_2^+$, or $[Tc(III)Cl_2(DMPE)_2]^+$

The chemistry of Tc(V) complexes offers many possibilities for the synthesis of coordination complexes. In fact, the availability of different central cores can be used to stabilize a considerable number of ligands with very different coordinating properties. The TcO^{3+}, *trans*-TcO_2^+ and *trans*-$XTcO_2^+$ (X = halogenide, O, alcoholate, N groups, etc.) cores are stabilized by various ligands, and the existence of one or the other core is attributed to the arrangement of the coordinating atoms in the horizontal plane perpendicular to Tc=O. Soft atoms stabilize a TcO^{3+} core, while harder ones produce *trans*-$XTcO_2^+$ or *trans*-TcO_2^+ cores (Davison 1983).

For example, a cationic *trans*-$TcO_2{}^+$ complex is produced with tetradentate cyclam (Zuckman et al. 1981). $[Tc(V)O_2(Cyclam)]^+$ (Fig. 2.1.13) was investigated as a transchelating compound rather than as a radiopharmaceutical because it has no useful biological properties.

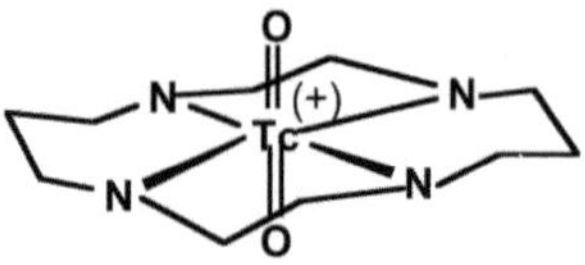

Fig. 2.1.13. TcO_2 (Cyclam)$^+$

2.1.2.11 TcN(NOEt) and Heterocomplexes with Metal–Nitrogen Multiple Bond

Nitrido Tc(V) complexes with a technetium–nitrogen triple bond were introduced by Baldas et al. (1978); structural verification of a stable TcN^{2+} core was also documented (Marchi et al. 1990). Some donors of nitrido nitrogen atom (N^{3-}) to yield the $Tc \equiv N^{2+}$ group have been evaluated; *N*-methyl-*S*-methyl dithiocarbazate $[H_2NN(CH_3)$–$C(=S)SCH_3]$ in acidic solution was found to be the most efficient ligand (Marchi et al. 1990; Pasqualini et al. 1992). In the presence of $Tc \equiv N^{2+}$, a high variability of the chelating set was observed. In comparison with $Tc{=}O^{3+}$ cores, softer coordinating atoms produce more stable complexes. Dithiocarbazate seems to produce prereduced intermediary Tc complexes containing a $Tc \equiv N$ core, which undergo facile substitution reactions with the final ligands.

The neutral $[TcN(Et(OEt)NCS_2]$ complex, called TcN(NOEt), was studied as a myocardial agent, demonstrating different biological properties with respect to the monocationic species (Pasqualini et al. 1994). The ^{99}Tc complex with an Et_2NCS_2 ligand was structurally defined (Bolzati et al. 2002) (Fig. 2.1.14), showing a square-pyramidal configuration with two dithiocarbamate groups bound in the equatorial plane and a $Tc \equiv N$ core; the same species is present at the NCA level.

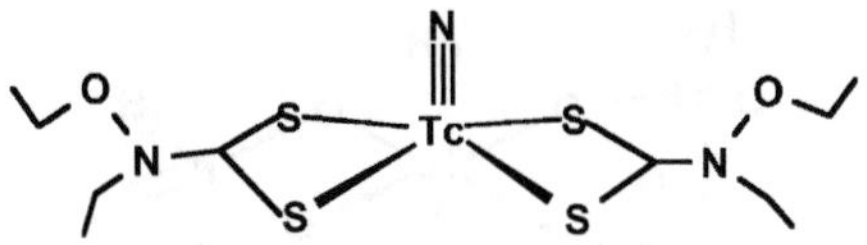

Fig 2.1.14.. The neutral [TcN(Et(OEt)NCS$_2$], also called Tc-NOEt

2.1.3 Technetium(IV), (III), and (I) complexes

Technetium is stabilized at low oxidation states by suitable ligands such as phosphines, isonitriles, carbon monoxide, and thiourea (Gorski and Koch 1970). Organometallic carbonyl (CO) complexes are interesting precursors for a new class of ^{99m}Tc(I) radiopharmaceuticals (Alberto et al. 2001; Schibli et al. 2000).

As reported previously, EDTA (Burgi et al. 1981) (Fig. 2.1.15) and many other ligands with no π-accepting groups can produce Tc(IV) or Tc(III) stable complexes, because other parameters such as chelating effect and metal–metal bonds contribute to their stabilization. Usually six-coordinated complexes in an octahedral configuration are obtained, but some exceptions are possible. In addition, the complex charge may vary in dependence of the ligand charge, deprotonation of the coordinating group; however, very exceptionally is the net charge more than one, negative, or positive.

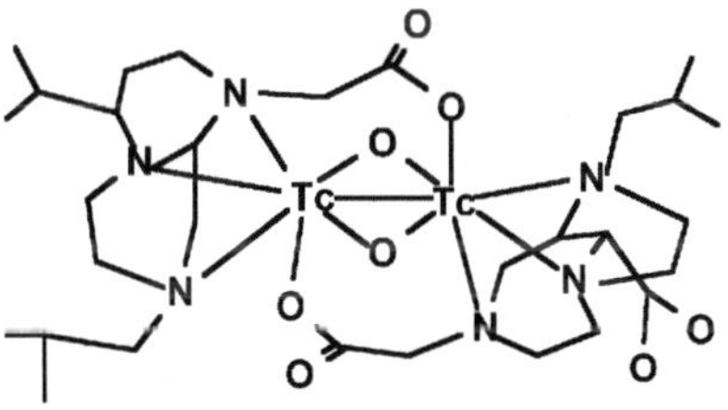

Fig. 2.1.15. Tc-ethylenediaminetetraacetic acid (EDTA)

2.1.3.1 Tc-Diphosphonates

^{99m}Tc-methylenediphosphonate (MDP) and polyphosphate complexes were studied as ^{99}Tc complexes, but only one x-ray structure was obtained with the diphosphonate ligand (Subramanian et al. 1975). ^{99m}Tc-MDP is a polymeric species in which tin is incorporated (Libson et al. 1980) (Fig. 2.1.16). The figure shows primarily the configuration at the central technetium.

Reduction of $^{99}TcO_4^-$ with $NaBH_4$ in the presence of hydroxyethylene diphosphonate (HEDP) produced seven different components detected by HPLC. It was demonstrated that the different species differ in molecular weight, depending on the size of the polymers (Van den Brand et al. 1981). Chemical yield of various components depends on the total technetium concentration, the polymerization reaction following high-order kinetics. As a single component, ^{99m}Tc-MDP and ^{99m}Tc-HMDP localize independently in the inorganic bone matrix. In order to obtain reproducible clinical results, bone agents must be pre-

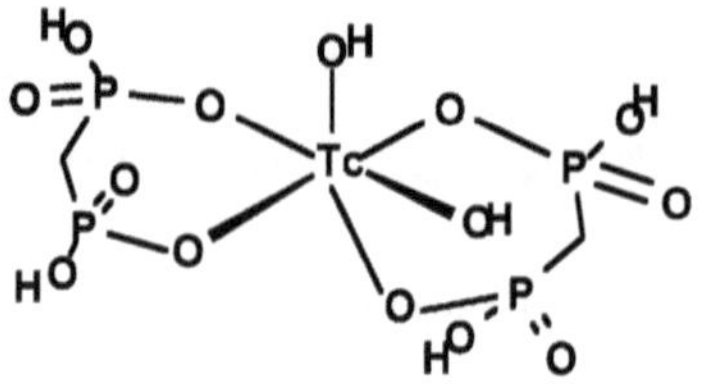

Fig. 2.1.16. Tc-diphosphonate

pared using fresh eluates obtained from a ^{99}Mo/^{99m}Tc generator that is eluted regularly. ^{99m}Tc diphosphonates show high skeletal uptake and are used for bone scintigraphy.

2.1.3.2 Tc-*N*-(2,6-dimethylphenylcarbamoylmethyl) iminodiacetic acid

N-(2,6-dimethylphenylcarbamoylmethyl)iminodiacetic acid (HIDA) and several ether derivatives have been evaluated as ligands for complexation, producing ^{99m}Tc complexes suitable as hepatobiliary agents. ^{99m}Tc-IDA complexes have a negative charge (Loberg and Fields 1978). Two molecules of ligand are coordinated to one Tc(III)-core (Nunn et al. 1983) (Fig. 2.1.17).

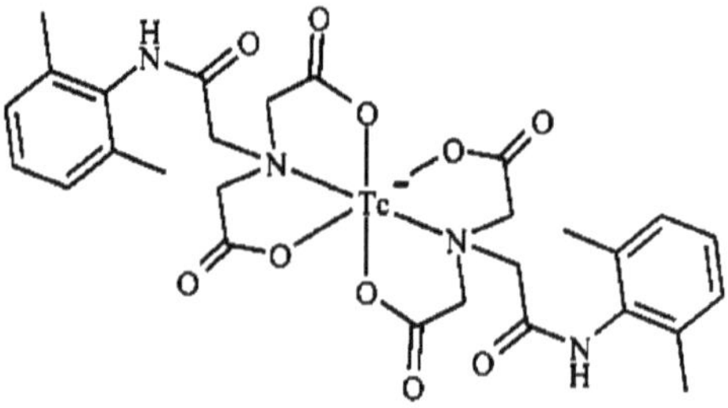

Fig. 2.1.17. Tc-*N*-(2,6-dimethylphenylcarbamoylmethyl)iminodiacetic acid)$_2$ – (HIDA)

2.1.3.3 Tc-Q12

Complexes of the Q series are defined by their structure belonging to the $[Tc^{III}P_2L]^+$ complexes, with polydentate Schiff bases stabilized at the +3 oxidation state by a tertiary phosphine ligand (Deutsch et al. 1987). In fact, ^{99}Tc(V)OCl-L-oxo complexes are easily reduced by a two-electron process to Tc(III). The final Tc(III) compound (Fig. 2.1.18) has an octahedral configuration with the two *trans* phosphines on the apexes and the tetradentate Schiff base on the equatorial plane (Jurisson et al. 1984). Tc-Q12 has a positive charge, the two hydroxyl groups being deprotonated. The ^{99}Tc complex has been prepared in two steps, with an intermediate Tc(V)-oxo complex (Abrams et al. 1982).

These complexes are well modified in the backbone, without decreasing the complex stability. Q12 is the best derivative in the series (Deutsch et al. 1987); however, none of these ligands has been used as a myocardial perfusion agent.

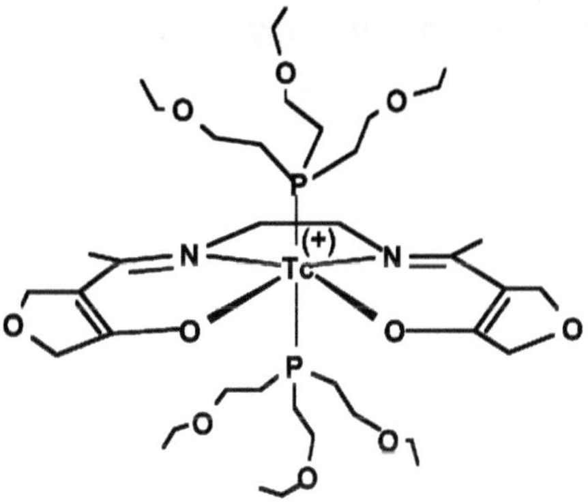

Fig. 2.1.18. Tc(Q12). *L* Equatorial tetracoordinate ligand

2.1.3.4 Tc-Boronic Adducts of Technetium Oximes

Dioxime type ligands can be considered as Schiff base bischelates (Deutsch et al. 1978; Bandoli et al. 1986). The first complexes with oxime ligands were described as mono-capped Tc(dioxime)$_3$(μOH)SnCl$_3$ (dioxime = dimethylglyoxime) complexes (Treher et al. 1989). The boronic adducts of technetium oximes – (BATOs) (Fig. 2.1.19) – were well characterized, and some could be used both as myocardial and cerebral perfusion agents. The complexes are neutral; technetium is coordinated to three N-bonded dioxime molecules and to one Cl or Br atom in an axial position (seven covalent bonds).

The three bidentate dioxime groups are joined through covalent B–O bonds to a tetrahedral boron cap derived from an alkyl boronic acid derivative. The six ligating nitrogen atoms form a monocapped distorted trigonal prism. It can be characterized by the geometry of the triangles of nitrogen or oxygen at the capped and uncapped ends of the complex.

Different oximes can be used, but the major structural modifications of the complex are achieved at the boronic side chain (R$_1$).

One BATO-derived radiopharmaceutical, ^{99m}Tc-teboroxime (Cardiotec), has been available in the United States.

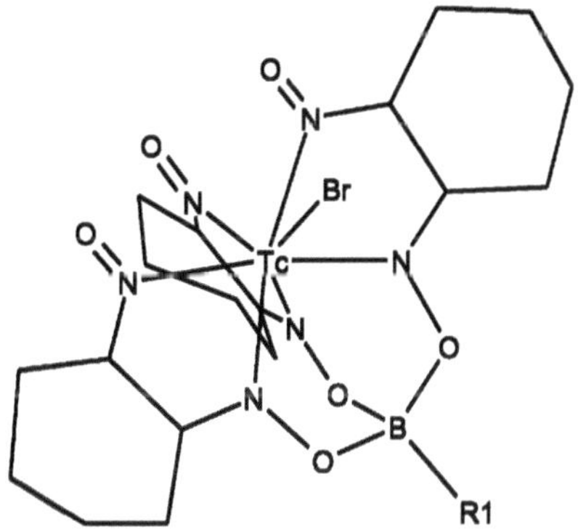

Fig. 2.1.19. Tc-boronic adducts of technetium oxime (BATO)

2.1.3.5 Tc-Methoxyisobutyl Isocyanide

Isonitriles, like carbon monoxide or phosphines, are ligands with high reducing properties together with a high capability of stabilizing low oxidation states. Tertiary butyl isonitrile (TBI) was the first ligand evaluated as a myocardial imaging agent (Holman et al. 1984). The positively charged Tc(I) complex showed high uptake in myocytes; however, clearance from the liver was slow. Introducing the 2-methoxy-derivative had a positive effect on the biodistribution, since the ether is metabolized and cleared faster.

Structural characterization of Tc(I) complexes was performed identifying ^{99m}Tc sestamibi as a complex with six monodentate methoxyisobutyl isocyanide (MIBI) ligands attached symmetrically to a central Tc(I) atom (Abrams et al. 1983; Jones et al. 1985; Fig. 2.1.20).

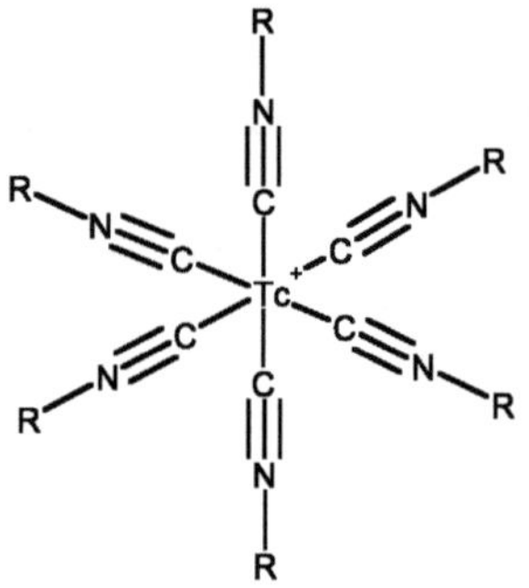

Fig. 2.1.20. Tc-sestamibi. $R = $2-methoxyisobutyl

Cationic Tc(I)-hexakis(2-methoxy-isobutyl-isonitrile) tetraflurorobate is labeled by reacting tetrakis(2-methoxy-isobutyl-isonitrile)-copper(I) tetraflurorobate adduct with ^{99m}Tc-pertechnetate, using the kit formulation. Heating the reaction vial in a boiling water bath further facilitates formation of ^{99m}Tc(I) sestamibi.

Clinical studies showed high myocardial extraction of ^{99m}Tc sestamibi and fast background clearance (Wackers et al 1989). Redistribution of the lipophilic complex is blocked by intracellular binding.

2.1.4 ^{99m}Tc Labeling

^{99m}Tc chemistry is primarily the chemistry of anionic pertechnetate. This ^{99m}Tc species is eluted from the ^{99}Mo/^{99m}Tc generator with high specific activity as an isotonic solution. Accordingly, ^{99m}Tc chemistry is aqueous solution chemistry in saline suitable to be injected intravenously. Also, ^{99m}Tc chemistry is an NCA chemistry because ^{99m}Tc activity is present in the radiopharmaceutical kit at 10^{-8} to 10^{-9} M.

Direct labeling. Generally, direct labeling is performed by adding ^{99m}Tc eluate in a suitable volume to a sterile kit. The kit contains all chemical components, including a reducing agent. The labeling reaction requires reduction of pertechnetate, which is reacting with the ligand forming the labeled product in high yield (>90%).

Exchange labeling. In a few exceptions, an intermediate ligand complex is formed (MAG_3) that is stabilized by ligand exchange during heating. In the case of MIBI, the kit contains a preformed copper(I) complex, a so-called adduct, which facilitates formation of hexacoordinated ^{99m}Tc(I)-MIBI.

Effect of formulation. Kits contain very low amounts of stannous ion for reduction of ^{99m}Tc-pertechnetate; nevertheless, $SnCl_2$ is usually in high excess. There are several reasons for using stannous salt in excess. Stannous salts are spontaneously oxidized in air. Also, oxidant species in the eluate may have been formed by radiolysis; the amount of Sn(II) available in solution is very low with respect to the total amount of lyophilized $SnCl_2$. In order to assure validity of kits beyond the expiration date, an excess of $SnCl_2$ is used in the kit formulation.

On the other hand, there are cases in which the amount of reductant must be strictly controlled. This is indicated when more than one oxidation state is favored with a certain ligand, or when hydrolysis products interfere with complex stability. This precaution is possible with Sn(II) complexes (Sn-tartrate, Sn-gluconate, Sn-citrate, Sn-EDTA, etc.), which release small amounts of stannous ion into solution. In addition, another reducing agent, including the ligand itself, might be considered.

Formation of colloidal TcO_2 is avoided in the presence of ligand, which competes for the reduced technetium species, producing the labeled ^{99m}Tc pharmaceutical. In the absence of ligand, a mixture of hydrolized, insoluble ^{99m}Tc species, $TcO_2 \cdot nH_2O$, is formed. To increase the rate of coordination, a high amount of the ligand is generally used. The kinetic mechanism of reduction-substitution is rather complicated, and sometimes it depends on the concentration of carrier $^{99m}TcO_4^-$. This is observed when ^{99}Tc carrier in the eluate is increased to CA level.

Kit components. Kit composition is optimized to ensure that the unique ^{99m}Tc-labeled complex is obtained in high yield. Several factors influence the reduction/coordination process; these are primarily the nature and the amounts of reductant and ligand, pH, and temperature. Generally, the rate of complex formation is a good indicator of complex stability, which is essential to avoid increased background activity in vivo. In order to provide a suitable pH environment for the formation of a specific ^{99m}Tc complex, buffers are important components in kit formulations.

Additives include antioxidants, catalysts, accelerators, solubilizing agents, and fillers (Nowotnik 1994).

Antioxidants are added to the formulation in order to increase the stability of the radiopharmaceutical. Antioxidants for ^{99m}Tc complexes that have been used are ascorbic acid (Tofe and Francis 1976), gentisic acid (Tofe et al. 1980), and *p*-aminobenzoic acid (Rimmer 1982).

A *catalyst* might be a ligand, which rapidly forms an intermediary coordination complex such as gluconate, DTPA, and citrate (Davison 1983). Ligand exchange is applied when complex formation with a certain ligand is slow relative to formation of reduced, hydrolized technetium, resulting in a poor radiochemical yield.

Accelerators increase the radiochemical yield and rate of complex formation (Tweedle 1983).

Surfactants might be required to solubilize lipophilic ^{99m}Tc complexes (MIBI) (Bergstein and Subramanyam 1986) and particulate preparations (macroaggregated albumin, microspheres).

Solubility of the product in aqueous solution is indispensable. Equally important is the dissolution of the lyophilized kit contents when ^{99m}Tc eluate is added, in order to assure proper chemistry during the vitally important first few seconds of reconstitution.

Inert fillers are added in order to achieve rapid solubilization of the vial contents through the control of particle size during the lyophilization process. The size of the lyophilizate plug and particle size are controlled by the freeze-dry cycle in kit production. Sodium chloride is added to D,L-HMPAO kits and mannitol to MIBI kits.

Many variables in kit formulation have to be explored during the developmental phase of a new product in addition to the documentation of compatibility with different generator eluates and the shelf-life of the kit.

References

Abrams MJ, Davison A, Brodack JW, Jones AG, Faggiani R, Lock CJ (1982) The preparation of technetium(III) compounds in aqueous media. J Labeled Comp Radiopharm 14:1596–1597

Abrams MJ, Davison A, Jones AG, Costello CE, Pang H (1983) Synthesis and characterization of hexakis(alkylisocyanide) and hexakis(arylisocyanide) complexes of technetium(I). Inorg Chem 22:2798–2800

Abrams MJ, Larsen S, Zubieta J (1991) Investigations of the technetium hydrazido core – synthesis and structural characterization of $[(N\text{-}C_4H_9)_4N][Tc_2(NNPH_2)_2(C_6Cl_4O_2)_4].CH_2Cl_2{\cdot}2CH_3OH$, a Tc(V)/Tc(IV) catecholate complex with the hydrazido ligands adopting the unusual eta-1 bridging mode. Inorg Chem 30:2031–2035

Alberto R, Ortner K, Wheatley N, Schibli R, Schubiger PA (2001) Synthesis and properties of boranocarbonates: a convenient in situ CO source for the aqueous preparation of $[^{99m}Tc(OH_2)_3(CO)_3]^+$. J Am Chem Soc 123:3135–3136

Andros G, Harper PV, Lathrop KA, McCardle RJ (1965) Pertechnetate-99m localization in man with application to thyroid scanning and the study of thyroid physiology. J Clin Endocrinol Metab 25:1067–1076

Baldas J, Bonnyman J, Poier PM, William GA, Mackay MF (1981) Synthesis and structure of bis(diethyldithiocarbamato)nitrido technetium(V): a technetium-nitrogen triple bond. J Chem Soc Dalton Trans 9:1798–1801

Bandoli G, Mazzi U, Moresco A, Nicolini M, Refosco F, Tisato F (1986) Technetium complexes containing tridentate and bidentate Schiff base type ligands. In: Nicolini M, Bandoli G, Mazzi U (eds) Technetium in chemistry and nuclear medicine 2. Cortina International, Verona, Italy, pp 73–80

Bandoli G, Mazzi U, Roncari E, Deutsch E (1982) Crystal structures of technetium compounds. Coord Chem Rev 44:210

Bandoli G, Nicolini M, Mazzi U, Spies H, Muenze R (1984) Synthesis and X-ray crystal structure of tetraethylammonium bis[1,2-di(carbomethoxy)ethane-1,2-dithiolato]oxotechnetate(V). Transition Met Chem 9:127–129

Bergstein PL, Subramanyam V (1986) Ether isonitriles and radiolabeled complexes thereof. Eur. Patent Appl. EP 86117847.3

Bolzati C, Boschi A, Uccelli L, Tisato F, Refosco F, Cagnolini A, Duatti A, Pracash S, Bandoli G, Vittadini A (2002) Chemistry of the strong electrophilic metal fragment $[^{99}Tc(N)(PXP)]_2+$ (PXP diphosphine ligand). A novel tool for the selective labeling of small molecules. J Am Chem Soc 124:11468–11479

Brenner D, Davison A, Lister-James J, Jones AG (1984) Synthesis and characterization of a series of isomeric oxotechnetium(V) diamino dithiolates. Inorg Chem 23:3793–3797

Burgi HB, Anderegg G, Blauenstein P (1981) Preparation, characterisation, crystal and molecular structure of $Na_2[N(CH_2COO)_3Tc^{(IV)}(m\text{-}O)_2Tc^{(IV)}(H_2EDTA){\cdot}5H_2O$. Inorg Chem 20:3829–3834

Costello CE, Brodack JW, Jones AG, Davison A, Johnson DL, Kasina S, Fritzberg AR (1983) The investigation of radiopharmaceutical components by fast atom bombardment mass spectroscopy: the identification of Tc-HIDA and the epimers of Tc-CO_2DADS. J Nucl Med 24:353–355

Coveney JR, Robbins MS (1987) Comparison of technetium-99m MAG_3 Kit with HPLC-purified technetium-99m MAG_3 and OIH in rats. J Nucl Med 28:1881–1887

Davison A (1983) The coordination chemistry of technetium In: Deutsch E, Nicolini M, Wagner HN Jr (eds) Technetium in chemistry and nuclear medicine 1. Cortina International, Verona, Italy, pp 3–14

Davison A, De Phamphilis BV, Jones AG, Franklin K, Lock CJL (1987) Synthesis and characterization of complexes containing the bis(1,2-thiolato)-oxotechnetium(V) core. Inorg Chim Acta 128:161–167

Davison A, Jones AG (1982) The chemistry of technetium(V). Int J Appl Radiat Isot 33: 875–881

Davison A, Jones AG, Orvig C, Sohn M. (1981) A new class of oxotechnetium(V) chelate complexes containing $TcON_2S_2$ core. Inorg Chem 20:1629–1632

Davison A, Sohn M, Orvig C, Jones AG, LaTegola MR (1979) A tetradentate ligand designed specifically to coordinate technetium. J Nucl Med 20:641

De Kieviet W (1981) Technetium radiopharmaceuticals: chemical characterization and tissue biodistribution of Tc-glucoheptonate using Tc-99m and carrier Tc-99. J Nucl Med 22:703–709

De Pamphilis BV, Jones AG, Davis MA (1974) Preparation and crystal structure of oxotechnetium bis(thiomercaptoacetate) and its relationship to radiopharmaceuticals labeled with ^{99m}Tc. J Am Chem Soc 78:5570–5571

Deutsch E, Elder RC, Laarge BA et al (1978) Structural characterization of a bridged technetium-99-tin-dimethylglyoxime complex: implication for the technetium-99m-labeled radiopharmaceuticals prepared by tin(II) reduction of pertechnetate. Proc Natl Acad Sci USA 73:653–660

Deutsch E, Glavan KA, Sodd VJ et al (1981) Cationic Tc-99m complexes as potential myocardial imaging agents. J Nucl Med 22:897–907

Deutsch E, Ketring AR, Libson K, Vanderheyden J-L, Hirth WJ (1989) The Noah's ark experiment: species-dependent biodistributions of cationic ^{99m}Tc complexes. Int J Rad Appl Instrum 16:191–232

Deutsch E, Libson K, Jurisson S, Lindoy LF (1983) Technetium chemistry and technetium radiopharmaceuticals In: Lippard SJ (ed) Progress in inorganic chemistry. Wiley, New York, pp 75–139

Deutsch E, Vanderheyden J-L, Gerundini P, Libson K et al (1987) Development of non reducible technetium-99m(III) cations as myocardial perfusion imaging agents: initial experience in humans. J Nucl Med 28:1870–1880

Deutsch E, Vanderheyden JL, Gerundini P, Libson K, Hirth W, Colombo F, Savi A, Fazio F (1987) Development of nonreducible technetium-99m(III) cations as myocardial perfusion imaging agents: initial experience in humans. J Nucl Med 28:1870–1880

Dewanjee MK (1990) The chemistry of ^{99m}Tc-labeled radiopharmaceuticals. In: Seminars in nuclear medicine XX. Saunders, Philadelphia

Eckelman WC, Meinken G, Richards P (1971) ^{99m}Tc-human serum albumin. J Nucl Med 12:707–710

Edwards D, Cheesman E,, Watson M, Maheu L, Nguyen S, Dimitre L, Nason T, Watson A, Walovitch R et al (1990) Synthesis and characterization of technetium and rhenium complexes of *N,N'*-1,2-ethylenediylbis-L-cysteine. Neurolite and its metabolites. In: Nicolini M, Bandoli G, Mazzi U (eds) Technetium and rhenium in chemistry and nuclear medicine 3. Cortina International, Verona, Italy, p 433

Epps LA, Burns HD, Lever SZ et al (1978) Brain imaging agents: synthesis and characterization of (*N*-piperidinylethyl)hexamethyl diaminodithiolate oxo-technetium(V) complexes. Int J Appl Radiat Isot 38:661–664

Fair CK, Troutner DE, Schlemper EO, Murmann RK, Hoppe ML (1984) Oxo[3,3'-(1,3-propanediyldiimino)bis(3-methyl-2-butanone oximato)(3$^-$)-*N,N',N'',N'''*]-technetium(V), [$TcO(C_{13}H_{25}N_4O_2)$]. Acta Cryst C40:1544–1546

Franklin KJ, Lock HE, Lock CJL (1982) Preparation, spectroscopic properties and structure of 1-oxo-2,3,6-(D-penicillaminato NSO)-4,5-(D-penicillaminato NS) technetium(V). Inorg Chem 21:1941

Fritzberg AR (1986) Advances in renal radiopharmaceuticals In: Fritzberg AR (ed) Radiopharmaceuticals: progress and clinical perspectives, vol. I. CRC Press, Boca Raton, pp 61–87

Fritzberg AR, Kasina S, Eshima D, Johnson DL (1986) Synthesis and biological evaluation of technetium-99m MAG_3 as a Hippuran replacement. J Nucl Med 27:111–116

Gerson M, Deutsch E, Nishiyma H et al (1983) Myocardial perfusion imaging with ^{9m}Tc-DMPE in man. Eur J Nucl Med 8:371–374

Gorski B, Koch H (1970) Technetium complex formation with chelate-forming ligands. II. J Inorg Nucl Chem 32:3831–3836

Harper PV, Lathrop KA, Gottschalk A (1966) Pharmacodynamics of some technetium-99m preparations. In: Andrews GA, Knisely RM, Wagner HN Jr (eds) Radioactive pharmaceuticals. AEC symposium series conf 651111 1966, pp 335–357

Holman BL, Jones AG, Lister-James J et al (1984) A new Tc-99m-labeled myocardial imaging agent, hexakis(*t*-butylisonitrile)technetium(I) (Tc-99mTBI). Initial experience in the human. J Nucl Med 25:13501355

Ikeda I, Inoue O, Kurata K. (1976) Chemical and biological studies on ^{99m}Tc-DMS-II: effect of Sn(II) on the formation of various Tc-DMS complexes. Int J Appl Radiat Isot 27:681–688

Ikeda I, Inoue O, Kurata K. (1977 a) Preparation of various Tc-99m dimercaptosuccinate complexes and their evaluation as radiotracers. J Nucl Med 18:1222–1229

Ikeda I, Inoue O, Kurata, K (1977 b) Chemical and biological studies on ^{99m}Tc-DMS-I: formation of complexes by four different methods. Int. J Nucl Med Biol 4:56–65

Johannsen B, Spies H (1988) Progress and problems in the chemistry of technetium-99m tracers. Isotopenpraxis 24:449–454

Johannsen B, Spies H, Syhre R (1979) Studies on complexation of ^{99m}Tc with dimercaptosuccinic acids with regard to organ specificity of ^{99m}Tc-radiopharmaceuticals. Eur J Nucl Med 4:148

Jones AG, Davison A (1982) The chemistry of technetium I, II, III, IV. Int Appl Radiat Isot 33:867–874

Jones AG, Dionauge GF, Davison A et al (1985) Biological distribution and structure function relationship of hexakis isonitrile Tc(I) complexes (abstract). J Nucl Med All Sci 29:200

Jurisson S, Dancey K, McPartlin M, Tasker P, Deutsch E (1984) Synthesis, characterization, and electrochemical properties of technetium complexes containing both tetradentate Schiff base ligands and monodentate tertiary phopshine ligands: single crystal structure of trans-(*N,N'*-ethylene bis(acetylacetone-iminato)bis(triphenylphos- phine)-technetium(III)-hexafluoro-phosphate. Inorg Chem 23:4743–4744

Jurisson S, Schlemper EO, Troutner DE, Canning LR, Nowotnik DP, Neirinckx RD (1987) Synthesis, characterization and X-ray structural determination of Technetium(V)-oxo-tetradentate amine oxime complexes. Inorg Chem 25:3576–3582

Katti KV, Singh PR, Barnes CL, Katti KK, Kopicka K, Ketring AR, Volkert WA (1993) Organometallic phosphinimines as building blocks for potential new radiopharmaceuticals. Z Naturforsch 486:1381–1386

Kelly JD, Forster AM, Higley B, Archer CM, Booker FS, Canning LR, Chiu KW, Edwards B, Gill HK, McPartlin M, Nagle KR, Latham IA, Pickett RD, Storey AE, Webbon PM (1993) Technetium-99m-tetrofosmin as a new radiopharmaceutical for myocardial perfusion imaging. J Nucl Med 34:222–227

Leveille J, Demonceau G, De Roo M, Rigo P, Taillefer R, Morgan R, Kupranick D, Walovitch RC (1989) Characterization of technetium-99m-L,L-ECD for brain perfusion imaging, part 1. Pharmacology of technetium-99m ECD in nonhuman primates. J Nucl Med 30:1892

Lever SZ (1995) Technetium and rhenium compounds. In: Wagner HN Jr, Szabo S, Buchanan JW (eds) Principles of nuclear medicine, 2nd edn. Saunders, Philadelphia, pp 213–220

Lever SZ, Burns HD, Kervitsky TM, Goldfarb HW, Woo DV, Wong DF, Epps LA, Kramer AV, Wagner HN Jr (1985) Design, preparation and biodistribution of a technetium-99 diaminodithiol complex to assess regional cerebral blood flow. J Nucl Med 26:1287–1294

Libson K, Deutsch E, Barnett BL (1980) Structural characterisation of a Tc-99 diphosphonate complex. Implication for the chemistry of technetium-99m skeletal imaging agents. J Am Chem Soc 102:2476–2478

Linder KE (1986) Aminocarboxylate complex of technetium. PhD thesis, Massachusettes Institute of Technology, June

Loberg MD, Fields AT (1978) Chemical structure of technetium-99m-labeled *N*-(2,6-dimethyl-phenylcarbamoylmethyl)-iminodiacetic acid (^{99m}Tc-HIDA). Int J Appl Radiat Isot 29:167–173

Marchi A, Garuti P, Duatti A et al (1990) Synthesis of technetium(V)-nitrido complexes with chelating amines: a novel class of monocationic, octahedral complexes containing the $[Tc{\equiv}N]^{2+}$ core. Crystal structures of $[TcN(en)_2Cl]^+$ (en = ethylenediamine) and $[TcN(tad)Cl]^+$ (tad = 1,5,8,12-tetraazadodecane). Inorg Chem 29:2091–2096

Mazzi U (1989) The coordination chemistry of technetium in its intermediate oxidation states. Polyhedron 8:1983–1688

McAfee JG, Fueger GF, Baggish MS, Holzman GB, Zolle I (1964) ^{99m}Tc-labeled serum albumin for scintillation scanning of the placenta. J Nucl Med 5:936–946

McAfee JG, Fueger GF, Stern HS, Wagner HN Jr, Migita T (1964) ^{99m}Tc-pertechnetate for brain scanning. J Nucl Med 5:811–827

Morgan G, Deblaton M, Hussein W, Thornback J, Evrard G, Durant F, Stach J, Abram U, Abram S (1991) Rhenium(V) and technetium(V) complexes with *N*-(2(1H-pyrolmethyl))-*N'*-(4-pentene-3-one-2)ethane-1,2-diaminate ($C_{12}H_{16}N_3O$, MRP20) – X-ray crystal-structures of H3MRP20 and TcO(MRP20). Inorg Chim Acta 190:257–264

Morgan GF, Abram U, Evrard G, Durant F, Deblaton M, Clemens P, Vandenbroeck P, Thornback JR (1990) Structural Characterization of the new brain imaging agent [^{99m}Tc][TcO(L)], H_3L=N-4-oxopentan-2-ylidene-*N'*-pyrrol-2-ylmethylen-ethane-1,2-diamine (MRP20). J Chem Soc Chem Comm 24:1772–1773

Neirinckx RD, Canning LR, Piper IM, Nowotnik DP, Pickett RD, Holmes RA, Volkert WA, Forster AM, Weisner PS, Mariott JA, Chaplin SB (1987) Technetium-99m D,L-HM-PAO: a new radiopharmaceutical for SPECT imaging of regional cerebral blood perfusion. J Nucl Med 28:191–202

Noll B, Seifert S, Muenze R (1980) New Tc(IV) compounds with nitrilo-triacetic acid. Radiochem Radioanal Lett 43:215–218

Nosco DL, Tofe AJ, Dunn TJ, Lyle LR et al (1989) New developments in radiopharmaceuticals at Mallinckrodt. In: Nicolini M, Bandoli G, Mazzi U (eds) Technetium and rhenium in chemistry and nuclear medicine 3. Cortina International, Verona, Italy, pp 381–392

Nowotnik DP (1994) Physico-chemical concepts in the preparation of radiopharmaceuticals In: Sampson CB (ed) Textbook of radiopharmacy: theory and practice, second enlarged edition. Gordon and Breach, Reading

Nunn AD, Loberg MD, Conley RA (1983) A structure–distribution–relationship approach leading to the development of Tc-99m-mebrofenin: an improved cholescintigraphic agent. J Nucl Med 24:423–430

Nunn AD, Treher EN, Feld T (1986) Boronic acid adducts of technetium oxime complexes (BATOs), a new class of neutral complexes with myocardial imaging capabilities. J Nucl Med 27:893

Pasqualini R, Comazzi V, Bellande E, Duatti A, Marchi A (1992) A new efficient method for the preparation of ^{99m}Tc-radiopharmaceuticals containing the Tc≡N multiple bond. Int J Appl Radiat Isot 43:1329–1333

Pasqualini R. Duatti A, Bellande E, Comazzi V, Brucato V et al (1994) Bis(dithiocarbamato)nitrido technetium-99m radiopharmaceuticals. A class of neutral myocardial imaging agents. J Nucl Med 35:334–341

Peacock RD (1966) The chemistry of technetium and rhenium. Elsevier, London

Perrier C, Segrè E (1937) Radioactive isotopes of element 43. Nature 140:193–194

Perrier C, Segrè E (1947) Technetium: the element of atomic number 43. Nature 159:24 (Letter)

Richards P (1966) Nuclide generators. In: Radioactive pharmaceuticals. USAEC symposium series, no. 6, (CONF-651111), Oak Ridge, Tenn., pp 155–163

Rimmer J (1982) Radiopharmaceutical composition based on technetium-99m and the reagent for making it. Eur Patent Appl EP 78,642

Schibli R, Labela R, Alberto R, Garcia-Garayoa E, Ortner K, Abram U, Schubiger PA (2000) Influence of the denticity of ligand systems on the in vitro and in vivo behavior of Tc-99m(I)-tricarbonyl complexes: a hint for the future functionalization of biomole-cules. Bioconjugate Chem 11:345–351

Schwochau K (1983) The present status of technetium chemistry. Radiochim Acta 32:139–152

Segrè E, Seaborg GT (1938) Nuclear isomerism in element 43. Phys Rev 54:772

Seifert S, Noll B, Muenze R (1982) Studies of the complex formation of technetium(IV) with aminopolycarboxylic acids in aqueous solution. Int J Appl Radiat Isot 33:1391–1398

Sharp PF, Smith FW, Gemmel HG, Lyall D, Evans NTS, Gvozdanovic D, Davidson J, Tyrrell DA, Pickett RD, Neirinckx RD (1986) Technetium-99m HM-PAO stereoisomers as potential agents for imaging regional cerebral blood flow: human volunteer studies. J Nucl Med 27:171–177

Singh PR, Ketring AR, Volkert WA, Katti KV (1996) Potential of phosphinimines and phosphinimine-containing polymers as scavenging agents for the extraction of $^{99}TcO_4^-$ from aqueous media. In: Bandoli G, Mazzi U, Nicolini M, SG Editoriali (eds) Technetium and rhenium in chemistry and nuclear medicine 4. Padova, Italy, pp 239–242

Spies H, Johannsen B, Muenze R (1980) Kinetics investigations on the reaction of technetium(V) - gluconate with meso-dimercaptosuccinic acid and meso-dimercapto-succinic acid dimethylester. Radiochemical Radioanal Lett 43:311–318

Steigman J, Eckelman WC (1992) The chemistry of technetium in medicine. National Academy Press, Washington, DC

Steigman J, Meinken G, Richards P (1975) Reduction of pertechnetate-99 by stannous chloride. I. Stoichiometry of the reaction in hydrochloric acid, in a citrate buffer, and in a DTPA buffer. Int J Appl Radiat Isot 26:601–609

Stern HS, McAfee JG, Subramanian G (1966) Preparation, distribution and utilization of technetium-99m-sulfur colloid. J. Nucl Med 7:655–675

Stern HS, Zolle I, McAfee JG (1965) Preparation of ^{99m}Tc-labeled serum albumin. Int J Appl Radiat Isot 16:283–288

Subramanian G, McAfee JG, Blair RG, Kallfelz FA, Thomas FD (1975) Technetium-99m-methylene-diphosphonate – a superior agent for skeletal imaging: comparison with other technetium complexes. J Nucl Med 16:744–755
Tofe AJ, Bevan JA, Fawzi MB, Francis MD, Silberstein EB, Alexander GA, Gunderson DE, Blair K (1980) Gentisic acid: a new stabilizer for low tin skeletal imaging agents: concise communication. J Nucl Med 21:366–370
Tofe AJ, Francis MD (1976) In vitro stabilization of a low tin bone imaging kit. J Nucl Med 16:414–422
Treher EN, Francesconi LC, Gougoutas JZ, Malley M, Nunn A (1989) Mono-capped tris dioxime complexes of technetium (III): synthesis and structural characterization of TCX (dioxime)$_3$, B–R (X=Cl, Br; dioximedimethylglyoxime cyclohexanedioxime; R=CH_3, C_4H_9). Inorg Chem 28:3411–3416
Troutner DE, Volkert WA, Hoffman TJ, Holmes RA (1984) A neutral lipophilic complex of ^{99m}Tc with a multidentate amine oxime. Int J Appl Radiat Isot 35:467–470
Tweedle MF (1983) Accelerators for forming cationic technetium complexes useful as diagnostic agents. Int. Patent Appl. PCT 83,02,615
Van den Brand JAGM, Das HA, Dekker B, De Ligny CL (1981) The gel chromatographic separation and identification of the Tc(Sn) HEDP complexes using the radiotracers ^{32}P, ^{99m}Tc, ^{18}Sn. Int J Appl Radiat Isot 32:637
Verbruggen A, Bormans G, Cleynhens B, Hoogmartens M, Vandecruys A, De Roo M (1989) Separation of the enantiomers of technetium-99m-MAG$_3$ and their renal excretion in baboons and a volunteer. Nuklearmedizin 25:436–439
Verbruggen A, Bormans G, Van Nerom C, Cleynhens B, Crombez D, De Roo M (1989) Isolation of the mono-ester mono-acid derivatives of ^{99m}Tc-ECD and their biodistribution in mice. In: Nicolini M, Bandoli G, Mazzi U (eds) Technetium and rhenium in chemistry and nuclear medicine 3. Cortina International, Verona, Italy, pp 445–452
Wackers FJTh, Berman DS, Maddahi J, Watson DD, Beller GA, Strauss HW, Boucher CA, Picard M, Holman BL, Fridrich R, Inglese E, Delaloye B, Bischof-Delaloye A, Camin L, McKusick K (1989) Technetium-99m hexakis 2-methoxyisobutyl isonitrile: human biodistribution, dosimetry, safety, and preliminary comparison to thallium-201 for myocardial perfusion imaging. J Nucl Med 30:301–311
Watson AD, Tulip TH, Roe DC (1987) The synthesis, characterization and multinuclear NMR studies of a technetium bisaminebisthiol complex: a new radiopharmaceutical precursor. In: Nicolini M, Bandoli G, Mazzi U (eds) Technetium in chemistry and nuclear medicine 2. Cortina International, Verona, Italy, 61–64
Yokoyama A, Hata N, Horiuchi K, Matsuda H, Saji H, Ohta H et al (1985) The design of a pentavalent ^{99m}Tc-dimercaptosuccinate complex as a tumor imaging agent. J Nucl Med 12:273–279
Yokoyama A, Horiuchi K, Hata N et al (1979) Technetium in technetium-99m radiopharmaceuticals. I. Tetravalent mononuclear technetium penicillamine complex. J Labeled Compd Radiopharm 16:80–81
Zuckman SA, Freeman GM, Troutner DE, Volkert WA, Holmes RA, Van Derveer DG, Barefield EK (1981) Preparation and X-ray structure of *trans*-dioxo(1,4,8,11-tetraazacyclotetradecane)technetium(V) perchlorate hydrate. Inorg Chem 20:2386–2389

Further Reading

The following is a list of proceedings of the International Symposia on Technetium in Chemistry and Nuclear Medicine, Academia Cusanus, Bressanone (Italy).

Deutsch E, Nicolini M, Wagner HN Jr, (1983) Technetium in chemistry and nuclear medicine 1. Cortina International, Verona, Italy
Nicolini M, Bandoli G, Mazzi U (1986) Technetium in chemistry and nuclear medicine 2. Cortina International, Verona, Italy
Nicolini M, Bandoli G, Mazzi U (1990) Technetium and rhenium in chemistry and nuclear medicine 3. Cortina International, Verona, Italy
Nicolini M, Bandoli G, Mazzi U (1996) Technetium and rhenium in chemistry and nuclear medicine 4. SG Editoriali, Padova, Italy
Nicolini M, Bandoli G, Mazzi U (1999) Technetium, rhenium and other metals in chemistry and nuclear medicine 5. SG Editoriali, Padova, Italy
Nicolini M, Mazzi U (2002) Technetium, rhenium and other metals in chemistry and nuclear medicine 6. SG Editoriali, Padova, Italy

2.2 The Technetium and Rhenium Tricarbonyl Core

R. Schibli

An aqua ion of technetium and rhenium, in analogy to, e.g. $[Cu(OH_2)_6]^{2+}$, would be most convenient for radiolabeling procedures. However, such an aqua ion presumably does not exist or is very unstable. In an effort to solve this dilemma, Alberto and coworkers have designed and developed an organometallic semiaqua ion of the general formula $[M(OH_2)_3(CO)_3]^+$ (M Tc, Re), useful as precursor for the radiolabeling of biomolecules for diagnostic and therapeutic purposes (Alberto et al. 1995, 1998, 1999; Egli et al. 1997). The metal centers are in the oxidation state +1 with a low-spin d^6 electronic configuration. The high electron density is stabilized by three strong π-acceptors (CO) facially arranged. The precursors are water stable and water soluble, and the water molecules readily undergo ligand exchange, whereas the carbonyl ligands are substitution stable.

The precursor $[M(OH_2)_3(CO)_3]^+$ is readily accessible directly from the corresponding sodium permetallate, $Na[MO_4]$ (Alberto et al. 1998; Schibli et al. 2002). The preparation comprises a six-electron reduction and concomitant coordination of three COs. The basis of Tc-99m kit formulation is disodium boronocarbonate (BC), $Na_2[H_3BCO_2]$, which serves as an in situ CO source and at the same time reduces the technetium center (Alberto et al. 2001). The kit is nowadays commercially available under the name Isolink (Mallinckrodt-Tyco Med) for research purposes. BC is stable in aqueous solution and can be lyophilized. The precursor $[^{99m}Tc(OH_2)_3(CO)_3]^+$ can be synthesized in quantitative yield by adding generator eluate to the vial and subsequent heating to 100 °C for 20 min (Alberto et al. 2001). The preparation of the rhenium homologue $[^{188}Re(OH_2)_3(CO)_3]^+$ deviated slightly from technetium, since rhenium is more difficult to reduce and reacts, in general, much slower. Therefore, $H_3B{\cdot}NH_3$ is the reducing agent (eventually in combination with another polymer bound reducing agent), and the reaction must be carried out in the presence of H_3PO_4 at acidic pH (Park et al. 2006; Schibli et al. 2002). This formulation presently excludes an instant kit formulation for $[^{188}Re(OH_2)_3(CO)_3]^+$.

As mentioned above, $[M(OH_2)_3(CO)_3]^+$ can be considered as a normal aqua ion with only three available coordination sites. Substitution of the water molecules with almost any type of chelator (classic/nonclassic) forms kinetically stable coordination compounds. This holds true for mono-, bi-, and tridentate ligands, regardless of their hardness or softness. This behavior also represents the distinct feature of $[M(OH_2)_3(CO)_3]^+$ as compared with other technetium and rhenium metal centers, and is one of its major advantages for the labeling of molecules for imaging and therapeutic purposes.

An enormous variety of mono-, bi-, and tridentate ligand systems comprising different donor atoms or groups have been developed and are still designed and optimized. Bifunctional chelating agents (BFCA) have been readily developed specifically for the purpose of functionalization of biomolecules and subsequent radiolabeling with the $M(CO)_3$ core (Alberto et al. 2004; Alves et al. 2005; Banerjee et al. 2002, 2005b; Correia et al. 2001; Garcia et al. 2000, 2002; He et al. 2005; Karagiorgou et al. 2005; Lazarova et al. 2005; Lipowska et al. 2004; Mandal et al. 1998; Mundwiler et al. 2005; Schibli et al. 2002; Stephenson et al. 2003; Stichelberger et al. 2003; van Staveren et al. 2004, 2005; Fig. 2.2.1). Bidentate ligands have proven to be very fast coordinating entities, in particular if they are of anionic nature. Tridentate ligands do not display significantly higher thermodynamic stability than bidentate chelates do. However, their reaction rate is

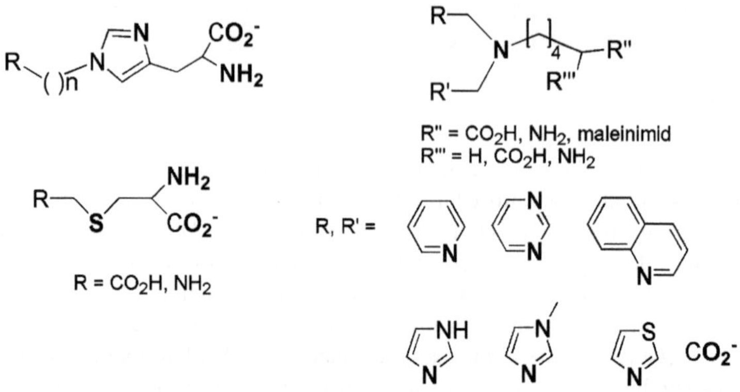

Fig. 2.2.1. Various ligand bifunctional, tridentate chelating systems designed for the coupling to biomolecules and subsequent radiolabeling with the $[M(OH_2)_3(CO)_3]^+$ (coordinative atoms in *boldface*). Single amino acid chelates (SAAC): *R″* CO_2H, *R‴* NH_2

much faster, which becomes the decisive point for radiopharmaceutical application. In that respect, tridentate ligands are favored. In addition, tridentate ligands shield the organometallic metal center from, e.g., in vivo-observed crossreactivity with serum proteins, as observed in the case for complexes of the general formula $[M(OH_2)(L^2)(CO)_3]$ (L^2 bidentate chelate) (Schibli et al. 2000). It has been observed that ^{99m}Tc-tricarbonyl complexes, which are coordinated with a tridentate chelating system, reveal good stability when challenged in human plasma and with excess cysteine, histidine, or glutathione. These complexes show also very good clearance form the blood pool and all tissue and organs when tested in BALB/c mice. In contrast, complexes, which are coordinated in a bidentate fashion, show significant aggregation with plasma proteins in vitro and in vivo. They are significantly retained in the blood and in the organs of excretion such as the liver and the kidneys. These differences may be related to the susceptibility of the third, nonchelating coligand (H_2O) to exchange with more reactive functional groups in vivo, allowing the ^{99m}Tc to be retained in tissues (Pietzsch et al. 2000; Schibli et al. 1999). For both reasons mentioned above, many groups are focusing on the development of novel, potent, tridentate chelates and tridentate BFCAs tailor-made for the $M(CO)_3$ fragment.

Egli et al. (1999) have investigated the ability of amino acids and amino acid fragments to react with the ^{99m}Tc-tricarbonyl core. The most important finding was that histidine reacts quantitatively with the organometallic precursor at very low concentrations (10^{-6} M). In an effort to create novel, bifunctional analogues of histidine, Alberto et al. recently derivatized histidine by introducing various functional groups at the ε-N of the imidazole ring (Alberto et al. 2004; van Staveren et al. 2004). Attachment of these histidine derivatives to the C or N terminus of peptides is an elegant approach, and at the same time liberates both the *α*-amino group and carboxyl group to participate in tridentate chelation along with the *δ*-N of the imidazole ring. A similar strategy was applied for *S*-functionalized cysteine BFCAs (van Staveren et al. 2005). Although the amino acid cysteine (and methionine) per se was found to be a rather "slow" coordinating ligand, the situation changed significantly if the sulfur group of cysteine was functionalized.

Valliant and Zubieta developed BFCAs for the $Tc(CO)_3$ core, based on a lysine backbone comprising pyridyl, imidazole, thiolate, carboxylate groups, etc., for the specific purpose of conjugation to small peptides by solid-phase synthetic methods (Banerjee et al. 2002, 2004, 2005 a, b; Stephenson et al. 2003, 2004; Wei et al. 2005). A whole library of such single amino acid chelates (SAAC) derivatives of lysine has been prepared and readily conjugated to small peptides.

The organometallic nature of the $M(CO)_3$ core also allows the introduction and combination with other nonclassic organometallic ligands, such as cyclopentadienes (cp) or cyclopentadienyls (cp^-). Cp^- is one of the smallest ligands with a low molecular weight, but is able to occupy three coordination sites. Complexes of the cymantrene type $[CpM(CO)_3]$ (M Tc, Re) are stable in physiological media (Wenzel 1992; Wenzel et al. 1993, 1994). Cps can also be further derivatized with, e.g., an acetyl group (Bernard et al. 2003) (Fig. 2.2.2). The acetyl group can act as an anchoring group for biomolecules, giving rise to cp biomolecule conjugates. Reaction of such cp derivatives with $[^{99m}Tc(OH_2)_3(CO)_3]^+$ in aqueous media formed the corresponding radiolabeled conjugates in high yields, but at relatively high ligand concentrations (10^{-4} to 10^{-3} M) (Bernard et al. 2003).

Valliant's group has recently built an interesting link between boron neutron capture therapy and diagnostic radiopharmacy (Fig. 2.2.2). The carborane 3-isocyano-1,2-dicarba-*closo*-dodecaborane and functionalized derivatives thereof react with $[^{99m}Tc(OH_2)_3(CO)_3]^+$ under basic conditions quantitatively (Sogbein et al. 2004, 2005 a, b). If the carboranes were coupled to targeting biomolecules, this approach would secure first the site-specific delivery of high quantities of boron atoms and second to quantify and to visualize the distribution of boron conjugates, a task that is not easily verified with nonradiolabeled boron compounds. Both types of nonclassic ligand systems, cps, and carboranes became only useful for radiopharmacy because of their reactivity with the novel synthon $[M(OH_2)_3(CO)_3]^+$.

The tricarbonyl technology not only provides new opportunities with respect to the use of "exotic" ligand systems, but it also opens perspectives for novel labeling strategies.

The group at the Paul Scherrer Institute and Plückthun et al. have successfully developed a direct labeling protocol employing $[^{99m}Tc(OH_2)_3(CO)_3]^+$ for scFvs and "mini-antibodies" (bi- and trivalent constructs of scFv) carrying an N- or a C-terminal His-tag (Deyev et al. 2003; Waibel et al. 2000; Willuda et al. 1999, 2001). The method is particu-

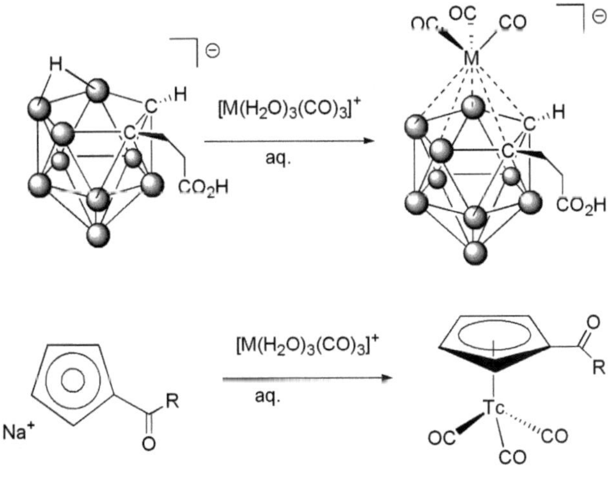

Fig. 2.2.2. Aqueous-base preparation of $M(CO)_3$ complexes comprising nonclassic ligand systems such as functionalized carboranes and cyclopentadienyls

larly elegant and versatile, because His-tags are frequently genetically expressed for ease of purification of the protein on a nickel affinity column. This His-tag can be considered as a multidentate ligand, since two or more imidazoles from histidine can coordinate the metal centre (Fig. 2.2.3). Mixing of such an His-tag protein with $[^{99m}Tc(OH_2)_3(CO)_3]^+$ in buffer at 37 °C for 15 min resulted in >90 % stable and specific incorporation of the total activity. This gentle procedure allows for the first time the radiolabeling of recombinant proteins "from the shelf", thus, without any chemical modification of the protein structure. The procedure is convenient for the quick and noninvasive evaluation of targeting proteins.

An approach that is particularly interesting for radiolabeling of receptor-targeting radiopharmaceuticals with high specific activity is based on a peculiarity of functionalized aliphatic amines. It was observed that ternary amines involved in the coordination of the $^{99m}Tc(CO)_3$ fragment are cleaved from a solid-phase support during the labeling reaction. It could be shown that metal-assisted cleavage allows the preparation of essentially carrier-free complexes or bioconjugates (Mundwiler et al. 2004). Cleavage occurs exclusively with technetium but not with rhenium. Typical yields of these processes varied between 10 and 50%, relative to the total activity of ^{99m}Tc (Fig. 2.2.4).

Mixed-ligand approaches are well documented for Tc and Re in higher oxidation states. The $M(CO)_3$ fragment allows a similar possibility. As mentioned earlier, the water ligand in complexes of the type $[M(OH_2)(L^2)(CO)_3]$ (L^2 bidentate chelate) is loosely bound and can be exchanged by a potent monodentate ligand (L^1), forming complexes of the general formula $[M(L^1)(L^2)(CO)_3]$. Either L^1 or L^2 can be readily coupled to biomolecules (Mundwiler et al. 2004; Fig. 2.2.5). Several problems have been addressed with this "2+1 approach": (1) the functionalization of biomolecules is minimized or simplified, (2) the resulting complexes/bioconjugates are coordinatively saturated, and (3) further flexibility with respect to the fine-tuning of the physicochemical properties of the radiopharmaceutical is possible.

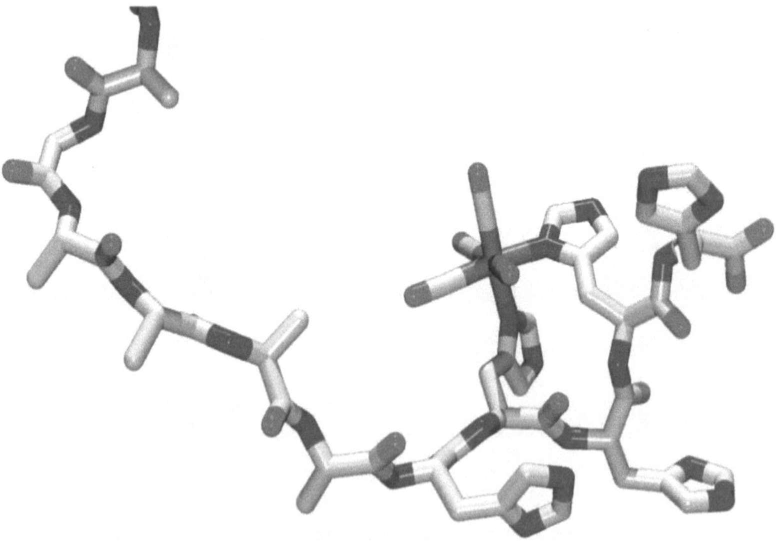

Fig. 2.2.3. Model of the potential coordination of the $M(CO)_3$ core to His_5-tag of a recombinantly produced protein. *Purple* technetium, *bright blue* nitrogen, *red* oxygen, *gray* carbon

Zubieta and coworkers have recently taken advantage of the fluorescent and luminescent properties of organometallic complexes of rhenium (and technetium) comprising certain aromatic ligand systems (Fig. 2.2.6). The nonradioactive $^{nat}Re(CO)_3$ bioconjugates with a formyl peptide receptor-targeting peptide (fMLF), enabeled visualization

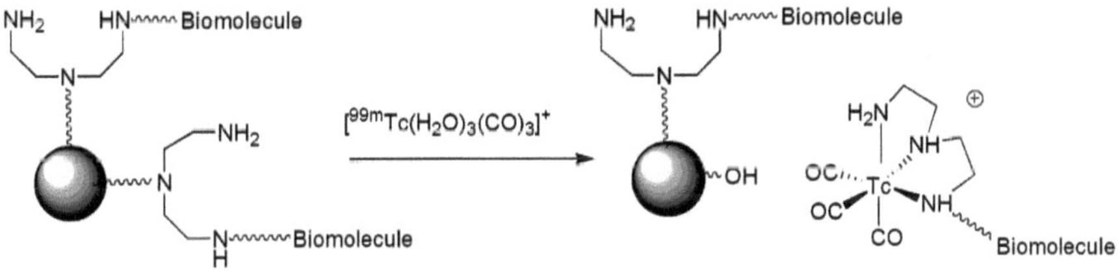

Fig. 2.2.4. $Tc(CO)_3$-assisted cleavage of solid-phase bound biomolecules functionalized with an aliphatic triamine chelate, leading to high specific activity of radiotracer

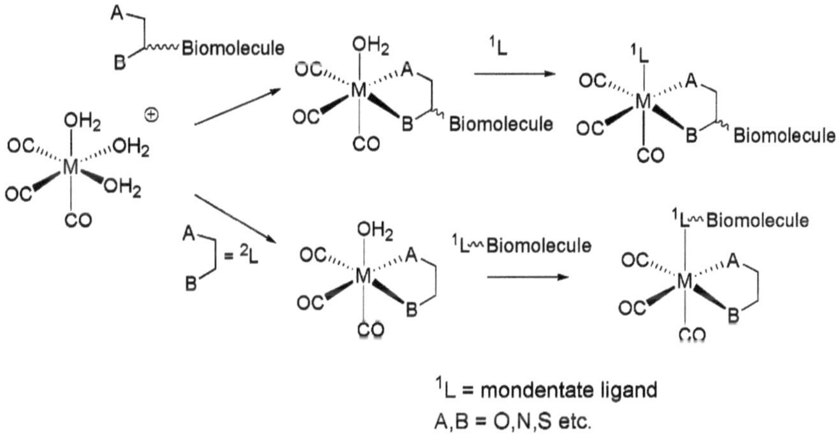

Fig. 2.2.5. Schematic drawing of mixed-ligand approaches using combination of mono- and bidentate ligand systems coupled to biomolecules

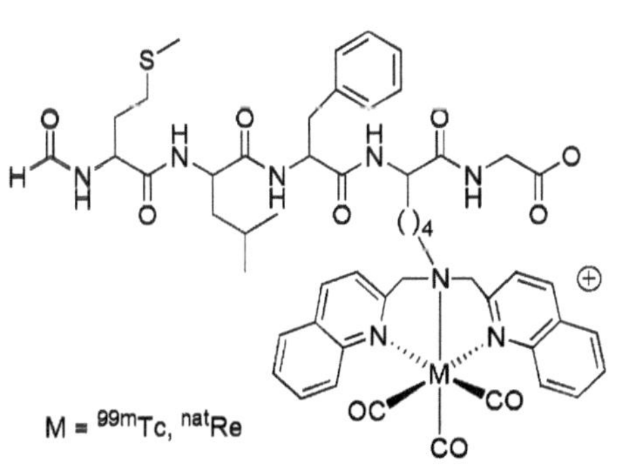

Fig. 2.2.6. Formyl peptide receptor-targeting peptide (fMLF)[(SAACQ-M(CO)$_3$)$^+$] conjugate useful for in vitro fluorescent microscopy (where M = ^{nat}Re) and in vivo single-photon emission computer tomography (SPECT) (where M = ^{99m}Tc) with isostructural technetium and rhenium complexes

of receptor targeting on the cellular level by means of fluorescent microscopy (Stephenson et al. 2004). The isostructural bioconjugate in the radiolabeled form (with the $^{99m}Tc(CO)_3$ core) allowed the noninvasive detection of corresponding cancer sites in vivo via single-photon emission tomography. Hence, the tricarbonyl technology allows bridging the intrinsic gap between in vitro and in vivo imaging.

2.2.1 Bioconjugates Comprising the $M(CO)_3$ Core

The number of technetium and rhenium tricarbonyl compounds in preclinical evaluation is remarkable. These efforts comprise small molecules as well as macromolecules useful in diagnostic and/or therapeutic nuclear medicine. There are also clinical data available with tumor affine peptides such as neurotensin receptor- and somatostatin receptor-targeting peptides radiolabeled with the $^{99m}Tc(CO)_3$ core.

Dopamine transporter ligand DAT and the 5-HT_{1A} serotonergic receptor ligand WAY100635 have been, and are still, subjects of intense investigation in conjunction with the carbonyl labeling technology (Fig. 2.2.7). WAY100635, has been functionalized with cyclopentadiene and bidentate Schiff-base chelates (Alberto et al. 1999; Arterburn et al. 2003; Bernard et al. 2003; Bigott et al. 2005). The conjugates revealed an IC_{50} value in the low-nanomolar range toward the 5-HT_{1A} receptor. For the preparation of the cp-arylpiperazine derivative, a one-pot, single-step synthesis was described (yields >95%), starting directly from aqueous $[^{99m}TcO_4]^-$, applying the strategy illustrated in Fig. 2.2.2 (Wald et al. 2001). In vitro the receptor affinity and the selectivity of the organometallic derivatives were preserved. However, in vivo the compounds displayed insufficient brain uptake.

Metal carbonyl complexes of steroids have been synthesized by the groups of Johannsen and Katzenellenbogen (Arterburn et al. 2003; Bigott et al. 2005; Luyt et al. 2003; Wust et al. 1998, 1999). Various 17*β*-progesterone and 7*α*-estradiole dithioether and cyclopentadiene complexes of technetium/rhenium(I) tricarbonyl have been pre-

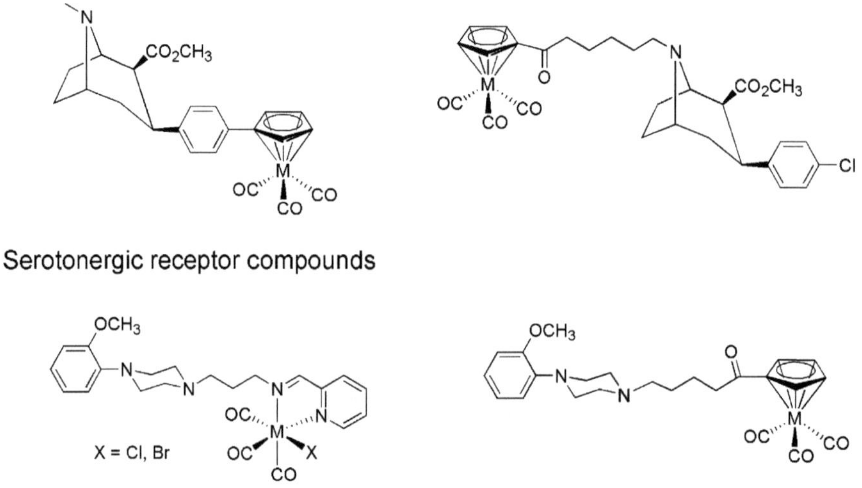

Fig. 2.2.7. Examples of organometallic Tc-99m central nervous system (CNS) complexes

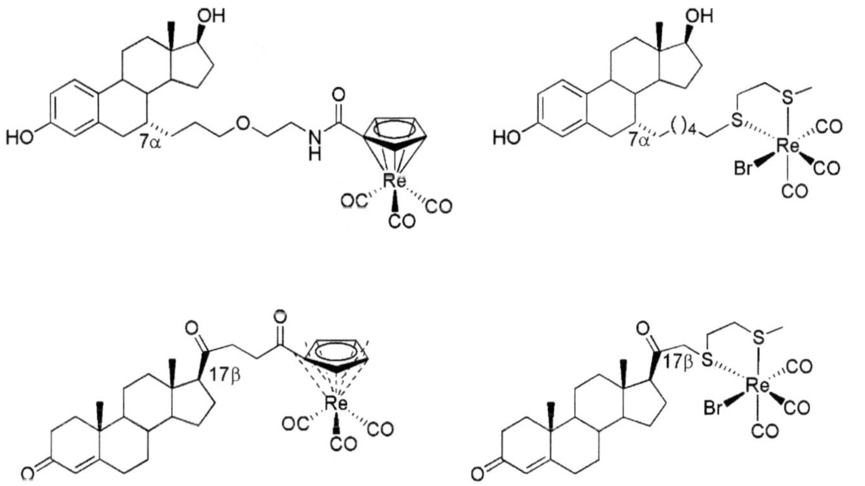

Fig. 2.2.8. Structure of various organometallic steroids for potential radiodiagnostic and radiotherapeutic targeting of progesterone receptor (PR)- and estrogen receptor (ER)-positive cancer

pared and tested (Fig. 2.2.8). The relative binding affinity (RBA) was found to depend on the nature of the spacer between the metal chelate and the steroid moiety. Similar observation and tendencies have been reported for the progestin complexes (Wust et al. 1999). For both examples, the organometallic cyclopentadienyl–tricarbonyl systems were superior to the dithioether-tricarbonyl in terms of RBA for the corresponding receptors. Synthesis and biodistribution studies of the corresponding Tc-99m and even Tc-94m analogues have been performed that suggested limited usefulness of these systems as effective imaging agents for progesterone receptor (PR)- and estrogen receptor (ER)-positive breast cancer.

Schibli and coworkers and other groups have recently published organometallic folate derivatives for targeting *α*-folate receptor over expressing cancer cells (Müller et al. 2004). Preclinical in vivo single-photon emission computer tomography (SPECT)/CT studies in tumor-bearing mice have revealed almost identical pharmacokinetics for both Tc-99m and the homologous Re-188 folate (Fig. 2.2.9). Based on these in vivo results and results of other organometallic Tc-99m/Re-188-labeled biomolecules (*vide infra*), it is reasonable to propose that for the tricarbonyl technology, the concept of the "matched pair" Tc/Re is indeed valid in various aspects.

Alberto and coworkers have demonstrated that vitamin B_{12}, essential for tumor growth, can be functionalized at several positions and radiolabeled with a $Tc(CO)_3$ core (Kunze et al. 2004; van Staveren et al. 2004). The in vivo assessment of several promising derivatives is currently under investigation.

The most thoroughly studied class of biomolecules that was tested with the tricarbonyl technology was the tumor affine peptides. Peptides reveal biological and pharmacological characteristics (e.g., biological half-life), which are very suited for the imaging and therapy with Tc-99m or Re-188. In fact, peptides have been among the first examples for the efficient labeling with the $M(CO)_3$ core. A number of other peptides have been studied in detail, such as neurotensin, bombesin, octreotide, annexin (Biechlin et al. 2005; Tait et al. 2002), and neuropeptide Y. Neurotensin and stabilized derivatives thereof were derivatized with histidine, either through an amide bond to the carboxylic

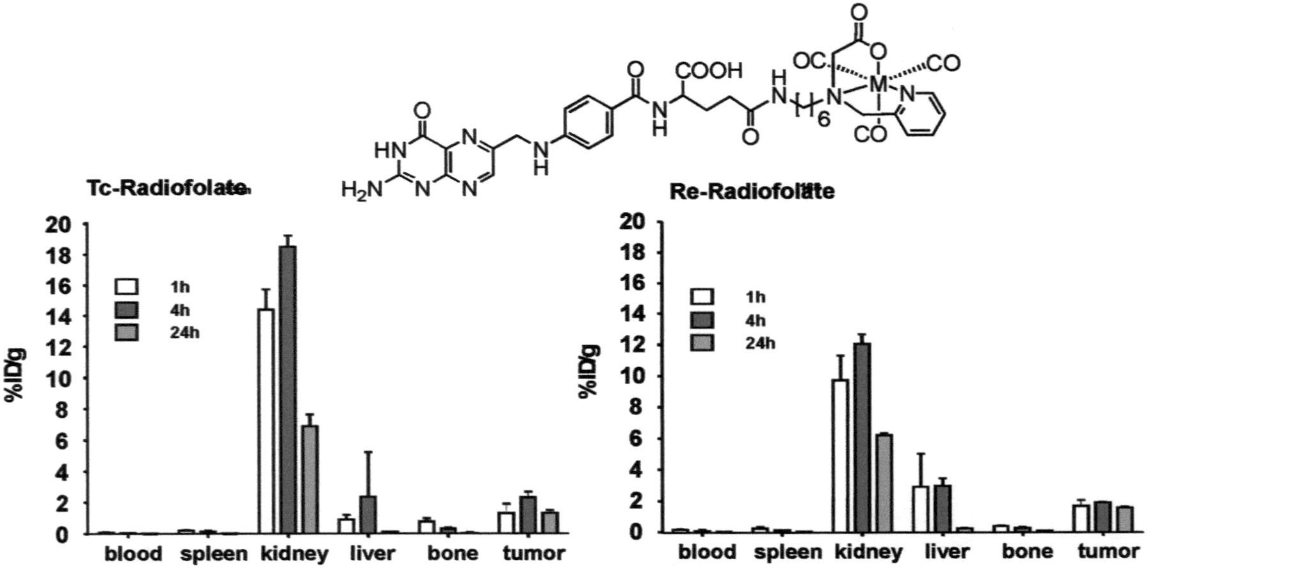

Fig. 2.2.9. Time-dependent biodistribution of novel $^{99m}Tc/^{188}Re$ organometallic folate derivatives in nude mice bearing folate receptor overexpressing KB-cell xenografts

acid to produce a bidentate NN chelator or through alkylation at the N-amino group in order to retain the tripodal coordinating feature (Blauenstein et al. 2004; Bruehlmeier et al. 2002; Egli et al. 1999; Garcia-Garayoa et al. 2001, 2002; Waibel et al. 2000). Biodistribution studies with Tc-99m showed that tridentate ligands are superior to bidentate ones, which is in agreement with the findings and preferences mentioned previously. A phase I clinical study is ongoing with ^{99m}Tc-labeled neurotensin derivatives. Neurotensin analogues with improved pharmacological profiles are currently employed in preclinical therapy studies with $^{188}Re(CO)_3$.

^{0}Tyr^{3}octreotate analogues functionalized with various BFCA have been tested (Marmion et al. 1999). The BFCA gave rise to complexes of different overall charge (+1 to −3). Wester et al. have coupled picoline aminoacetic acid to a carbohydrated octreotide. The carbohydrate makes the conjugate much more hydrophilic, and an excellent biodistribution in humans was observed (Wester 2003; Wester et al. 2001).

Bombesin was derivatized at the C terminus with bidentate chelators (Smith et al. 2003 a, b). The labeled peptide fully retained the biological activity and was stable in vitro and in vivo. Since the bidentate coordination is not optimal with respect to stability (pharmacokinetics), the coordination sphere of the metal tricarbonyl core has been saturated with a highly hydrophilic phosphine. This additional coordination is an example of the 2+1 approach mentioned in the previous section. The mixed-ligand approach resulted in significantly higher hydrophilicity of the radioconjugates and an improved biodistribution labeled with Tc-99m and also with Re-188 (Smith et al. 2003).

In the case of other receptor avid peptides and proteins, which express an endogenous histidine such as, e.g., bombesin or neuropeptide Y, the pronounced avidity of the tricarbonyl core for histidine can create a problem with unspecific binding (Langer et al. 2001; La Bella et al. 2002 a, b). Prelabeling procedures can circumvent these problems (Langer et al. 2001). However, Garcia et al. could show that a site-specific postlabeling of bombesin is possible by introduction of a potent tridentate ligand such as, e.g., the N_α-Ac-histidine at the N terminus of the peptide (La Bella et al. 2002). As a result, a single, stable species was formed, and unspecific labeling was negligible.

The high efficiency combined with the mild reaction conditions applicable with $[M(OH_2)_3(CO)_3]^+$ is very attractive for radiolabeling of sensitive proteins. This has been recognized by several groups. MUC1 mucin is upregulated and abnormally glycosylated in bladder cancer, and is a promising target for intravesical radioimmunotherapy. The in vivo results in tumor mice have clearly revealed a better retention of immunoreactivity of the $^{188}Re(CO)_3$-labeled monoclonal antibody (mAb) as compared with the 2-mercaptoethanol-reduced and Re(V) labeled mAb (Murray et al. 2001). The surfactant protein B was nonspecifically labeled with $[^{99m}Tc(OH_2)_3(CO)_3]^+$. The highly lipophilic protein has potential in the diagnosis of acute respiratory disease syndrome (Amann et al. 2001). Waibel et al. and Deyev et al. have pioneered the use of site-specific labeling of recombinant proteins via a multi–His-tag (Willuda et al. 2001). The ease of radiolabeling is remarkable and unmet with any other technetium methodology (Deyev et al. 2003).

In conclusion, it is apparent that organometallic compounds are a valuable and realistic alternative for the labeling of biomolecules in, e.g., radiopharmacy. The encouraging results of preclinical and clinical studies with organometallic-labeled tumor affine peptides and vitamins build the scaffold for further investigations. The tricarbonyl technology is the creative precedent of such novel techniques. However, the perspective and potential of organometallic labeling techniques in nuclear medicine will also depend on the success of new compounds for therapeutic use and the availability of appropriate radionuclides. In the future, chemists and radiopharmacists will be equally challenged to exploit the aqueous organometallic chemistry of potential radionuclides to develop novel techniques and compounds for diagnostic and therapeutic application.

References

Alberto R, Schibli R, Egli A, Schubiger PA, Herrmann WA, Artus G, Abram U, Kaden TA (1995) Metal carbonyl syntheses XXII. Low-pressure carbonylation of $[MOCl_4]^-$ and $[MO_4]^-$: the technetium(I) and rhenium(I) complexes $[NEt_4]_2[MCl_3(CO)_3]$. J Organomet Chem 493:119–127

Alberto R, Schibli R, Egli A, Schubiger AP, Abram U, Kaden TA (1998) A novel organometallic aqua complex of technetium for the labeling of biomolecules: synthesis of $[^{99m}Tc(OH_2)_3(CO)_3]^+$ from $(TcO_4)^-$ in aqueous solution and its reaction with a bifunctional ligand. J Am Chem Soc 120:7987–7988

Alberto R, Schibli R, Schubiger AP, Abram U, Pietzsch HJ, Johannsen B (1999) First application of *fac*-$[^{99m}Tc(OH_2)_3(CO)_3]^+$ in bioorganometallic chemistry: design, structure, and in vitro affinity of a 5-HT1A receptor ligand labeled with Tc-99m. J Am Chem Soc 121:6076–6077

Alberto R, Schibli R, Waibel R, Abram U, Schubiger AP (1999) Basic aqueous chemistry of $M(OH_2)_3(CO)_3^+$ (M Re, Tc) directed towards radiopharmaceutical application. Coord Chem Rev 192:901–919

Alberto R, Ortner K, Wheatley N, Schibli R, Schubiger AP (2001) Synthesis and properties of boranocarbonate: A convenient in situ CO source for the aqueous preparation of $[^{99m}Tc(OH_2)_3(CO)_3]^+$. J Am Chem Soc 123:3135–3136

Alberto R, Pak JK, van Staveren D, Mundwiler S, Benny P (2004) Mono-, bi-, or tridentate ligands? The labeling of peptides with Tc-99m-carbonyls. Biopolymers 76:324–333

Alves S, Paulo A, Correia JDG, Gano L, Smith CJ, Hoffman TJ, Santos I (2005) Pyrazolyl derivatives as bifunctional chelators for labeling tumour-seeking peptides with the *fac*-$M(CO)_3^+$ moiety (M = Tc-99m, Re): synthesis, characterization, and biological behavior. Bioconjugate Chem 16:438–449

Amann A, Decristoforo C, Ott I, Wenger M, Bader D, Alberto R, Putz G (2001) Surfactant protein B labelled with $^{99m}Tc(CO)_3(H2O)_3^+$ retains biological activity in vitro. Nucl. Med. Biol. 28:243–250

Arterburn JB, Corona C, Rao KV, Carlson KE, Katzenellenbogen JA (2003) Synthesis of 17-alpha-substituted estradiol-pyridin-2-yl hydrazine conjugates as effective ligands for labeling with Alberto's complex *fac*-$Re(OH_2)_3(CO)_3^+$ in water. J Org Chem 68:7063–7070

Banerjee SR, Levadala MK, Lazarova N, Wei LH, Valliant JF, Stephenson KA, Babich JW, Maresca KP, Zubieta J (2002) Bifunctional single amino acid chelates for labeling of biomolecules with the $[Tc(CO)_3]^+$ and $[Re(CO)_3]^+$ cores. Crystal and molecular structures of $ReBr(CO)_3(H_2NCH_2C_5H_4N)$, $Re(CO)_3[(C_5H_4NCH_2)_2NH]$ Br, $Re(CO)_3[(C_5H_4NCH_2)_2NCH_2CO_2H]$ Br, $Re(CO)_3[X(Y)NCH_2CO_2CH_2CH_3]$ Br (XY=2-pyridylmethyl; X=2-pyridyimethyl, Y=2-(1-methylimidazolyl)methyl; XY=2-(1-methylimidazolyl)methyl), $ReBr(CO)3[(C5H_4NCH_2)NH(CH_2C_4H_3S)]$, and $Re(CO)_3[(C_5H_4NCH_2)N(CH_2C_4H_3S)(CH_2CO_2)]$. Inorg Chem 41:6417–6425

Banerjee SR, Babich JW, Zubieta J (2004) Bifunctional chelates with aliphatic amine donors for labeling of biomolecules with the $[Tc(CO)_3]^+$ and $[Re(CO)_3]^+$ cores: the crystal and molecular structure of $Re(CO)_3[(H_2NCH_2CH_2)_2N(CH_2)_4CO_2Me]$. Br Inorg Chem Commun 7:481–484

Banerjee SR, Babich JW, Zubieta J (2005a) Site directed maleimide bifunctional chelators for the $M(CO)_3^+$ core (M=Tc-99m, Re). Chemical Commun 13:1784–1786

Banerjee SR, Schaffer P, Babich JW, Valliant JF, Zubieta J (2005b) Design and synthesis of site directed maleimide bifunctional chelators for technetium and rhenium. Dalton Transactions 3886–3897

Bernard J, Ortner K, Spingler B, Pietzsch HJ, Alberto R (2003) Aqueous synthesis of derivatized cyclopentadienyl complexes of technetium and rhenium directed toward radiopharmaceutical application. Inorg Chem 42:1014–1022

Biechlin ML, d'Hardemare AD, Fraysse M, Gilly FN, Bonmartin A (2005) Improvement in radiolabelling proteins with the Tc-99m-tricarbonyl-core $^{99m}Tc(CO)_3^+$, by thiol-derivatization with iminothiolane: application to gamma-globulins and annexin V. J Labelled Compd Radiopharm 48:873–885

Bigott HM, Parent E, Luyt LG, Katzenellenbogen JA, Welch MJ (2005) Design and synthesis of functionalized cyclopentadienyl tricarbonylmetal complexes for technetium-94m PET imaging of estrogen receptors. Bioconjugate Chem 16:255–264

Blauenstein P, Garayoa EG, Ruegg D, Blanc A, Tourwe D, Beck-Sickinger A, Schubiger PA (2004) Improving the tumor uptake of Tc-99m-labeled neuropeptides using stabilized peptide analogues. Cancer Biother Radiopharm19:181–188

Bruehlmeier M, Garayoa EG, Blanc A, Holzer B, Gergely S, Tourwe D, Schubiger PA, Blauenstein P (2002) Stabilization of neurotensin analogues: effect on peptide catabolism, biodistribution and tumor binding. Nucl Med Biol 29:321–327

Correia JDG, Domingos A, Santos I, Alberto R, Ortner K (2001) Re tricarbonyl complexes with ligands containing P,N,N and P,N,O donor atom sets: synthesis and structural characterization. Inorg Chem 40:5147–5151

Deyev SM, Waibel R, Lebedenko EN, Schubiger AP, Pluckthun A (2003) Design of multivalent complexes using the barnase-barstar module. Nat Biotechnol 21:1486–1492

Egli A, Hegetschweiler K, Alberto R, Abram U, Schibli R, Hedinger R, Gramlich V, Kissner R, Schubiger PA (1997) Hydrolysis of the organometallic aqua ion *fac*-triaquatricarbonylrhenium(I). Mechanism, pKa, and formation constants of the polynuclear hydrolysis products. Organometallics 16:1833–1840

Egli A, Alberto R, Tannahill L, Schibli R, Abram U, Schaffland A, Waibel R, Tourwe D, Jeannin L, Iterbeke K, Schubiger PA (1999) Organometallic Tc-99m-aquaion labels peptide to an unprecedented high specific activity. J Nuc Med 40:1913–1917

Garcia R, Paulo A, Domingos A, Santos I, Ortner K, Alberto R (2000) Re and Tc complexes containing B–H·M agostic interactions as building blocks for the design of radiopharmaceuticals. J Am Chem Soc 122:11240–11241

Garcia-Garayoa E, Allemann-Tannahill L, Blauenstein P, Willmann M, Carrel-Remy N, Tourwe D, Iterbeke K, Conrath P, Schubiger PA (2001) In vitro and in vivo evaluation of new radiolabeled neurotensin(8-13) analogues with high affinity for NT1 receptors. Nucl Med Biol 28:75–84

Garcia R, Xing YH, Paulo A, Domingos A, Santos I (2002) Rhenium(I) tricarbonyl complexes with mercaptoimidazolylborate ligands bearing piperazine fragments. Dalton Transactions 22:4236–4241

Garcia-Garayoa E, Blauenstein P, Bruehlmeier M, Blanc A, Iterbeke K, Conrath P, Tourwe D, Schubiger PA (2002) Preclinical evaluation of a new, stabilized neurotensin(8-13) pseudopeptide radiolabeled with Tc-99m. J Nuc Med 43:374–383

He HY, Lipowska M, Xu XL, Taylor AT, Carlone M, Marzilli LG (2005) $Re(CO)_3$ complexes synthesized via an improved preparation of aqueous *fac*-$Re(CO)_3(H_2O)_3^+$ as an aid in assessing Tc-99m imaging agents. Structural characterization and solution behavior of complexes with thioether-bearing amino acids as tridentate ligands. Inorg Chem 44:5437–5446

Karagiorgou O, Patsis G, Pelecanou M, Raptopoulou CP, Terzis A, Siatra-Papastaikoudi T, Alberto R, Pirmettis I, Papadopoulos M (2005) *S* (2 (2′ pyridyl)ethyl)cysteamine and *S*(2 (2 ′ pyridyl)ethyl) D,L-homocysteine as ligands for the "*fac*-$M(CO)_3^+$" (M = Re, ^{99m}Tc) Core Inorg Chem 44:4118–4120

Kunze S, Zobi T, Kurz P, Spingler B, Alberto R (2004) Vitamin B_{12} as a ligand for technetium and rhenium complexes. Angew Chem Int Edit 43:5025–5029

La Bella R, Garcia-Garayoa E, Bahler M, Blauenstein P, Schibli R, Conrath P, Tourwe D, Schubiger PA (2002a) A Tc-99m(I)-postlabeled high affinity bombesin analogue as a potential tumor imaging agent. Bioconjugate Chem 13:599–604

La Bella R, Garcia-Garayoa E, Langer M, Blauenstein P, Beck-Sickinger AG, Schubiger PA (2002b) In vitro and in vivo evaluation of a Tc-99m(I)-labeled bombesin analogue for imaging of gastrin releasing peptide receptor-positive tumors. Nucl Med Biol 29:553–560

Langer M, La Bella R, Garcia-Garayoa E, Beck-Sickinger AG (2001) Tc-99m-labeled neuropeptide Y analogues as potential tumor imaging agents. Bioconjugate Chem 12:1028–1034

Lazarova N, Babich J, Valliant J, Schaffer P, James S, Zubieta J (2005) Thiol- and thioether-based bifunctional chelates for the $[M(CO)_3^+$ core (M Tc, Re). Inorg Chem 44:6763–6770

Lipowska M, Cini R, Tamasi G, Xu XL, Taylor AT, Marzilli LG (2004) Complexes having the *fac*-$[M(CO)_3]^+$ core (M = Tc, Re) useful in radiopharmaceuticals: X-ray and NMR structural characterization and density functional calculations of species containing two sp3 N donors and one sp3 O donor. Inorg Chem 43:7774–7783

Luyt LG, Bigott HM, Welch MJ, Katzenellenbogen JA (2003) 7 alpha- and 17 alpha-substituted estrogens containing tridentate tricarbonyl rhenium/technetium complexes: Synthesis of estrogen receptor imaging agents and evaluation using MicroPET with technetium-94m. Bioorg Med Chem 11:4977–4989

Mandal SK, Ho DM, Qing LG, Orchin M (1998) The preparation and crystal structure of (dppe) $(CO)_3$Re-OC(O)O-Re$(CO)_3$(dppe). Polyhedron 17:607–611

Marmion ME, Alberto R, Bugaj J, Chinen L, Schmidt M, Srinivasan A (1999) Preparation and biodistribution of [^{99m}Tc$(CO)_3$His^0,Tyr3]octreotate. J Labelled Compd Radiopharm 42:S231–S233

Müller C, Hofmann U, Schubiger AP, Schibli R (2004) Organometallic 99mTc-technetium(I)- and Re-rhenium(I) folate derivatives for potential use in nuclear medicine. J Organomet Chem 289:4712–4721

Mundwiler S, Candreia L, Hafliger P, Ortner K, Alberto R (2004) Preparation of no-carrier-added technetium-99m complexes via metal-assisted cleavage from a solid phase. Bioconjugate Chem 15:195–202

Mundwiler S, Kundig M, Ortner K, Alberto R (2004) A new 2+1 mixed ligand concept based on 99mTc $(OH_2)_3(CO)_3^+$: a basic study. Dalton Transactions 1320–1328

Mundwiler S, Waibel R, Spingler B, Kunze S, Alberto R (2005) Picolylamine-methylphosphonic acid esters as tridentate ligands for the labeling of alcohols with the fac-$M(CO)_3^+$ core (M=Tc-99m, Re): synthesis and biodistribution of model compounds and of a Tc-99m-labeled cobinamide. Nucl Med Biol 32:473–484

Murray A, Simms MS, Scholfield DP, Vincent RM, Denton G, Bishop MC, Price MR, Perkins AC (2001) Production and characterization of Re-188-C595 antibody for radioimmunotherapy of transitional cell bladder cancer. J Nucl Med 42:726–732

Park SH, Seifert S, Pietzsch HJ (2006) Novel and efficient preparation of precursor Re-[188Re $(H_2O)_3(CO)_3]^+$ for the labeling of biomolecules. Bioconjugate Chem 17:223–225

Pietzsch HJ, Gupta A, Reisgys M, Drews A, Seifert S, Syhre R, Spies H, Alberto R, Abram U, Schubiger PA, Johannsen B (2000) Chemical and biological characterization of technetium(I) and rhenium(I) tricarbonyl complexes with dithioether ligands serving as linkers for coupling the $Tc(CO)_3$ and $Re(CO)_3$ moieties to biologically active molecules. Bioconjugate Chem 11:414–424

Schibli R, Katti KV, Higginbotham C, Volkert WA, Alberto R (1999) In vitro and in vivo evaluation of bidentate, water-soluble phosphine ligands as anchor groups for the organometallic *fac*-[99mTc(CO)3]+core. Nucl Med Biol 26:711–716

Schibli R, La Bella R, Alberto R, Garcia-Garayoa E, Ortner K, Abram U, Schubiger PA (2000) Influence of the denticity of ligand systems on the in vitro and in vivo behavior of Tc-99m(I)-tricarbonyl complexes: a hint for the future functionalization of biomolecules. Bioconjugate Chem 11:345–351

Schibli R, Schwarzbach R, Alberto R, Ortner K, Schmalle H, Dumas C, Egli A, Schubiger AP (2002) Steps toward high specific activity labeling of biomolecules for therapeutic application: preparation of precursor $[^{188}Re(H_2O)_3(CO)_3]^+$ and synthesis of tailor-made bifunctional ligand systems. Bioconjugate Chem 13:750–756

Smith CJ, Sieckman GL, Owen NK, Hayes DL, Mazuru DG, Kannan R, Volkert WA, Hoffman TJ (2003 a) Radiochemical investigations of gastrin-releasing peptide receptor-specific $99mTc(X)(CO)_3$-Dpr-Ser-Ser-Ser-Gln-Trp-Ala-Val-Gly-His-Leu-Met-(NH_2) in PC-3, tumor-bearing, rodent models: syntheses, radiolabeling, and in vitro/in vivo studies where Dpr = 2,3-diaminopropionic acid and X H_2O or $P(CH_2OH)_3$. Cancer Res 63:4082–4088

Smith CJ, Sieckman GL, Owen NK, Hayes DL, Mazuru D, Volkert WA, Hoffman TJ (2003 b) Radiochemical investigations of $^{188}Re(H_2O)(CO)_3$-diaminopropionic acid-SSS-bombesin(7-14)NH_2: syntheses, radiolabeling and in vitro/In vivo GRP receptor targeting studies. Anticancer Res 23:63–70

Sogbein OO, Merdy P, Morel P, Valliant JF (2004) Preparation of Re(I)- and Tc-99m(I)-metallocarboranes in water under weakly basic reaction conditions. Inorg Chem 43:3032–3034

Sogbein OO, Green AEC, Schaffer P, Chankalal R, Lee E, Healy BD, Morel P, Valliant JF (2005a) Synthesis of ortho- and meta-Re(I)-metallocarboranes in water. Inorg Chem 44:9574–9584

Sogbein OO, Green AEC, Valliant JF (2005b) Aqueous fluoride and the preparation of $^{99m}Tc(CO)_3(OH_2)_3^+$ and Tc-99m-carborane complexes. Inorg Chem 44:9585–9591

Maresca KP (2003) Bifunctional single amino acid chelates (SAAC) as synthons for the solid phase synthesis of Tc(I) and Re(I) radiopharmaceuticals. J Nucl Med 44:48P–48P

Staveren DR van, Benny PD, Waibel R, Kurz P, Pak JK, Alberto R (2005) I-functionalized cysteine: powerful ligands for the labelling of bioactive molecules with triaquatricarbonyltechnetium-99m $^{99m}Tc(CO)_3(H_2O)_3^+$. Helv Chim Acta 88:447–460

Staveren DR van, Mundwiler S, Hoffmanns U, Pak JK, Spingler B, Metzler-Nolte N, Alberto R (2004) Conjugation of a novel histidine derivative to biomolecules and labelling with $^{99m}Tc(CO)_3(H_2O)_3^+$. Org Biomol Chem 2:2593–2603

Staveren DR van, Waibel R, Mundwiler S, Schubiger PA, Alberto R (2004) Conjugates of vitamin B12 with N-epsilon-functionalized histidine for labeling with $[^{99m}Tc(OH_2)_3(CO)_3]^+$: synthesis and biodistribution studies in tumor bearing mice. J Organomet Chem 689:4803–4810

Stephenson KA, Banerjee SR, Besanger T, Sogbein OO, Levadala MK, McFarlane N, Lemon JA, Boreham DR, Maresca KP, Brennan JD, Babich JW, Zubieta J, Valliant JF (2004) Bridging the gap between in vitro and in vivo imaging: isostructural Re and Tc-99m complexes for correlating fluorescence and radioimaging studies. J Am Chem Soc 126:8598–8599

Stephenson KA, Valliant JF, Zubieta J, Banerjee SR, Levadala MK, Taggart L, Ryan L, McFarlane N, Boreham DR, Babich JW, Stephenson KA, Zubieta J, Banerjee SR, Levadala MK, Taggart L, Ryan L, McFarlane N, Boreham DR, Maresca KP, Babich JW, Valliant JF (2004) A new strategy, for the preparation of peptide-targeted radiopharmaceuticals based on ion of peptide-targeted an Fmoc-lysine-derived single amino acid chelate (SAAC). Automated solid-phase synthesis, NMR characterization, and in vitro screening of fMLF(SAAC)G and fMLF $(SAAC\text{-}Re(CO)_3)^{+}$G. Bioconjugate Chem 15:128–136

Stichelberger A, Waibel R, Dumas C, Schubiger PA, Schibli R (2003) Versatile synthetic approach to new bifunctional chelating agents tailor made for labeling with the *fac*-$[M(CO)_3]^{+}$ core (M = Tc, ^{99m}Tc, Re): synthesis, in vitro, and in vivo behavior of the model complex $[M(APPA)(CO)_3]$ (appa [(5-amino-pentyl)-pyridin-2-yl-methyl-amino]-acetic acid). Nucl Med Biol 30:465–470

Tait JF, Smith C, Gibson DF (2002) Development of annexin V mutants suitable for labeling with Tc(I)-carbonyl complex. Bioconjugate Chem 13:1119–1123

Waibel R, Novak-Hofer I, Schibli R, Blauenstein P, Garcia-Garayoa E, Schwarzbach R, Zimmermann K, Pellikka R, Gasser O, Blanc A, Bruhlmeier M, Schubiger PA (2000) Radiopharmaceuticals for targeted tumor diagnosis and therapy. Chimia 54:683–688

Waibel R, Stichelberger R, Alberto R, Schubiger PA, Chester KA, Begent RHJ (2000) Site-directed labelling of single chain antibodies with $^{99(m)}$technetium and $^{(188)}$rhenium. Eur J Nucl Med 27:15

Wald J, Alberto R, Ortner K, Candreia L (2001) Aqueous one-pot synthesis of derivatized cyclopentadienyl-tricarbonyl complexes of Tc-99m with an in situ CO source: application to a serotonergic receptor ligand. Angew Chem Int Edit 40:3062–3066

Wei LH, Babich J, Zubieta J (2005) Bifunctional chelates with mixed aromatic and aliphatic amine donors for labeling of biomolecules with the $[Tc(CO)_3]^{+}$ and $[Re(CO)_3]^{+}$-cores. Inorg Chim Acta 358:3691–3700

Wenzel M (1992) Tc-99m Labeling of cymantrene-analogs with different substituents - a new approach to Tc-99m radiodiagnostics. J Labelled Compd Radiopharm 31:641–650

Wenzel M, Saidi M (1993) Esters of Tc-99m labeled cytectrenecarboxylic acid with alcohols of cyclic amines as cerebral radiodiagniostic agents. J Labelled Compd Radiopharm 33:77–80

Wenzel M, Klinge C (1994) Tc-99m-labeled estradiol derivatives - synthesis, organ distribution and tumour affinity. J Labelled Compd Radiopharm 34:981–987

Wester HJ (2003) Carbohydrated peptides. Cancer Biother Radiopharm18:277–277

Wester HJ, Schottelius M, Schwaiger M (2001) $^{99m}Tc(CO)_3$-labeled carbohydrated SSTR-ligands: synthesis, internalization kinetics and biodistribution on a rat pancreatic tumor model. J Nucl Med 42:115P–115P

Willuda J, Honegger A, Waibel R, Schubiger AP, Stahel R, Zangemeister-Wittke U, Pluckthun A (1999) High thermal stability is essential for tumour targeting of antibody fragments: Engineering of a humanized anti-epitelial glycoprotein-2 (epitelial cell adhesion molecule) single-chain Fv fragment. Cancer Res 59:5758–5767

Willuda J, Kubetzko S, Waibel R, Schubiger PA, Zangemeister-Wittke U, Pluckthun A (2001) Tumour targeting of mono-, di-, and tetravalent Anti-p185(HER-2) miniantibodies multimerized by self-associating peptides. J Biol Chem 276:14385–14392

Wust F, Carlson KE, Katzenellenbogen JA, Spies H, Johannsen B (1998) Synthesis and binding affinities of new 17 alpha-substituted estradiol-rhenium "n+1" mixed-ligand and thioethercarbonyl complexes. Steroids 63:665–671

Wust F, Skaddan MB, Leibnitz P, Spies H, Katzenellenbogen JA, Johannsen B (1999) Synthesis of novel progestin-rhenium conjugates as potential ligands for the progesterone receptor. Bioorg Med Chem 7:1827–1835

2.3 Technetium Coupled to Biologically Active Molecules

H.-J. Pietzsch, J.-U. Künstler and H. Spies

2.3.1 Introduction

Many ^{99m}Tc pharmaceuticals were designed for the measurement of organ function, based on regional blood flows, ion transport, and cellular retention. Organ specificity is governed by molecular characteristics (e.g., size, shape, charge) and physiological factors.

Primarily, these radiotracers are coordination complexes of technetium leaving either a positive or a negative charge; neutral, lipophilic complexes pass the blood-brain barrier. Organ function is related to regional perfusion (e.g., brain, heart). Hepatocyte function is measured by the excretion of iminodiacetic acid (IDA) derivatives into bile, simulating the active transport of bilirubin. Increased osteogenic activity correlates with increased regional uptake of ^{99m}Tc-diphosphonate complexes in bone structures, delineating tumor and metastatic growth. The functional state of the kidneys as measured by active tubular secretion requires a negatively charged complex with a carboxylate anion.

^{99m}Tc pharmaceuticals based on coordination complexes with functionalized ligands are also known as "Tc essentials"; those concerning labeled particles and macromolecules are called "Tc-tagged" radiopharmaceuticals. A variety of chelating agents have been developed for complex formation with certain oxidation states of technetium, providing the structural requirements for uptake and retention (Schwochau 2000). Examples of Tc essentials are shown in Fig. 2.3.1.

The outstanding interest in the development of novel ^{99m}Tc pharmaceuticals is documented in recent reviews (Hom and Katzenellenbogen 1997; Johannsen and Pietzsch 2002a; Jurisson and Lydon 1999).

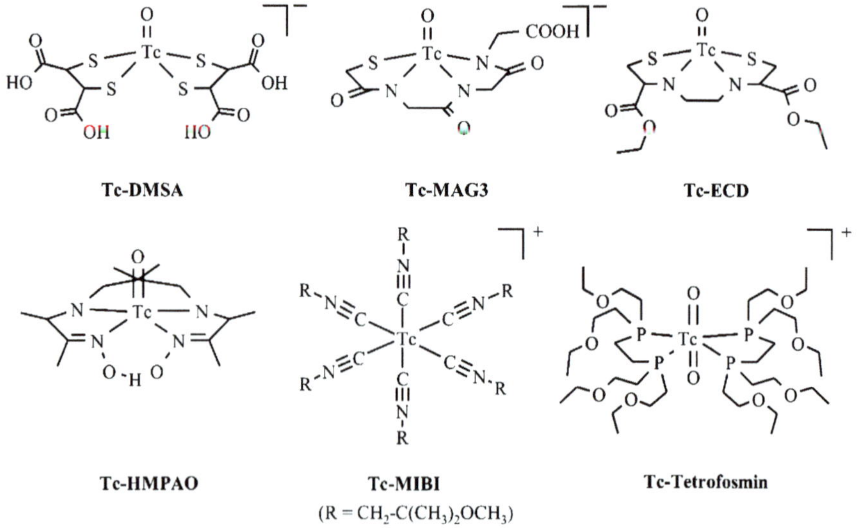

Fig. 2.3.1. "Tc essential" radiopharmaceuticals in clinical use

2.3.1.1 Target-Specific ^{99m}Tc Pharmaceuticals

Besides the merits of coordination complexes for diagnostic imaging, few applications of tumor diagnosis are in clinical use. The need for radiotracers binding specifically to epitopes expressed on tumor cells has grown over the past decade, promoting new labeling techniques, by which technetium is attached to biomolecules.

Direct or random labeling of biologically active molecules with reduced technetium did not produce pharmacologically acceptable radiotracers. Therefore, some known ^{99m}Tc complexes were specifically evaluated as potential chelating units, such as mercaptoacetyltriglycine (MAG_3) and diaminodithiol (DADT).

Prerequisites of an optimal chelator:

- The ligand used as a chelator should not alter the in vivo characteristics of a biomolecule.
- The chelate unit containing technetium should preferably be an integral part of the biomolecule (Johannsen and Pietzsch 2002b).
- The chelating unit should not affect the potency of the biomolecule.

The design of site-specific technetium molecules may complete the quest for the optimal chelator in accordance with the target-specific biomolecule, combining the chemical and biological requirements for tumor imaging.

2.3.2 Factors Affecting In Vivo Performance

Unlike ^{99m}Tc complexes, which are symmetric, small molecules (Schwochau 2000), labeled biomolecules might be asymmetric, integrating building blocks with distinct functions, such as a linker and the technetium chelating unit (Fig. 2.3.2).

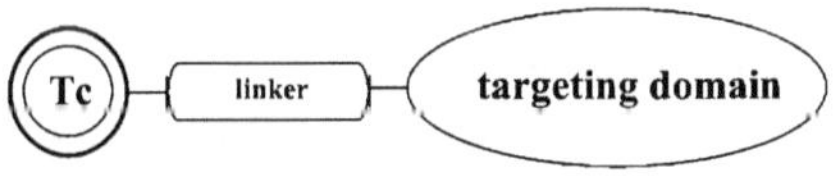

Fig. 2.3.2. Schematic representation of a specific ^{99m}Tc biomolecule

The in vivo distribution of a ^{99m}Tc biomolecule is influenced by its chemistry and by biological factors. The chemical nature of the targeting domain determines uptake and retention in the biological system. Structural integrity includes both chemical and metabolic stability of the labeled conjugate. Thus, the ligand that is used as a chelator, the type of linker, and the biomolecule will determine the bioavailability of the radiotracer.

Biological factors are related to the specific recognition of metal-based molecules, to membrane transport (particularly crossing the blood–brain barrier), clearance from nontarget sites, and high-affinity binding to tumor cell epitopes, permitting imaging and disease assessment (Ballinger 2002).

In Vivo Stability versus Reactivity. Both the chelate unit and the organic moiety may undergo transformations in vivo. In the case of "3+1" mixed-ligand complexes, the monodentate thiol ligand is exchanged by other SH-containing compounds such as glutathione (Nock et al. 1999) or reacts with proteins (Seifert et al. 2001). In vivo transchelation has been observed with certain ^{99m}Tc complexes (methylenediphosphonate [MDP]/gluconate).

Complexes with robust tetradentate chelate units have shown metabolic degradation, splitting off the whole chelate as observed with ^{99m}Tc-Trodat-1 (Kushner et al. 1999; Mu et al. 1999).

Furthermore, the carbon bond between the linker and the tertiary nitrogen of the coordination shell may break, even during the labeling procedure, as recently reported for ^{99m}Tc tricarbonyl-labeled glucose (Pak and Alberto 2001).

Transport across Cell Membranes. In the body, many interactions of the radiotracer with biological components exist, affecting regional uptake. Transport across membranes and binding affinity have to be verified in suitable models. Specific radiotracers are designed to use transporter-mediated processes for unidirectional uptake; generally, uptake into cells across membranes should be rapid.

Lipophilicity. Lipophilicity of a radiotracer facilitates diffusion across membranes, particularly passing the blood–brain barrier, which is required for brain uptake.

Receptor Binding. Receptor binding is based on high-affinity binding of the radiotracer molecule, which is an antagonist. Since receptor density in the brain is generally in the picomolar range, high specific activity is a prerequisite for receptor ligands.

Displacement Studies. Enzyme-inhibition studies have similar requirements if the target molecule is an enzyme and the radiotracer used for quantification is an inhibitor. An alternate mechanism is based on substrate analogs, like ^{18}F-FDG, which block enzymatic degradation, thus facilitating quantification.

2.3.3 Chelate Units in the Design of Target-Specific ^{99m}Tc Pharmaceuticals

Tc chelates suitable for conjugate formation with biomolecules are derived from Tc in oxidation states V, III, and I. Organometallic carbonyl (CO) complexes serve as precursors for the synthesis of ^{99m}Tc(I) pharmaceuticals (Sect. 2.2).

Oxotechnetium(V) Complexes. The dominant structural element is the oxotechnetium core, TcO^{3+}. The presence of the oxo ligand has a significant effect on the structure and stability of these complexes. DADT-derived ligands have found application in nuclear medicine because of the thiophilic nature of technetium for the thiolate donor group (Schwochau 2000) (Fig. 2.3.3). Tetradentate diaminodithiol (N_2S_2) forms neutral, lipid-soluble technetium complexes – the most prominent is the ethylcysteinate dimer (ECD) – used for the measurement of brain perfusion. The combination of an amine-N or amide-N in an N_2S_2 arrangement (monoamine, monoamine [MAMA]) results in a more polar derivative than the diaminedithiol system. This might be preferable when less lipophilicity is required, e.g., for labeling of proteins. Tripeptides combine donor atoms of different reactivity (e.g., N_3S, N_4). The prototypic N_3S chelator MAG_3 forms

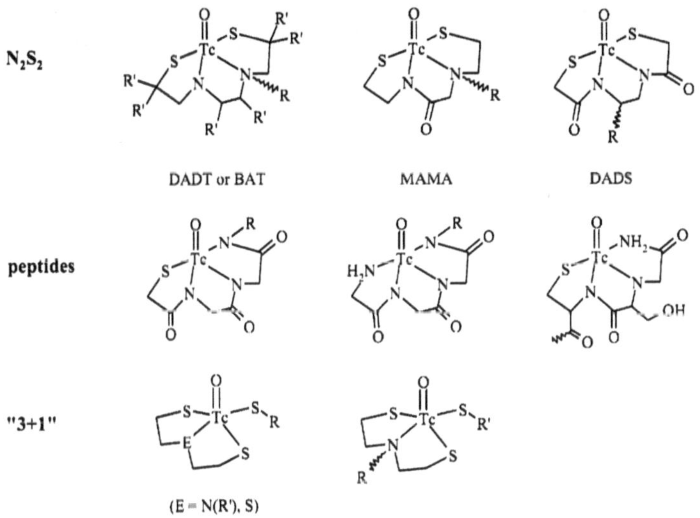

Fig. 2.3.3. Different types of oxotechnetium(V) chelates derived from N,S-ligands for radiotracer design. *R* spacer + targeting domain, *R'* H, alkyl, aryl

the radiopharmaceutical $[TcO(MAG_3)]^-$, which is used for studies of renal tubular function (Fig. 2.3.1).

Mixed-ligand complexes were synthesized in order to reduce the synthetic expenditure necessary for tetradentate compounds. A combination of tridentate (S_3 or NS_2) and monodentate (thiol) ligands is employed in the so-called "3+1" complexes (Spies et al. 1999). While the oxotechnetium/tridentate unit is very stable, exchange of the monodentate ligand has been observed in vivo (Nock et al. 1999; Syhre et al. 1998).

Tc(V) Hydrazino Nicotinamide (HYNIC) Derivatives. The introduction of the Tc(V)-HYNIC system (Schwartz et al. 1991) represents a milestone in the development of Tc-99m radiopharmaceuticals. Particularly, peptides have been labeled with very high specific activity (Edwards et al. 1999a; Harris et al. 1999; Rose et al. 1998). Since the HYNIC linker occupies only one coordination site, coligands such as tricine, ethylenediamine diacetic acid (EDDA), etc., may complete the coordination sphere of the metal (Babich et al. 2000; Edwards et al. 1999b; Ono et al. 2000) (Fig. 2.3.4).

Nitridotechnetium(V) Heterocomplexes. An asymmetric nitridotechnetium(V) heteromoiety has been proposed for radiolabeling bioactive molecules (Bolzati et al. 2000, 2002; Boschi et al. 2001; Pasqualini et al. 1992, 1994; Refosco et al. 2000). The metal fragment $[Tc(N)(PXP)]^{2+}$ can be used as an efficient synthon for the preparation of a series of nitrido heterocomplexes containing bidentate chelators such as dithiocarbamates, dithiocarbazates, cysteine, and dithiolates (Bolzati et al. 2004; Boschi et al. 2005; Tisato et al. 2004) (Fig. 2.3.5).

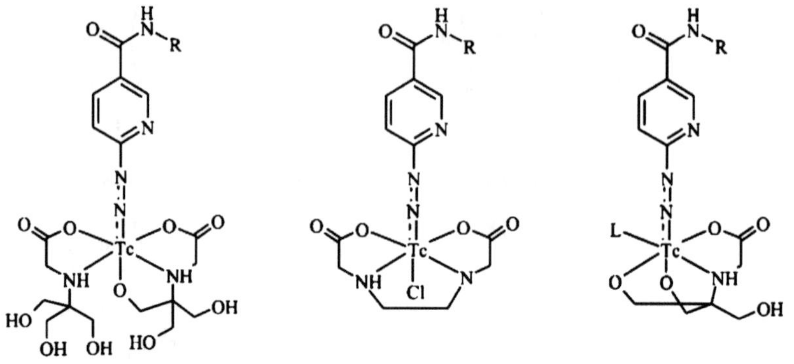

Fig. 2.3.4. Proposed structures of ^{99m}Tc-hydrazino nicotinamide (*HYNIC*) tricine derivatives with various coligands. *R* biomolecule

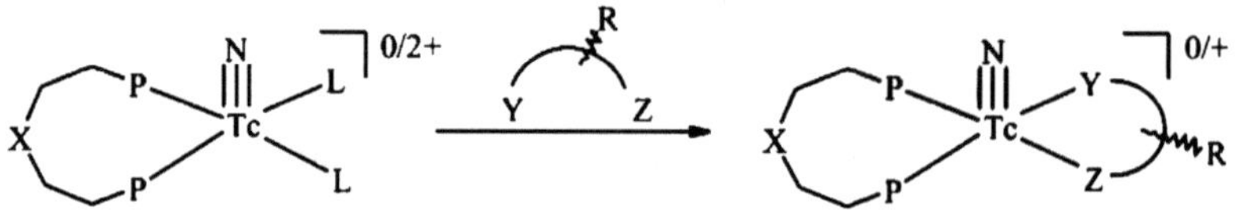

Fig. 2.3.5. Schematic representation of the labeling approach using the novel $[Tc(N)(PXP)]^{2+}$ moiety

Tc(III) Complexes. A novel type of Tc(III) chelate formed by the tripodal chelator 2,2′,2″-nitrilotris(ethanethiol) and a tertiary phosphine or an isocyanide as coligands contains sterically well-shielded oxo-free Tc(III) (Fig. 2.3.6) (Pietzsch et al. 2001 a; Seifert et al. 2004; Spies et al. 1999). This moiety fulfils the requirements of a nonpolar building block stable against ligand exchange reactions in vivo.

Another type of neutral Tc(III) complexes derived from the reaction of oxotechnetium(V) "3+1" precursors with tertiary phosphines, namely compounds of the general formula $[M(PR_3)(SES)(SR)]$ (SES = tridentate dithiol ligand; E = S, NR, O), suffers from instability against cysteine and glutathione (Pietzsch et al. 2001 b; Seifert et al. 2000). Stability of this class of compounds can be enhanced when a bidentate P,S phosphinothiol ligand is used instead of the monodentate ligand. The resulting "3+2" coordinated Tc(III) mixed-ligand complexes have the general formula $[Tc(SES)(R_2PS)]$ (Pietzsch et al. 2003) (Fig. 2.3.7).

Tc(I) Complexes. The organometallic ligand cyclopentadienyl (cp) offers advantages because of its small size and low molecular weight (Wenzel and Klinge 1994). Stable Re(cp) and Tc(cp) complexes have been prepared that were conjugated to octreotide (Spradau et al. 1999), piperidine (Fig. 2.3.8) (Saidi et al. 2001), tropane (Cesati et al. 2002), and steroid hormones (LeBideau et al. 2001; Mull et al. 2002). However, this approach still suffers from unacceptable reaction conditions for routine use of technetium-99m.

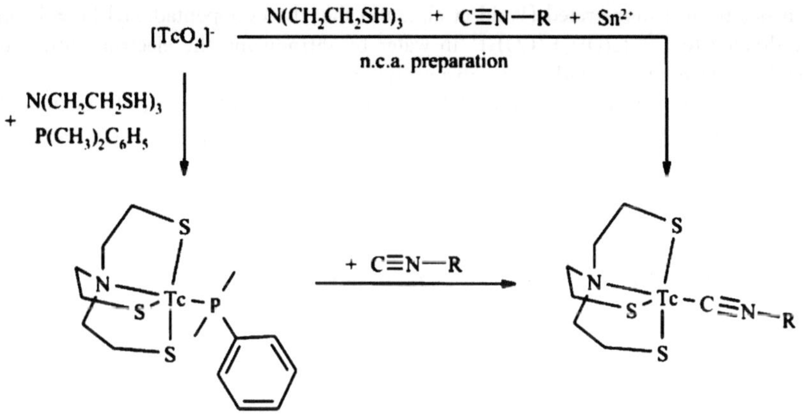

Fig. 2.3.6. Formation of technetium(III) complexes with tetradentate/monodentate coordination

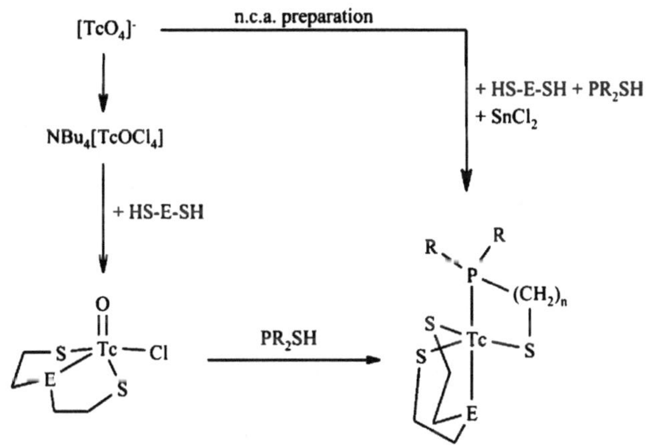

Fig. 2.3.7. Reaction routes to Tc(III) complexes with "3+2" coordination. $E = N(CH_3)$, S

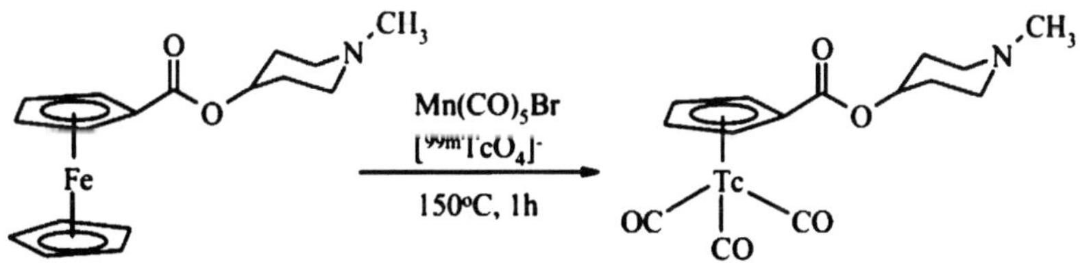

Fig. 2.3.8. Preparation of $^{99m}Tc(CO)_3$ cyclopentadienyl (cp) carboxylate derivative illustrating the double-ligand transfer approach (Saidi 2001; Wenzel 1994)

It has been demonstrated (Wald et al. 2001) that the cyclopentadienyl ligand can be coordinated to $[^{99m}Tc(OH_2)_3(CO)_3]^+$ in water by introducing the electron withdrawing acetyl group in cyclopentadiene to give acetyl-cp.

Technetium(I) chemistry initiated by (Alberto et al. 2001) is greatly facilitated by the available Tc(I)-tricarbonyl synthon. Recent developments and investigations are presented in Sect. 2.2.

2.3.4 Search for Novel Tc Pharmaceuticals

Peptides. Peptides with low molecular weight consisting of 5–15 amino acids have attracted much attention in radiopharmaceutical design because of their low immunogenicity, suitable pharmacokinetic properties, and high binding affinities. They are easier to synthesize and to modify than are larger molecules (Signore 1995). To be suitable as a tumor-imaging peptide, the density of the peptide affine receptor on tumors must be considerably higher than in other regions of the body. The metabolic stability and affinity for the receptor should be high. Many excellent reviews have been published discussing different aspects of technetium radiopharmaceuticals based on peptides (Aloj and Morelli 2004; Eberle et al. 2004; Fichna and Janecka 2003; Giblin et al. 2005; Langer and Beck-Sickinger 2001; Liu 1999; Liu and Edwards 2002; Maecke 2005; Okarvi 2004; Signore et al. 2001). Naturally occurring peptides that can be used for tumor imaging are listed in Table 2.3.1.

Modified derivatives of somatostatine have been synthesized to prolong the biological half-life of native somatostatine. The most important derivative is octreotide (sandostatin), a cyclic peptide with 8 amino acids, unlike the 14 in somatostatine.

The successful use of octreoscan in the diagnosis of somatostatine receptor-positive tumors has intensified the search for improved or new peptide-based agents for imaging thrombi, infection/inflammation, and different tumors.

Among the chelate units used for peptide labeling, the Tc-HYNIC and Tc-tricarbonyl cores have gained importance. A freeze-dried kit formulation for the preparation of ^{99m}Tc-EDDA-HYNIC-D-Phe(1), Tyr(3)-octreotide, another somatostatin analog for tumor diagnosis, has recently been published (von Guggenberg et al. 2004).

Ongoing research on ^{99m}Tc-HYNIC somatostatin analogs has further clarified the effect of labeling methods and peptide sequence on bioperformance (Bangard et al. 2000; Decristoforo and Mather 1999 a, b; Decristoforo et al. 2000). A variety of coligands used for labeling HYNIC-derivatized peptides has been explored, e.g., 2-mercaptopyridines and 2-mercaptopyrimidines (Babich et al. 2001).

Recently described labeled HYNIC-conjugated peptides also involve RGD (Arg-Gly-Asp) peptides targeting the integrin $\alpha_v\beta_3$ (vitronectin) receptor. Tertiary ligand complexes of HYNIC-conjugated peptide, tricine and trisodium triphenylphosphine-3,3′,3″-trisulfonate (TPPTS) have been published (Liu et al. 2001; Su et al. 2002).

Interleukin-8, a chemotactic cytokine involved in activation of neutrophils to areas of infection, can be labeled with ^{99m}Tc-HYNIC with preservation of its leukocyte receptor-binding capacity (Rennen et al. 2001).

After the introduction of the Tc(I) tricarbonyl approach, its application to peptide labeling has been pursued (Egli et al. 1999).

Other chelating frameworks have been studied such as the novel dithia-bisphosphine chelator (Gali et al. 2001), or further employed such as tripeptide N_3S chelators for the

Table 2.3.1. List of naturally occurring peptides that can be used for tumor imaging

Ligand	Selectivity
Somatostatine	Neuroendocrinic tumors, non-Hodgkin's lymphoma,
Derivatives	Melanomas, breast tumors
Alpha-MSH	Melanomas
LHRH	Prostata tumors, breast tumors
VIP/PACAP	SCLC, tumors of colon, stomach, pancreas
RGD	Blood vessels of tumors
CCK-B/gastrine	MTC, SCLC, pancreas tumors, astrocytomes
Neutrotensin	SCLC, colon tumors, exocrinic pancreas tumors
Bombesin/GRP	SCLC, colon tumors, glioblastomas, prostata tumors
Substance P	Glioblastomas, astrocytomas, MTC, breast tumors, Peritoneal blood vessels

SCLC small cell lung carcinoma, *MTC* medullary thyroid cancer

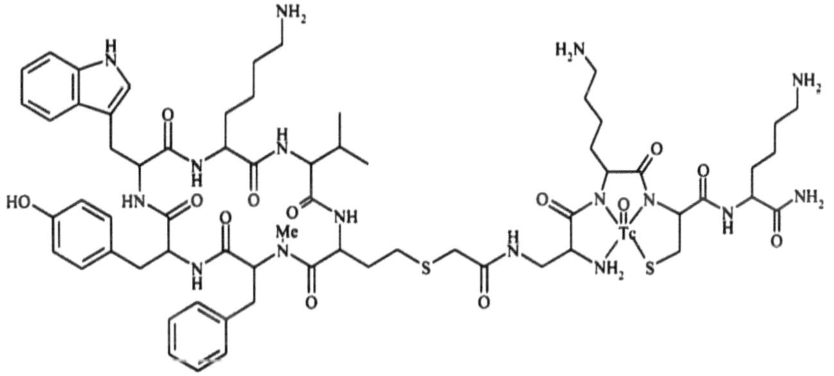

Fig. 2.3.9. Structure of Tc-labeled depreotide (NeoTect^{TM})

inflammation imaging agent ^{99m}Tc-RP128 (Caveliers et al. 2001), a tuftsin receptor-binding peptide (Wong et al. 2001) and melanocortin receptor-1 specific ligands for targeting melanoma (Sharma et al. 2000).

The commercial kit NeoTect^{TM} (Diatide) was designed as a radiopharmaceutical for somatostatin-receptor imaging of lung tumors (Virgolini et al. 1998). It is based on the peptide P829 (depreotide), a structural modification of octreotide, with the technetium binding N_3S sequence diaminopropionic acid-lysine-cysteine built into the molecule (Cyr et al. 1999) (Fig. 2.3.9). This modification is an alternative to octreotide, where the labeling process leads to the reduction of the disulfide bond, resulting in a loss of receptor-binding affinity (Blum et al. 1999; Vallabhajosula et al. 1996).

Novel ^{99m}Tc-based tetra-amine-functionalized [Tyr^3]octreotate analogues (Fig. 2.3.10) have been developed for imaging of somatostatin receptor-positive tumors (Maina et al. 2002; Nikolopoulou et al. 2006).

An intrapatient comparison of ^{99m}Tc-N_4-[Tyr^3]octreotate with ^{99m}Tc-EDDA/HYNIC-[Tyr^3]octreotide showed that ^{99m}Tc-Demotate is a promising agent for somatostatin receptor scintigraphy (Gabriel et al. 2004).

The same open-chain tetra-amine ligand has been conjugated to various bombesin derivatives. First studies in mice showed high and specific accumulation of ^{99m}Tc-De-

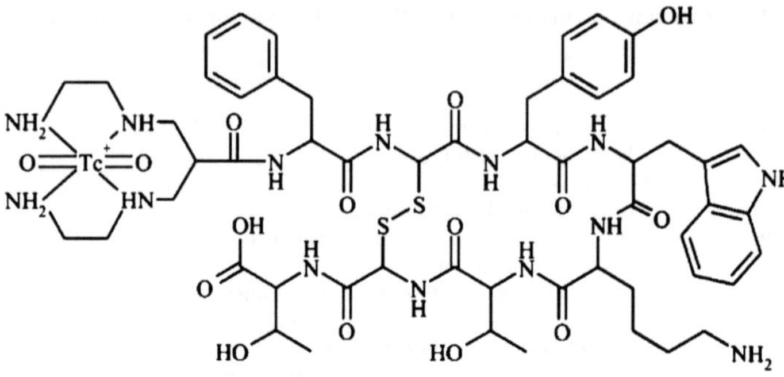

Fig. 2.3.10. Structure of tetra-amine-functionalized ^{99m}Tc-Demotate

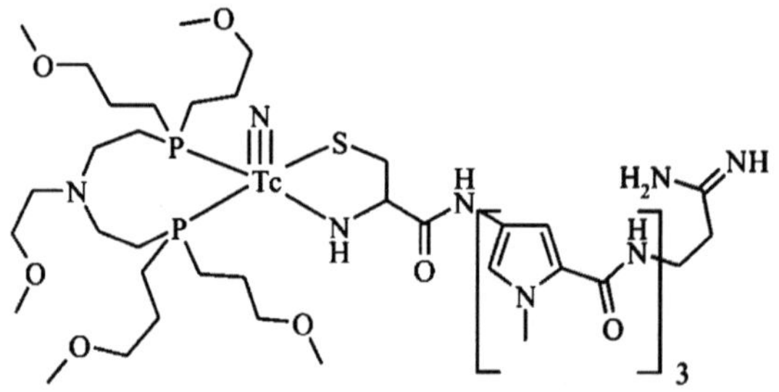

Fig. 2.3.11. Hybrid distamycin-cysteine conjugated with a $[^{99m}Tc(N)(PP)]^{2+}$ fragment

mobesin 1 in gastrin releasing peptide receptor (GRP-R)-positive regions (pancreas, gastrointestinal tract) (Nock 2003).

A new high-affinity technetium-99m-bombesin analogue with low abdominal accumulation has been recently published (Lin et al. 2005).

^{99m}Tc-UBI 29-41, a technetium-99m-labeled peptide derived from ubiquicidine, targets bacterial and fungal infections in experimental animals. Welling et al. reported on the radiochemical and biological features of this radioactive agent and the importance of the amino acid sequence of UBI 29-41 for imaging of infections (Lupetti et al. 2002; Welling et al. 2002, 2005).

An attempt to exploit the chemistry of nitridotechnetium(V) complexes for labeling small biomolecules has been described (Baraldi et al. 2000). The tripyrrole peptide distamycin A, an antibiotic agent that binds to DNA, was functionalized with cysteine to obtain a bidentate ligand, which forms a mixed-ligand complex with a $[^{99m}Tc(N)(PP)]^{2+}$ fragment (Fig. 2.3.11).

Proteins and Antibodies. In the past, considerable work has been focused on the development of ^{99m}Tc-labeled monoclonal antibodies and their fragments. Three main strategies for labeling can be distinguished: direct labeling, bifunctional chelating agent (BFCA)-based prelabeling, and BFCA-based postlabeling.

Among the direct labeling methods, reduction of the antibody by a thiol reagent, such as mercaptoethanol or dithiothreitol, results in high labeling yields (Reilly 1993; Schwarz et al. 1988; Thakur et al. 1991). Table 2.3.2 compiles ^{99m}Tc-labeled antibodies approved as radiopharmaceuticals in the United States and the European Union.

Table 2.3.2. ^{99m}Tc-labeled antibodies approved as radiopharmaceutical (2005)

Drug	Indication	Antibody	Target	^{99m}Tc-binding	Year of approval
Neutrospec	Equivocal signs and appendicitis (infection/ inflammation)	Fanolesomab (IgM, murine)	CD15	Reduced protein	2004 (US)
Humaspect	Colorectal cancer	Votumumab (IgG, human)	CTAA16.88	Reduced protein	1998 (EU)
Leukoscan	Osteomyelitis (infection/ inflammation in bone)	Sulesomab (Fab′, murine)	CEA and NCA90	Reduced protein	1997 (EU)
CEA-Scan	Colorectal cancer	Arcitumomab (Fab′, murine)	CEA	Reduced protein	1996 (US); 1996, withdrawn 2005 (EU)
Verluma	Small cell lung cancer	Nofetumomab (Fab′, murine)	CD20	N_2S_2 chelate	1996 (US)
Tecnemab-K-1	Melanoma	Antimelanoma mAb fragments (Fab′ and $F(ab')_2$, murine)	HMW-MAA	Reduced protein	1996, withdrawn 2000 (EU)

Sources: pharmacos.eudra.org; www.fda.gov; www.biopharma.com
CEA carcinoembryonic antigen, *CD* cluster of differentiation, *mAb* monoclonal antibody, *CTAA* cytokeratine tumor-associated complex of antigens, *NCA* granulocyte nonspecific crossreacting antigen, *HMW-MAA* high-molecular-weight melanoma-associated antigen

Table 2.3.3. ^{99m}Tc-labeled monoclonal antibodies and antibody fragments for potential application

Antigen	Potential imaging application	^{99m}Tc-binding method	References
CA125	Ovarian cancer	Direct labeling	Kobayashi et al. 1993
CD4	Rheumatoid arthritis	Direct labeling	Kinne et al. 1995; Becker et al. 1990
CD22	Non-Hodgkin's lymphoma	MAG_3	Postema et al. 2003
CD44v6	Head and neck squamous cell carcinoma	MAG_3	Stroomer et al. 2000; Colnot et al. 2003
CD62E (E-Selectin)	Infection/inflammation	Direct labeling	Jamar et al. 2002
EGFR	EGFR-expressing tumors	EC, direct labeling	Schechter et al. 2003; Meenakshi et al. 2003
G250	Renal cell carcinoma	HYNIC, MAG_3, direct labeling	Steffens et al. 1999
MUC1	Bladder cancer, breast cancer	Tricarbonyl, direct labeling	Waibel et al. 1999; Simms et al. 2001
Myosin	Myocardial infraction	Direct labeling	Iwasaki et al. 2001; Taillefer et al. 1995
$P185^{HER-2}$	Breast cancer	Tricarbonyl	Willuda et al. 2001
TAG-72	Adenocarcinomas	HYNIC, introduced SH-group	Goel et al. 2001; Ranadive et al. 1993

CA cancer antigen, *CD* cluster of differentiation, *MUC* mucin, *TAG* tumor-associated glycoprotein, *EGFR* epidermal growth factor receptor, *MAG_3* mercaptoacetyltriglycine, *EC* ethylcysteinate, *HYNIC* hydrazino nicotinamide

Table 2.3.4. ^{99m}Tc-labeled proteins (excluding antibodies)

Protein	Imaging application	^{99m}Tc-binding unit	References
Polyclonal IgG	Infection/inflammation, Blood pool	HYNIC, direct labeling	Abrams et al. 1990; Pieri et al. 1991; Claessens et al. 1996; Dams et al. 2000
HSA	Blood pool	HYNIC, MAG_3, direct labeling	Verbeke et al. 1995; Pieri et al. 1991
Annexin V	Apoptotic cells	N_2S_2, HYNIC, MAG_3, EC, tricarbonyl, endogenous peptide sequences, direct labeling	Lahorte et al. 2004; Boersma et al. 2005
Interleukins	Infection/inflammation	HYNIC, N_3S-chelate	Rennen et al. 2001, 2003a; Signore et al. 2004; Chianelli et al. 1997
NGA	Liver disease	Direct labeling	Stadalnik et al. 2001
GSA	Liver disease	DTPA	Kokudo et al. 2003
Aprotinin	Amyloidosis	Direct labeling	Schaadt et al. 2003; Aprile et al. 1995
FGF-1	FGF-1 receptor	HYNIC	Zinn et al. 2000
EGF	EGF-receptor expressing tumors	Introduced thiol group (direct labeling)	Capala et al. 1997
Anaphylatoxin C5a, C5adR	Infection	HYNIC	Rennen et al. 2003b
NAP-2 (CXCL-7)	Infection	HYNIC	Rennen et al. 2004
Ubiquicidin	Infection	Direct labeling	Welling et al. 2000
Lactoferrin	Infection	Direct labeling	Welling et al. 2000
HuS (adapter protein)	Target with a docking tag	HYNIC	Blankenberg et al. 2004

IgG immunoglobulin G, *HSA* human serum albumin, *NGA* galactosyl neoglycoalbumin, *GSA* galactosyl human serum albumin, *FGF-1* acidic fibroblast growth factor, *EGF* epidermal growth factor, *NAP* neutrophil-activating peptide, *HuS* 110-amino acid fragment of human ribonuclease I, *MAG_3* mercaptoacetyltriglycine, *HYNIC* hydrazino nicotinamide, *EC* ethylcysteinate, *DTPA* diethylene triamine pentaacetate

Some further examples for current search in antibody labeling are given by Tang et al. (2005), Francis et al. (2004), and Jeong et al. (2004).

^{99m}Tc-labeled antibodies in experimental evaluation are summarized in Table 2.3.3.

A simple liquid formulation for the preparation of ^{99m}Tc-HYNIC-annexin V has been developed. Biodistribution studies in mice indicated that the target organs were the kidneys (Vanderheyden et al. 2002).

^{99m}Tc-HYNIC annexin V conjugates have been used for detection of apoptotic tumor response in vivo after a single dose of chemotherapy (Mochizuki et al. 2003), and for the evaluation of inflammation and apoptosis in rats with autoimmune myocarditis (Tokita 2003).

A selection of ^{99m}Tc-labeled proteins (excluding antibodies) is summarized in Table 2.3.4.

Oligonucleotides. Small oligonucleotide sequences that are complementary to a small mRNA segment could potentially target any specific mRNA molecule, and be used to image endogenous gene expression at the transcription level (Duatti 2004; Younes et al. 2002). Low in vivo stability continues to be a serious drawback. However, modifications may increase resistance to nucleases (Borkowski and Dinkelborg 2006; Usman and Blatt 2000).

Several authors reported the use of a so-called morpholino (MORF), a commercially available synthetic oligomer for pretargeting application (Liu et al. 2002, 2004). A construct of MAG_3 and cMORF was found to be effective in a mouse tumor model. Biodistribution data indicated high uptake in the tumor and low uptake in the normal tissues (Liu et al. 2002).

Recently, Qin et al. (2005) reported on molecular imaging of atherosclerotic plaques with ^{99m}Tc-labeled antisense oligonucleotides.

A review on recent progress in antisense targeting with radiolabeled DNA derivatives was given by Hnatowich and Nakamura (2004).

Central Nervous System (CNS) Receptor Imaging Agents. The development of ^{99m}Tc-based imaging agents selective for CNS receptors has been an area of considerable research endeavor. Progress has been made in the development of a dopamine transporter (DAT) imaging agent ^{99m}Tc-TRODAT-1 (Kung et al 1997) (Fig. 2.3.12), the development of another DAT ligand, ^{99m}Tc-O(15)O5T (Callahan et al. 2001), and synthesis of ^{99m}Tc complexes with nanomolar in vitro affinity for dopamine (D_1, D_2), serotonin (5-HT_{1A}, 5-HT_{2A}) and muscarinic acetylcholine receptors. The state-of-the-art of technetium-based CNS receptor ligands have been recently reviewed (Johannsen and Pietzsch 2002 a).

Molecular recognition of technetium complexes and their fit into the receptor-binding pocket is achievable. This is indicated by the high in vitro affinities of manifold Tc complexes to the serotonin-5-HT_{1A} receptor in the nanomolar and subnanomolar range (Alberto et al. 1999; Bernard et al. 2003; Bolzati et al. 2003; Boschi et al. 2003; Drews et al. 2002; Heimbold et al. 2002 a, b; Kara 2004; Leon et al. 2002; Papagianopoulou et al. 2002; Saidi et al. 2004; Samnick et al. 2004) (Fig. 2.3.13). Therefore, receptor binding would be high if the ligand would also demonstrate high uptake in brain; however, very low or absent brain uptake is the main issue in the development of receptor-binding imaging agents. A suitable combination of a high receptor affinity with a sufficient brain uptake was achieved only with the DAT ligands.

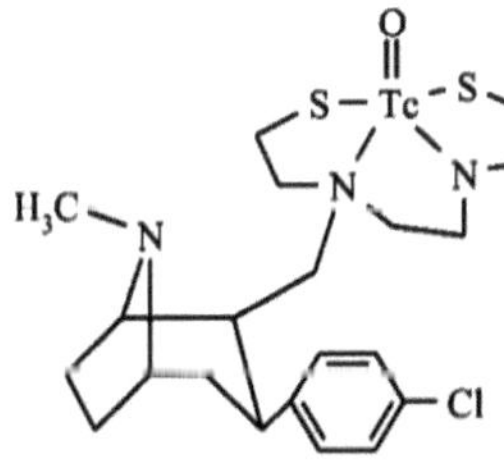

Fig. 2.3.12. Dopamine transporter (DAT) imaging agent ^{99m}Tc-TRODAT-1

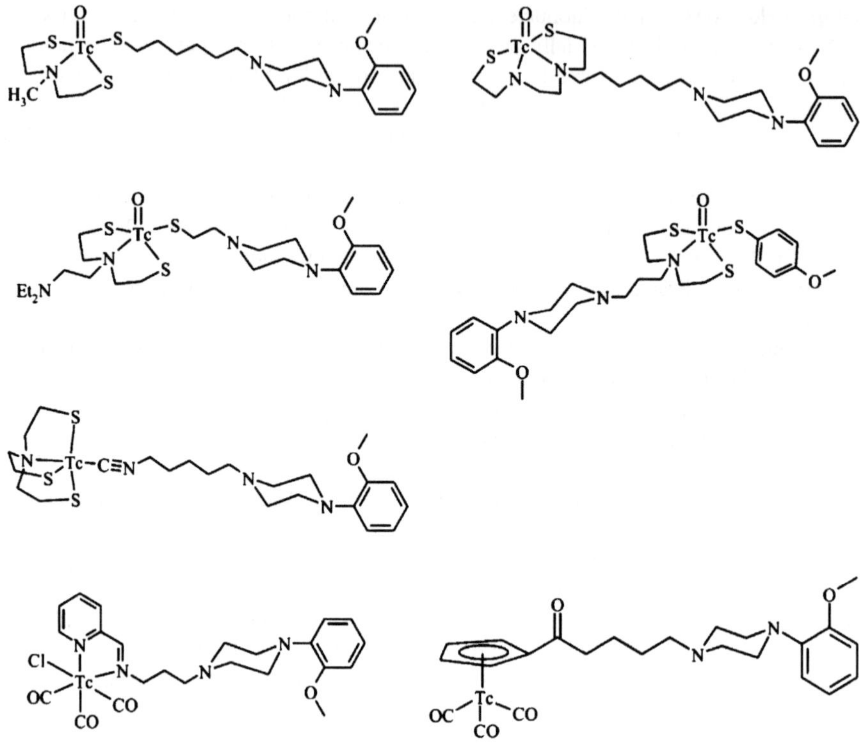

Fig. 2.3.13. ^{99m}Tc receptor ligands with nanomolar and subnanomolar affinities for the 5-HT_{1A} receptor (in vitro)

References

Abrams MJ, Juweid M, tenKate CI et al (1990) Technetium-99m-human polyclonal IgG radiolabelled via the hydrazino nicotinamide derivative for imaging focal sites of infection in rats. J Nucl Med 31:2022–2028

Alberto R, Schibli R, Schubiger PA et al (1999) First application of *fac*-[$^{99m}Tc(OH_2)_3(CO)_3]^+$ in bioorganometallic chemistry: design, structure, and in vitro affinity of a 5-HT_{1A} receptor ligand labelled with ^{99m}Tc. J Am Chem Soc 121:6076–6077

Alberto R, Ortner K, WheatleyN, Schibli R, Schubiger PA (2001) Synthesis and properties of boranocarbonate: a convenient in situ CO source for the aqueous preparation of [$^{99m}Tc(OH_2)_3(CO)_3]^+$. J Am Chem Soc 123:3135–3136

Aloj L, Morelli G (2004) Design, synthesis and preclinical evaluation of radiolabeled peptides for diagnosis and therapy. Curr Pharm Design 10:3009–3031

Aprile C, Marinone G, Saponaro R et al (1995) Cardiac and pleuropulmonary AL amyloid imaging with technetium-99m labelled aprotinin. Eur J Nucl Med 22:1393–1401

Babich JW, Coco WG, Barrow S et al (2000) ^{99m}Tc-labelled chemotactic peptides: influence of coligand on distibution of molecular species and infection imaging properties. Synthesis and structural characterization of model complexes with the [$Re(\eta^2$-$HNNC_5H_4N)(\eta^1$-$NNC_5H_4N)$] core. Inorg Chim Acta 309:123–136

Babich JW, Graham W, Femia FJ et al (2001) 6-Mercaptomethylpyridine-3-carboxylic acid (MEMNIC): a new reagent for peptide labelling with Tc-99m. Inorg Chim Acta 323:23–36

Ballinger JR (2002) The influence of carrier on ^{99m}Tc radiopharmaceuticals. Q J Nucl Med 46: 224–232

Bangard M, Behe M, Guhlke S et al (2000) Detection of somastotatin receptor-positive tumors using the new ^{99m}Tc-tricine-HYNIC-d-Phe^1-Tyr^3-octreotide: first results in patients and comparison with ^{111}In-DTPA-d-Phe^1-octreotide. Eur J Nucl Med 27:628–637

Baraldi PG, Romagnoli R, Duatti A et al (2000) Synthesis of hybrid distamycin-cysteine labelled with ^{99m}Tc: a model for a novel class of cancer imaging agents. Bioorg Med Chem Lett 10:1397–1400

Becker W, Emmrich F, Horneff G et al (1990) Imaging rheumatoid arthritis specifically with technetium 99m CD4-specific (T-helper lymphocytes) antibodies. Eur J Nucl Med 17:156–159

Bernard J, Ortner K, Spingler B et al (2003) Aqueous synthesis of derivatized cyclopentadienyl complexes of technetium and rhenium directed toward radiopharmaceutical application. Inorg Chem 42:1014–1022

Blankenberg FG, Mandl S, Cao YA et al (2004) Tumor imaging using a standardized radiolabelled adapter protein docked to vascular endothelial growth factor. J Nucl Med 45:1373–1380

Blum JE, Handmaker H, Rinne NA (1999) The utility of a somatostatin-type receptor binding peptide radiopharmaceutical (P829) in the evaluation of solitary pulmonary nodules. Chest 115:224–232

Boersma HH, Kietselaer BLJH, Stolk LML et al (2005) Past, present, and future of annexin A5: from protein discovery to clinical applications. J Nucl Med 46:2035–2050

Bolzati C, Boschi A, Duatti A et al (2000) Geometrically controlled selective formation of nitrido technetium(V) asymmetrical heterocomplexes with bidentate ligands. J Am Chem Soc 122:4510–4511

Bolzati C, Boschi A, Uccelli L et al (2002) Chemistry of the strong electrophilic metal fragment $[^{99m}Tc(N)(PXP)]^{2+}$ (PXP = diphosphine ligand). A novel tool for the selective labelling of small molecules. J Am Chem Soc 124:11468–11479

Bolzati C, Mahmood A, Malago E et al (2003) The $[Tc\text{-}99m(N)(PNP)]^{2+}$ metal fragment: a technetium-nitrido synthon for use with biologically active molecules the *N*-(2-methoxyphenyl)piperazyl-cysteine analogues as examples. Bioconjug Chem 14:1231–1242

Bolzati C, Benini E, Cazzola E et al (2004) Synthesis, characterization, and biological evaluation of neutral nitrido technetium(V) mixed ligand complexes containing dithiolates and aminodiphosphines. A novel system for linking technetium to biomolecules Bioconjug Chem 15:628–637

Borkowski S, Dinkelborg L (2006) Aptamers for in vivo imaging. In: Klussmann S (ed) The aptamer handbook. Wiley, Weinheim

Boschi A, Bolzati C, Benini E et al (2001) A novel approach to the high-specific-activity labelling of small peptides with the technetium-99m fragment $[^{99m}Tc(N)(PXP)]^{2+}$ (PXP = diphosphine ligand). Bioconjug Chem 12:1035–1042

Boschi A, Uccelli L, Duatti A et al (2003) Asymmetrical nitrido Tc-99m heterocomplexes as potential imaging agents for benzodiazepine receptors. Bioconjug Chem 14:1279–1288

Boschi A, Duatti A, Uccelli L (2005) Development of technetium-99m and rhenium-188 radiopharmaceuticals containing a terminal metal-nitrido multiple bond for diagnosis and therapy. Top Curr Chem 252:85–115

Callahan RJ, Dragotakes SC, Barrow SA et al (2001) A phase I clinical trial of the DAT ligand Tc-99m-O(15)O5T. J Nucl Med 42(Suppl):112S

Capala J, Barth RF, Bailey MQ et al (1997) Radiolabelling of epidermal growth factor with ^{99m}Tc and in vivo localization following intracerebral injection into normal and glioma-bearing rats. Bioconjug Chem 8:289–295

Caveliers V, Goodbody AE, Tran LL et al (2001) Evaluation of Tc-99m-RP128 as a potential inflammation imaging agent: human dosimetry and first clinical results. J Nucl Med 2001 42:154–161

Cesati RR, Tamagnan G, Baldwin RM et al (2002) Synthesis of cyclopentadienyltricarbonyl rhenium phenyltropanes by double ligand transfer: organometallic ligands for the dopamine transporter. Bioconjug Chem 13:29–39

Chianelli M, Signore A, Fritzberg AR et al (1997) The development of technetium-99m-labelled interleukin-2: a new radiopharmaceutical for the in vivo detection of mononuclear cell infiltrates in immune-mediated diseases. Nucl Med Biol 24:579–586

Claessens RAMJ, Boerman OC, Koenders EB et al (1996) Technetium-99m labelled hydrazinonicotinamido human non-specific polyclonal immunoglobulin G for detection of infectious foci: a comparison with two other technetium-labelled immunoglobulin preparations. Eur J Nucl Med 23:414–421

Colnot DR, Roos JC, de Bree R et al (2003) Safety, biodistribution, pharmacokinetics, and immunogenicity of ^{99m}Tc-labeled humanized monoclonal antibody BIWA 4 (bivatuzumab) in patients with squamous cell carcinoma of the head and the neck. Cancer Immunol Immunother 52:576–582

Cyr JE, Pearson DA, Manchanda R et al (1999) Characterization and radiolabelling chemistry of Tc-99m depreotide: a somatostatin receptor binding tumor imaging agent. J Nucl Med 40(Suppl):321

Dams ETM, Nijhof MW, Boerman OC et al (2000) Scintigraphic evaluation of experimental chronic osteomyelitis. J Nucl Med 41:896–902

Decristoforo C, Mather SJ (1999a) 99m-Technetium-labelled peptide-HYNIC conjugates: effects of lipophilicity and stability on biodistribution. Nucl Med Biol 26:389–396

Decristoforo C, Mather SJ (1999b) Technetium-99m somatostatin analogues: effect of labelling methods and peptide sequence. Eur J Nucl Med 26:869–876

Decristoforo C, Melendez-Alafort L, Sosabowski JK et al (2000) ^{99m}Tc-HYNIC-[Tyr^{3}]-octreotide for imaging somastotatin-receptor-positive tumors: preclinical evaluation and comparison with ^{111}In-octreotide. J Nucl Med 41:1114–1119

Drews A, Pietzsch HJ, Syhre R et al (2002) Synthesis and biological evaluation of technetium(III) mixed-ligand complexes with high affinity for the cerebral 5-HT1A receptor and the alpha1-adrenergic receptor. Nucl Med Biol 29:389–398

Duatti A (2004) In vivo Imaging of oligonucleotides with nuclear tomography. Curr Drug Targets 5:753–760

Eberle AN, Mild G, Froidevaux S (2004) Receptor-mediated tumor targeting with radiopeptides. Part 1. General concepts and methods: applications to somatostatin receptor-expressing tumors. J Rec Sig Trans 24:319–455

Edwards DS, Liu S, Harris AR et al (1999a) ^{99m}Tc-labelling of hydrazones of a hydrazinonicotinamide conjugated cyclic peptide. Bioconjug Chem 10:803–807

Edwards DS, Liu S, Ziegler MC et al (1999b) RP463: a stabilized technetium-99m complex of a hydrazinonicotinamide-derivatized chemotactic peptide for infection imaging. Bioconjug Chem 10:884–891

Egli A, Alberto R, Tannahill L et al (1999) Organometallic ^{99m}Tc-aquaion labels peptide to an unprecedented high specific activity. J Nucl Med 40:1913–1917

Fichna J, Janecka A (2003) Synthesis of target-specific radiolabelled peptides for diagnostic imaging. Bioconjug Chem 14:3–17

Francis RJ, Mather SJ, Chester K et al (2004) Radiolabelling of glycosylated MFE-23 CPG2 fusion protein (MFECP1) with Tc-99m for quantitation of tumour antibody-enzyme localisation in antibody-directed enzyme pro-drug therapy (ADEPT). Eur J Nucl Med Mol Imaging 31:1090–1096

Gabriel M, Decristoforo C, Maina T et al (2004) ^{99m}Tc-N_4-[Tyr^{3}]octreotate versus ^{99m}Tc-EDDA/HYNIC-[Tyr^{3}]octreotide: an intrapatient comparison of two novel technetium-99m labelled tracers for somatostatin receptor scintigraphy. Cancer Biother Radiopharm 19:73–79

Gali H, Hoffman TJ, Sieckman GL et al (2001) Synthesis, characterization, and labelling with Tc-99m/Re-188 of peptide conjugates containing a dithia-bisphosphine chelating agent. Bioconjug Chem 12:354–363

Giblin MF, Veerendra B, Smith CJ (2005) Radiometallation of receptor-specific peptides for diagnosis and treatment of human cancer. In Vivo 19:9–30

Goel A, Baranowska-Kortylewicz J, Hinrichs SH et al (2001) ^{99m}Tc-labelled divalent and tetravalent CC49 single-chain fv's: novel imaging agents for rapid in vivo localization of human colon carcinoma. J Nucl Med 42:1519–1527

Guggenberg von E, Mikolajczak R, Janota B et al (2004) Radiopharmaceutical development of a freeze-dried kit formulation for the preparation of [Tc-99m-EDDA-HYNIC-D-Phe(1),Tyr(3)]-octreotide, a somatostatin analog for tumor diagnosis. J Pharm Sciences 93:2497–2506

Harris TD, Sworin M, Williams N et al (1999) Synthesis of stable hydrazones of a hydrazinonicotinyl-modified peptide for the preparation of ^{99m}Tc-labelled radiopharmaceuticals. Bioconjug Chem 10:808

Heimbold I, Drews A, Syhre R et al (2002a) A novel technetium-99m radioligand for the 5-HT1A receptor derived from desmethyl-WAY-100635 (DWAY). Eur J Nucl Med 29:82–87

Heimbold I, Drews A, Kretzschmar M et al (2002b) Synthesis, biological and autoradiographic evaluation of a novel Tc-99m radioligand derived from WAY 100635 with high affinity for the 5-HT_{1A} receptor and the alpha1-adrenergic receptor. Nucl Med Biol 29:375–387

Hnatowich DJ, Nakamura K (2004) Antisense targeting in cell culture with radiolabelled DNAs – a brief review of recent progress. Ann of Nucl Med 18:363–368

Hom RK, Katzenellenbogen JA (1997) Technetium-99m-labelled receptor-specific small-molecule radiopharmaceuticals: recent developments and encouraging results. Nucl Med Biol 24:485–498

Iwasaki T, Iwasaki I, Aihara Y et al (2001) Immunoscintigraphy of aortic dissection with ^{99m}Tc-labelled murine anti-smooth muscle myosin monoclonal antibody in rats. J Nucl Med 42:130–137

Jamar F, Houssiau FA, Devogelaer JP et al (2002) Scintigraphy using a technetium 99m-labelled anti-E-selectin Fab fragment in rheumatoid arthritis. Rheumatology 41:53–61

Jeong JM, Hong MK, Lee J et al (2004) Tc-99m-neolactosylated human serum albumin for imaging the hepatic asialoglycoprotein receptor. Bioconjug Chem 15:850–855

Johannsen B, Pietzsch HJ (2002a) Development of technetium-99m-based CNS receptor ligands: have there been any advances? Eur J Nucl Med Mol Imaging 29:263

Johannsen B, Pietzsch H-J (2002b) Bioactivity of small technetium complexes In: Nicolini M., Mazzi U (eds) Technetium, rhenium and other metals in chemistry and nuclear medicine. SG Editoriali, Padova, Italy, pp 273–283

Jurisson SS, Lydon JD (1999) Potential technetium small molecule radiopharmaceuticals. Chem Rev 99:2205–2218

Kara G (2004) A novel mechanism for guanidino succinic acid (GSA) and technetium-99m-GSA, a novel agent for muscarinic acetylcholine receptor imaging. Eur J Nucl Med Mol Imaging 31(Suppl): 359–359

Kinne RW, Becker W, Koscheck T et al (1995) Rat adjuvant arthritis: imaging with technetium-99m-anti-CD4 fab fragments. J Nucl Med 36:2268–2275

Kobayashi H, Sakahara H, Saga T et al (1993) A human/mouse chimeric monoclonal antibody against CA125 for radioimmunoimaging of ovarian cancer. Cancer Immunol Immunother 37:143–149

Kokudo N, Vera DR, Makuuchi M (2003) Clinical application of TcGSA. Nucl Med Biol 30:845–849

Kung MP, Stevenson DA, Plössl K et al (1997) [Tc-99m]TRODAT-1: a novel technetium-99m complex as a dopamine transporter imaging agent. Eur J Nucl Med 24:372–380

Kushner SA, McElgin WT, Kung MP et al (1999) Kinetic modeling of [Tc-99m]TRODAT-1: a dopamine transporter imaging agent. J Nucl Med 40:150–158

Lahorte CMM, Vanderheyden JL, Steinmetz N et al (2004) Apoptosis-detecting radioligands: current state of the art and future perspectives. Eur J Nucl Med 31:887–919

Langer M, Beck-Sickinger AG (2001) Peptides as carrier for tumor diagnosis and treatment. Curr Med Chem Anticancer Agents 1:71–93

Le Bideau F, Salmain M, Top S et al (2001) New and efficient routes to biomolecules substituted with cyclopentadienyltricarbonylrhenium and -technetium derivatives. Chem Eur J 7:2289–2294

Leon A, Rey A, Mallo L et al (2002) Novel mixed ligand technetium complexes as 5-HT_{1A} receptor imaging agents. Nucl Med Biol 29:217–226

Lin KS, Luu A, Baidoo KE et al (2005) A new high affinity technetium-99m-bombesin analogue with low abdominal accumulation. Bioconjug Chem 16:43–50

Liu G, Mangera K, Liu N et al (2002) Tumor pretargeting in mice using ^{99m}Tc-labelled morpholino, a DNA analog. J Nucl Med 43:384–391

Liu GZ, He J, Dou SP et al (2004) Pretargeting in tumored mice with radiolabeled morpholino oligomer showing low kidney uptake. Eur J Nucl Med Mol Imaging 31:417–424

Liu S, Edwards DS (1999) Tc-99m-labelled small peptides as diagnostic radiopharmaceuticals. Chem Rev 99:2235–2268

Liu S, Edwards DS, Ziegler MC et al (2001) Tc-99m-Labelling of a hydrazinonicotinamide-conjugated vitronectin receptor antagonist useful for imaging tumors. Bioconjug Chem 12:624–629

Liu S, Edwards DS (2002) Fundamentals of receptor-based diagnostic metalloradiopharmaceuticals. In: Topics in current chemistry: contrast agents II, optical, ultrasound, x-ray and radiopharmaceutical imaging. Springer, Berlin Heidelberg New York, pp 259–278

Lupetti A, Welling MM, Mazzi U et al (2002) Technetium-99m labelled fluconazole and antimicrobial peptides for imaging of Candida albicans and Aspergillus fumigatus infections. Eur J Nucl Med Mol Imaging 29:674–679

Maecke HR (2005) Radiolabelled peptides in nuclear oncology: influence of peptide structure and labelling strategy on pharmacology. Ernst Schering Res Found Workshop 49:43–79

Maina T, Nock B, Nikolopoulou A et al (2002) ^{99m}Tc-Demotate, a new ^{99m}Tc-based [Tyr^3]octreotate analogue for the detection of somastotatin receptor-positive tumors: synthesis and preclinical results. Eur J Nucl Med 29:742–753

Meenakshi A, Ganesh V, Suresh Kumar R et al (2003) Radioimmuno targetting technetium-labeled anti epidermal growth factor receptor monoclonal antibodies in experimental tumor models. Q J Nucl Med 47:139–144

Mochizuki T, Kuge Y, Zhao S et al (2003) Detection of apoptotic tumour response in vivo after a single dose of chemotherapy with ^{99m}Tc-annexin V. J Nucl Med 44:92–97

Mu M, Kung MP, Plössl K et al (1999) Quantitation of Tc-TRODAT in human plasma samples by a simple extraction method. J Labelled Comp Radiopharm 42:213–215

Mull ES, Sattigeri VJ, Rodriguez AL et al (2002) Aryl cyclopentadienyl tricarbonyl rhenium complexes: novel ligands for the estrogen receptor with potential use as estrogen radiopharmaceuticals. Bioorg Med Chem 10:1381–1398

Nikolopoulou A, Maina T, Sotiriou P et al (2006) Tetraamine-modified octreotide and octreotate: labelling with Tc-99m and preclinical comparison in AR4-2J cells and AR4-2J tumor-bearing mice. J Peptide Science 12:124–131

Nock BA, Maina T, Yannoukakos D et al (1999) Glutathione-mediated metabolism of technetium-99m SNS/S mixed ligand complexes: a proposed mechanism of brain retention. J Med Chem 42:1066–1075

Nock B, Nikolopoulou A, Chiotellis E et al (2003) ^{99m}Tc-Demobesin 1, a novel potent bombesin analogue for GRP receptor-targeted tumour imaging. Eur J Nucl Med Mol Imaging 30:247–258

Okarvi SM (2004) Peptide-based radiopharmaceuticals: future tools for diagnostic imaging of cancers and other diseases. Med Res Rev 24:357–397

Ono M, Arano Y, Mukai T et al (2000) Control of radioactivity pharmacokinetics of ^{99m}Tc-HYNIC-labelled polypeptides derivatized with ternary ligand complexes. Bioconjug Chem 13:491–501

Pak JK, Alberto R (2001) Coordination reactions of glucose derivatives with the $[Tc(CO)_3]^+$ moiety for radiopharmaceutical application. J Labelled Comp Radiopharm 44:498–500

Papagiannopoulou D, Pirmettis I, Tsoukalas C et al (2002) Oxotechnetium ^{99m}TcO[SN(R)S][S] complexes as potential 5-HT_{1A} receptor imaging agents. Nucl Med Biol 29:825–832

Pasqualini R, Comazzi V, Bellande E et al (1992) A new efficient method for the preparation of Tc-99m-radiopharmaceuticals containing the TCN multiple bond. J Appl Radiat Isot 43:1329–1333

Pasqualini R, Duatti A, Bellande E et al (1994) Bis(dithiocarbamato)-nitrido Tc-99m-radiopharmaceuticals – a class of neutral myocardial imaging agents. J Nucl Med 35:334–341

Pieri P, Fischman AJ, Ahmad M et al (1991) Cardiac blood-pool scintigraphy in rats and hamsters: comparison of five radiopharmaceuticals and three pinhole collimator apertures. J Nucl Med 32:851–855

Pietzsch HJ, Gupta A, Syhre R et al (2001 a) Mixed-ligand technetium(III) complexes with tetradentate/monodentate NS3/isocyanide coordination: a new nonpolar technetium chelate system for the design of neutral and lipophilic complexes stable in vivo. Bioconjug Chem 12:538–544

Pietzsch HJ, Tisato F, Refosco F et al (2001 b) Synthesis and characterization of novel trigonal bipyramidal technetium(III) mixed-ligand complexes with SES/S/P coordination (E=O, N(CH3), S). Inorg Chem 40:59–64

Pietzsch HJ, Seifert S, Syhre R et al (2003) Synthesis, characterization, and biological evaluation of technetium(III) complexes with tridentate/bidentate S,E,S/P,S coordination (E=O, $N(CH_3)$, S): a novel approach to robust technetium chelates suitable for linking the metal to biomolecules. Bioconjug Chem 14:136–143

Postema EJ, Raemaekers JMM, Oyen WJG et al (2003) Final results of the phase I radioimmunotherapy trial using ^{186}Re-Epratuzumab for the treatment of patients with non–Hodgkin's lymphoma. Clin Cancer Res 9:3995s–4002s

Qin GM, Zhang YX, Cao W et al(2005) Molecular imaging of atherosclerotic plaques with technetium-99m-labelled antisense oligonucleotides. Eur J Nucl Med Mol Imaging 32:6–14

Ranadive GN, Rosenzweig HS, Epperly MW et al (1993) A new method of technetium-99m labelling of monoclonal antibodies through sugar residues. A study with TAG-72 specific CC-49 antibody. Nucl Med Biol 20:719–726

Refosco F, Bolzati C, Duatti A et al (2000) Mixed ligand Tc- and Re-nitrido complexes for radiolabelling bioactive molecules. Recent Res Devel Inorganic Chem 2:89–98

Reilly RM (1993) Immunoscintigraphy of tumors using tc-99m-labelled monoclonal-antibodies – a review. Nucl Med Commun 14:347–359

Rennen HJJM, Boerman OC, Oyen WJG et al (2001) Specific and rapid scintigraphic detection of infection with Tc-99m-labeled interleukin-8. J Nucl Med 42:117–123

Rennen HJJM, Boerman OC, Oyen WJG et al (2003 a) Kinetics of ^{99m}Tc-labelled interleukin-8 in experimental inflammation and infection. J Nucl Med 44:1502–1509

Rennen HJ, Oyen WJ, Cain SA et al (2003 b) Tc-99m-labelled C5a and C5a des Arg^{74} for infection imaging. Nucl Med Biol 30:267–272

Rennen HJJM, Frielink C, Brandt E et al (2004) Relationship between neutrophil-binding affinity and suitability for infection imaging: comparison of ^{99m}Tc-labelled NAP-2 (CXCL-7) and 3C-terminally truncated isoforms. J Nucl Med 45:1217–1223

Rose DJ, Maresca KP, Nicholson T et al (1998) Synthesis and characterization of organohydrazino-complexes of technetium, rhenium and molybdenum with the $\{M(\eta^1\text{-}H_xNNR)(\eta^2\text{-}H_yNNR)\}$ core and their relationship to radiolabelled organohydrazine-derivatized chemotactic peptides with diagnostic applications. Inorg Chem 37:2701–2716

Saidi M, Kothari K, Pillai MRA et al (2001) Cyclopentadienyl 99m-technetium tricarbonyl piperidin conjugate: biodistribution and imaging studies. J Labelled Cpd Radiopharm 44:603–618
Saidi M, Seifert S, Kretzschmar M et al (2004) Cyclopentadienyl tricarbonyl complexes of 99m-Tc for the in vivo imaging of the serotonin 5-HT$_{1A}$ receptor in the brain. J Organomet Chem 689:4739–4744
Samnick S, Scheuer C, Munks S et al (2004) Technetium-99m labelled 1-(4-fluorobenzyl)-4-(2-mercapto-2-methyl-4-azapentyl)-4-(2-mercapto-2-methylpropylamino)-pipe ridine and iodine-123 metaiodobenzylguanidine for studying cardiac adrenergic function: a comparison of the uptake characteristics in vascular smooth muscle cells and neonatal cardiac myocytes, and an investigation in rats. Nucl Med Biol 31:511–522
Schaadt BK, Hendel HW, Gimsing P et al (2003) ^{99m}Tc-aprotinin scintigraphy in amyloidosis. J Nucl Med 44:177–183
Schechter NR, Yang DJ, Azhdarinia et al (2003) Assessment of epidermal growth factor receptor with 99mTc-ethylenedicysteine-C225 monoclonal antibody. Anticancer Drugs 14:49–56
Schwarz A, Steinstrasser A, Bosslet K (1988) A simple procedure of Tc-99m labelling for monoclonal-antibodies. Eur J Nucl Med 14:C8–C8
Schwartz DA, Abrams MJ, Hauser MM et al (1991) Preparation of hydrazino-modified proteins and their use for the synthesis of Tc-99m-protein conjugates. Bioconjug Chem 2:333–336
Schwochau K (2000) Technetium chemistry and radiopharmaceutical applications. Wiley, New York
Seifert S, Drews A, Gupta A et al (2000) Stability studies on 99m-technetium(III) complexes with tridentate/monodentate thiol ligands and phosphine ("3+1+1" complexes). Appl Radiat Isot 53:431–438
Seifert S, Gupta A, Syhre R et al (2001) Ligand-exchange reaction of labile "3+1" Tc-99m(V) complexes with SH group-containing proteins. Int J Appl Radiat Isot 54:637–644
Seifert S, Künstler JU, Schiller E (2004) Novel procedures for preparing ^{99m}Tc(III) complexes with tetradentate/monodentate coordination of varying lipophilicity and adaptation to ^{188}Re analogues. Bioconjug Chem 15: 856–863
Sharma SD, Cai HZ, Yang W et al (2000) Melanocortin receptor-1 specific Tc-99m-metallopeptides for targeting melanoma. J Nucl Med 41:1021S
Signore A (1995) Receptor ligands. Q J Nucl Med 39:83–85
Signore A, Annovazzi A, Chianelli M et al (2001) Peptide radiopharmaceuticals for diagnosis and therapy. Eur J Nucl Med 28:1555–1565
Signore A, Annovazzi A, Barone R et al (2004) ^{99m}Tc-interleukin-2 scintigraphy as a potential tool for evaluating tumor-infiltrating lymphocytes in melanoma lesions: a validation study. J Nucl Med 45:1647–1652
Simms MS, Perkins AC, Price MR et al (2001) ^{99m}Technetium-C595 radioimmunoscintigraphy: a potential staging tool for bladder cancer. BJU Int 88:686–691
Spies H, Pietzsch HJ, Johannsen B (1999) The "n+1" mixed-ligand approach in the design of specific technetium radiopharmaceuticals: potentials and problems. In: Nicolini M, Mazzi U (eds) Tc, Re and other metals in chemistry and nuclear medicine, vol. 5. SG Editoriali, Padova, Italy, pp 101–108
Spradau TW, Edwards WB, Anderson CJ et al (1999) Synthesis and biological evaluation of Tc-99m-cyclopentadienyltricarbonyltechnetium-labelled octreotide. J Nucl Med Biol 26:1–7
Stadalnik RC, Vera DR (2001) The evolution of ^{99m}Tc-NGA as a clinically useful receptor-binding radiopharmaceutical. Nucl Med Biol 28:499–503
Steffens MG, Oosterwijk E, Kranenborg MHGC et al (1999) In vivo and in vitro characterization of three ^{99m}Tc-labeled monoclonal antibody G250 preparations. J Nucl Med 40:829–836
Stroomer JWG, Roos JC, Sproll M et al (2000) Safety and biodistribution of ^{99m}Technetium-labeled anti-CD44v6 monoclonal antibody BIWA 1 in head and neck cancer patients. Clin Cancer Res 6:3046–3055
Su ZF, Liu G, Gupta S et al (2002) In vitro and in vivo evaluation of a technetium-99m labelled cyclic RGD petide as a specific marker of $\alpha_v\beta_3$ integrin for tumor imaging. Bioconjug Chem 13:561–570
Syhre R, Seifert S, Spies H, et al (1998) Stability versus reactivity of "3+1" mixed-ligand technetium-99m complexes in vitro and in vivo. Eur J Nucl Med 25:793–796
Taillefer R, Boucher L, Lambert R et al (1995) Technetium-99m antimyosin antibody (3–48) myocardial imaging: human biodistribution, safety and clinical results in detection of acute myocardial infarction. Eur J Nucl Med 22:453–464
Tait JF, Brown DS, Gibson DF et al (2000) Development and characterization of annexin V mutants with endogenous chelation sites for ^{99m}Tc. Bioconjug Chem 11:918–925

Tait JF, Smith C, Gibson DF (2002) Development of annexin V mutants suitable for labelling with Tc(I)-carbonyl complex. Bioconjug Chem 13:1119–1123

Tang Y, Scollard D, Chen P et al (2005) Imaging of HER2/neu expression in BT-474 human breast cancer xenografts in athymic mice using [Tc-99m]-HYNIC-trastuzumab (Herceptin) Fab fragments. Nucl Med Commun 26:427–432

Thakur ML, DeFulvio J, Richard MD et al (1991) Tc-99m labelled monoclonal-antibodies – evaluation of reducing agents. Nucl Med Biology 18:227–233

Tisato F, Refosco F, Porchia M, et al (2004) The crucial role of the diphosphine heteroatom X in the stereochemistry and stabilization of the substitution-inert [M(N)(PXP)](2+) metal fragments (M Tc, Re; PXP diphosphine ligand). Inorg Chem 43:8617–8625

Tokita N, Hasegawa S, Murayama K et al (2003) 99m-Tc-HYNIC-annexin V imaging to evaluate inflammation and apoptosis in rats with autoimmune myocarditis. Eur J Nucl Med Mol Imaging 30:232–238

Usman N, Blatt LRM (2000) Nuclease-resistant synthetic ribozymes: developing a new class of therapeutics. J Clin Invest 106:1197–1202

Vallabhajosula S, Moyer BR, Lister-James J et al (1996) Preclinical evaluation of technetium-99m-labelled somatostatin receptor-binding peptides. J Nucl Med 37:1016–1022

Vanderheyden JL, Verbeke K, Kieffer D et al (2002) Product development and formulation of ^{99m}Tc NYNIC-rh-annexin V. In: Nicolini M, Mazzi U (eds) Tc, Re and other metals in chemistry and nuclear medicine, vol. 6. SG Editoriali, Padova, Italy, pp 335–338

Verbeke K, Hjelstuen O, Debrock E et al (1995) Comparative evaluation of ^{99m}Tc-HYNIC-HSA and ^{99}Tc^{m}-MAG3-HSA as possible blood pool agents. Nucl Med Comm 16:942–957

Virgolini I, Leimer M, Handmaker H et al (1998) Somatostatin receptor subtype specific and in vivo binding of a novel tumor tracer, ^{99m}Tc-P829. Cancer Res 58:1850–1859

Waibel R, Alberto R, Willuda J et al (1999) Stable one-step technetium-99m labelling of His-tagged recombinant proteins with a novel Tc(I)-carbonyl complex. Nature Biotechnol 17:897–901

Wald J, Alberto R, Ortner K et al (2001) Aqueous one-pot synthesis of derivatized cyclopentadienyl-tricarbonyl complexes of Tc-99m with an in situ CO source: application to a serotonergic receptor ligand. Angew Chem Int Ed 40:3062–3066

Welling MM, Paulusma-Annema A, Balter HS et al (2000) Technetium-99m labelled antimicrobial peptides discriminate between bacterial infections and sterile inflammation. Eur J Nucl Med 27:292–301

Welling MM, Mongera S, Lupetti A et al (2002) Radiochemical and biological characteristics of 99m Tc-UBI 29-41 for imaging of bacterial infections. Nucl Med Biol 29:413–422

Welling MM, Korsak A, Gorska B et al (2005) Kit with technetium-99m labelled antimicrobial peptide UBI 29-41 for specific infection detection. J Labelled Comp Radiopharm 48:683–691

Wenzel M, Klinge C (1994) Tc-99m-labelled estradiol derivatives – synthesis, organ distribution and tumor affinity. J Labelled Cpd Radiopharm 34:981–987

Willuda J, Kubetzko S, Waibel R et al (2001) Tumor targeting of mono-, di-, and tetravalent anti-p185^{HER-2} miniantibodies multimerized by self-associating peptides. J Biol Chem 17:14385–14392

Wong E, Bennett S, Lawrence B et al (2001) Tuftsin receptor-binding peptide labeled with technetium: chemistry and preliminary in vitro receptor-binding study. Inorg Chem 40:5695–5700

Younes CK, Boisgard R, Tavitian B (2002) Labelled oligonucleotides as radiopharmaceuticals: pitfalls, problems and perspectives. Curr Pharm Des 8:1451–1466

Zinn KR, Kelpke S, Chaudhuri TR et al (2000) Imaging Tc-99m-Labeled FGF-1 targeting in rats. Nucl Med Biol 27:407–414

Chapter 3

Stannous Chloride in the Preparation of ^{99m}Tc Pharmaceuticals

3

H. Spies and H.-J. Pietzsch

3.1 Introduction

^{99m}Tc pharmaceuticals need to be labeled by a simple procedure shortly before use. Because there is no effective chemistry available to attach a pertechnetate ion to an organic moiety, reduction of Tc(VII) in TcO_4^- to a lower oxidation state is a prerequisite for ^{99m}Tc complex formation in high yield and purity. Experience with various ligands has shown that the oxidation state of technetium is affected by the nature of the reducing agent, the chelator, and the reaction conditions. The choice of a suitable reducing agent for one-step labeling at mild pH conditions has been a major research effort.

Requirements of an "ideal" reducing agent for kit preparation:

- Effective reduction at mild pH conditions
- Formation of a single-component complex with distinct oxidation state
- No interference with the complexation process
- Not included in the final complex
- Stable during storage of the kit (long shelf-life)

Mild reaction conditions mean a neutral or weakly acidic pH, exclusion of toxic substances, and labeling at room temperature. Certain ligand systems favor distinct oxidation states; therefore, in kit formulations the nature and amount of reducing agent should be in balance with the ligand to ensure quantitative conversion of pertechnetate for complex formation, without further reduction to lower oxidation states. The reductant should not participate or interfere with the complexation process. Neither the reductant itself nor its oxidized form is part of the tracer molecule.

Several reducing agents with a reduction potential below that of pertechnetate (+0.747 V, Schwochau 2000) are capable of reducing pertechnetate in aqueous solution. In the absence of an appropriate ligand, reduction proceeds to insoluble technetium(IV) oxide $TcO_2 \cdot xH_2O$. In order to avoid colloid formation, reductions are generally carried out in the presence of a ligand that will stabilize a lower valence state of a technetium complex, thus limiting colloid formation (Srivastava and Richards 1983; Fig. 3.1).

Labeling and, in particular pertechnetate reduction, have been the topic of a series of articles (Alvarez 1975; Clarke and Podbielski 1987; Eckelman and Steigman 1991; Noronha 1978; Novotnik 1990; Rhodes 1991; Srivastava et al. 1977; Steigman and Richards 1974).

From the beginning of technetium-99m chemistry, a wide range of reducing agents have been used for pertechnetate, examples are given below (Srivastava and Richards 1983):

- Ferric chloride and ascorbic acid has been used as a reducing mixture to prepare ^{99m}Tc-labeled albumin (Persson and Liden 1969; Stern et al. 1965; Yokoyama et al. 1975). Ascorbate ion itself does not reduce technetium efficiently. However, ferric salt is reduced at acidic pH, generating ferrous ion, which upon elevation of pH can

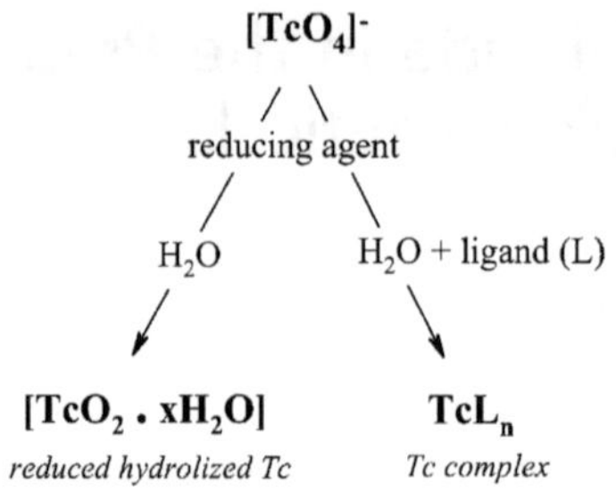

Fig. 3.1. Reduction and complex formation of pertechnetate

reduce pertechnetate; labeling of albumin occurred after adjustment to acidic pH. The reaction required strict control of the reaction conditions, several pH adjustments, and was therefore unsuitable for kit formulation.

- The use of borohydride (Deutsch et al. 1980; Smith et al. 1978) is presently limited to the synthesis of the *fac*-$[^{99m}Tc^{(I)}(H_2O)_3(CO)_3]^+$ as a precursor for technetium(I)-based radiopharmaceuticals (Alberto and Abram 2003). The typical feature of this reagent is production of an inhomogeneous mixture of various Tc species, even under well-controlled reaction conditions, including metallic technetium.
- Certain ligands, preferably those with phosphine (Deutsch et al. 1981; Vanderheyden et al. 1984) and thiol groups (Spies et al. 1978) also act as reductants. Although the reduction rate is low and an excess of ligand is required for quantitative yield, P and S ligands may have some advantages. Thus, water-soluble triphenylphosphine sulfonates suitably reduce disulfide bonds and pertechnetate simultaneously, which might offer advantages for direct labeling of peptides (Greenland et al. 2002).
- Hydrohalic acids (Hal = Cl, Br, I) (Thomas et al. 1979) are effective reducing agents for technetium, but their use requires harsh reaction conditions that lie outside the radiopharmaceutical milieu.
- Metallic reducing agents are convenient and efficient. Their disadvantage lies in the formation of colloids due to hydrolysis in aqueous solution, e.g., tin (II) and iron(II) (Lin et al. 1971), zinc metal (Kremer et al. 1989) or salts of titanium(III) (Kalincak et al. 1982), antimony(III) (Vilcek et al. 1982), molybdenum(III) (Vilcek et al. 1984), or tungsten(III) (Vilcek et al. 1985).
- Electrolytical reduction of ^{99m}Tc-pertechnetate has been investigated (Benjamin 1969, 1970; Dworkin and Gutkowski 1971; Eckelman et al. 1971 a; Gil et al. 1976). When using zirconium or tin electrodes, anodic dissolution of metal ions produced in situ reduction of pertechnetate (Steigman et al. 1974). Electrolysis has been used as a reliable method for laboratory production of ^{99m}Tc pharmaceuticals.
- Ferrous salt and tin(II) were identified as active reductants (Lin et al. 1971), producing labeling of human serum albumin (HSA) at pH 2.5. Ferrous salt needed pH elevation to increase its reduction potential (Zolle et al. 1975).

The advantages of stannous salts for kit formulation have been demonstrated (Eckelman and Richards 1970, 1972), and the stability of kits was increased by lyophilization (Deutsch and Redmond 1972), suggesting commercial production.

3.2 Stannous Chloride: the Preferred Reducing Agent for Tc Pharmaceuticals

The need for simple labeling methods for ^{99m}Tc pharmaceuticals had been expressed by scientists and physicians as early as 1965. The introduction of stannous ion as a reductant for kit preparation offered new perspectives in the development of ^{99m}Tc radiopharmaceuticals and attracted interest of all parties (Alvarez 1975; Eckelman et al. 1971b).

Kits are a powerful tool in ^{99m}Tc chemistry, offering labeling in isotonic solution at room temperature simply by adding the ^{99m}Tc activity in a suitable volume. Stannous salts are nontoxic, and stable when lyophilized and kept in a nitrogen atmosphere. Stannous salts are a reliable reductant used in all kit formulations.

Redox Chemistry. Tin forms compounds in the oxidation states +II and +IV. Potentials for the sequence $Sn^0 \rightarrow Sn^{II} \rightarrow Sn^{IV}$ in acidic and basic media are shown below (Wardell 1994):

Acidic solution:	Sn^0, 0.136 V	Sn^{II}, −0.15 V	Sn^{IV}
Basic solution:	Sn^0, 0.91 V	$Sn^{II}(OH)_3^-$, 0.93 V	$[Sn^{IV}(OH)_6]^{2-}$

The feature that makes tin(II) so interesting for Tc pharmaceutical preparation is the ease of the stannous ion to be oxidized to tin(IV) according to the reaction

$$Sn^{2+} \rightarrow Sn^{4+} + 2e^-$$

There were many investigations to explain the mechanism of reduction. Since direct chemical measurements are out of question at carrier-free concentrations of ^{99m}Tc (10^{-9} M), carrier technetium (^{99}Tc) in hydrochloric acid was used to determine the oxidation state of technetium in diethylene triamine pentaacetate (DTPA) and in citrate solution. Polarographic and iodometric techniques were used to analyze for unreacted stannous ion and to perform direct potentiometric titrations of pertechnetate-99 with stannous chloride (Münze 1980; Steigman et al. 1975). No quantitative kinetic studies had been made, but qualitative conclusions have been drawn for the reduction mechanism. Most probably, the first step is the reduction to Tc(V). Reduction to Tc(III) proceeds in two successive complementary reactions, both of which should be rapid in the low concentrations at radiopharmaceutical level:

$$\mathrm{Tc(VII)} + \mathrm{Sn(II)} \rightarrow \mathrm{Tc(V)} + \mathrm{Sn(IV)} \quad (1)$$

$$\mathrm{Tc(V)} + \mathrm{Sn(II)} \rightarrow \mathrm{Tc(III)} + \mathrm{Sn(IV)} \quad (2)$$

Whether the reaction stops at the Tc(V) (reaction a) or Tc (III) (reaction b) oxidation state, or subsequent reactions occur, e.g., to Tc(IV), is dependent primarily on the nature of the ligand applied. Incidentally, the first established case of the existence of a Tc(V) compound in water came from a study of Tc(V) citrate (Steigman et al. 1975).

Table 3.1. Content of stannous chloride and calculated Sn-to-Tc ratios in selected commercially available cold kits

Technetium kit	$SnCl_2 \times 2H_2O$ (mg)	Sn/Tc ratio [a]
MDP	0.5 [b]	1.2×10^5
HSA microspheres B1	0.4 [b]	9×10^4
DMSA	0.4 [b]	9×10^4
DTPA	0.3 [b]	7×10^4
HIDA	0.2 [b]	5×10^4
EC	0.2 [b]	5×10^4
HSA microspheres B20	0.1 [b]	2×10^4
MIBI (Cardiolite)	0.075	1.7×10^4
MAG_3	0.06 [a]	1.4×10^4
Tetrofosmin (Myoview)	0.03	7×10^3
ECD (Neurolite)	0.008	2×10^3
HMPAO (Ceretec)	0.0076	2×10^3

MDP methylenediphosphonate, *HSA* human serum albumin, *DMSA* dimercaptosuccinic acid, *DTPA* diethylene triamine pentaacetate, *HIDA* hepatoiminodiacetic acid, *EC* ethylene dicysteine, *HSA* human serum albumin, *MIBI* monodentate methoxyisobutyl isocyanide, MAG_3 mercaptoacetyltriglycine, *ECD* ethylene dicysteine dimer, *HMPAO* hexamethylpropylene amine oxime

[a] ^{99m}Tc eluate (370 MBq), ^{99}Tc is not considered

[b] Average of different kit formulations

Tin in the Labeling Process. Although stannous ion has become the reducing agent of choice, some inherent problems have to be considered:

- Complicated solution chemistry of stannous compounds
- Product contains Sn(IV)
- Easily oxidized to Sn(IV)
- Shelf-life

Stannous compounds have a complicated solution chemistry. $SnCl_2{\cdot}2H_2O$ is very difficult to purify; the purest commercially available product contains at least 5% of Sn(IV) (Donaldson and Moser 1960). Stannous ion is readily oxidized by various oxidants, such as the oxygen in air. Oxidation proceeds already on standing, but is more critical in both the freeze-dried kit formulation, and in particular in solution during reconstitution because of the low Sn concentration (Table 3.1). A minimum concentration of stannous ion must be maintained to guarantee reduction during the shelf-life. First, the amount of stannous ion must ensure the Sn(II) capacity required for reduction of both ^{99m}Tc and excess of long-lived ^{99}Tc-pertechnetate present in generator eluates. Second, since the reduction potential of a solution of $SnCl_2$ depends on the ratio of activities of Sn(IV) and Sn(II), the ratio of Sn(II) to Sn(IV) must not be too low. Oxygen and other oxidizing agents have to be carefully excluded from stannous chloride preparations. Due to oxidation on standing and side oxidation reactions, Sn(IV) is an unavoidable, relatively concentrated impurity in radiopharmaceutical preparations.

The amount of stannous chloride is empirically optimized for each individual kit formulation, maintaining the balance between two parameters: A large excess of stannous chloride should be used with respect to the added pertechnetate activity and the amount of stannous chloride kept as low as possible in order to avoid further reduction of pertechnetate to a lower oxidation state. In addition, the level of tin(IV) impurity in the radiopharmaceutical should be kept as low as possible. The calculated optimal content of stannous chloride in a series of commercial kits is shown in Table 3.1, where the stannous chloride amount per vial covers the range from 0.0076–0.5 mg, corresponding to a ratio of Sn to Tc in the range of 10^3 to 10^5.

The amount of reductant must be strictly controlled, in particular when the ligand used stabilizes technetium in more than one oxidation states.

In direct labeling of proteins, stannous chloride is administered for "pretinning". By incubating the protein with stannous ion, reactive disulfide bonds are reduced in addition to pertechnetate (Eckelman and Steigman 1991; Rhodes 1991).

The reaction kinetics of reduction and complexation are actually affected by the "usable" tin(II) and the ligand/tin ratio, summarized in the subsequent recommendations (Srivastava et al. 1977):

- Stannous solution used for formulation should be prepared with great caution to avoid oxidation and hydrolysis.
- A minimum quantity of Sn(II) and an excess of the complexing ligand (as optimized for a particular kit system) should be used.
- If the kit contains very little usable tin(II), the carrier content of $^{99m}TcO_4^-$ solutions should be evaluated. Total technetium sometimes may exceed the reductive capacity of tin.

Problems Associated with the Use of Stannous Ion. The kits could fail in the way that only a fraction of the original tin may be available in the desired form at reconstitution. Situations may occur that the kit contains very little usable tin(II), and the carrier content of the eluate may exceed the reductive capacity of tin. This is critical with kits containing a very small quantity of Sn(II) (see Table 3.1) (Srivastava et al. 1977).

Furthermore, an undesirable side reaction between tin and technetium may occur. As outlined above, there is a high excess of tin – as Sn(II) and Sn(IV) – over technetium and this fact leads to the idea that mixed-metal complexes may be formed in radiopharmaceutical preparations. Interest in the question, whether stannous or stannic tin could be involved in the radiopharmaceutical was further stimulated by the formation of a tin-capped ^{99}Tc-dimethylglyoxime complex, $^{99}Tc(oxime)_3(\mu\text{-}OH)SnCl_3$ (Deutsch et al. 1976). However, this compound was prepared under the condition of carrier-added technetium; its ready conversion to uncapped species gives no evidence for the existence of mixed-metal type compounds.

The behavior of stannous ion in kit preparations has been studied in a limited number of compounds, and the conclusion reached so far indicates that, apparently, tin ions only reduce TcO_4^- and indeed, apart from some tin-essential preparations such as stannous oxide colloid labeled with ^{99m}Tc (Subramanian and McAfee 1970), mixed Tc–Sn complexes have not been observed in low-molecular-weight radiopharmaceutical preparations.

Another aspect is hydrolysis and colloid formation. During the reduction of pertechnetate with stannous salts, tin is oxidized and hydrolyzed to form highly polydispersed colloidal particles. In some cases, mainly when weak or unsuitable ligands are used in the ^{99m}Tc labeling, interference of colloidal tin oxides on the biodistribution of ^{99m}Tc-radiolabeled tracers may occur. Such effects and the biodistribution of ^{99m}Tc-Sn colloid in dependence of the preparation conditions were subject of detailed studies.

Some authors hope to prevent complications related to the use of stannous chloride by using other stannous salts such as stannous fluoride, oxalate, tartrate, citrate, and phosphates.

Some efforts were made to use stannous ion fixed to a carrier to improve the labeling procedure. Albumin was labeled using stannous ion adsorbed to an ion-exchange resin in the presence of albumin and pertechnetate (Dreyer and Münze 1969); an insoluble macromolecular Sn(II) complex (R-Sn) was proposed in order to avoid the disadvantages of stannous chloride being hydrolized and oxidized (Nakayama et al. 1995).

Reduced, hydrolized ^{99m}Tc is a colloidal impurity in kit preparations. Depending on the amount present, it may distort the biodistribution pattern and limit the diagnostic value of the ^{99m}Tc pharmaceutical. The biodistribution of ^{99m}Tc-Sn colloid itself was studied separately (Syhre et al. 1976). The authors observed that the liver-to-kidney ratio is highly dependent on the pH of the preparation, reflecting the hydrolytic properties of tin hydroxide.

Furthermore, it has been shown that tin participates in the chemical binding process between technetium and tetracycline and similarly, it might very well explain how some other complexes are produced (Alvarez 1975).

A more recent example is the biodistribution of ^{99m}Tc-radiolabeled chitosan nanoparticles, using two different methods for reduction, namely, stannous chloride or sodium borohydride. The authors reported a considerable dependence on the method used. They concluded that nanoparticles are adsorbed on the surface of colloidal tin oxide particles that are generated during the labeling process (Banerjee et al. 2005).

Tin as a Foreign Element in the Body. The toxicity of tin compounds is related to the chemical form. The LD_{50} of stannous chloride in dogs (intravenous) is 20–50 mg/kg. The toxicology of colloidal tin oxide was thoroughly investigated (Fisher 1957). Data indicated that four doses as high as 350 mg/kg of tin produced no toxicity. Because of the low doses of tin applied in ^{99m}Tc pharmaceuticals, no toxic effects are expected.

Determination of Sn(II) (Quality Control). Determination of the Sn(II) content in radiopharmaceutical kits is an important aspect of quality control for commercial producers and registration authorities (Rakias and Zolle 1997). Classical methods for the determination of Sn(II) involve titrimetric, electrochemical, spectrophotometric, and chromatographic methods (Lejeune et al. 1996).

Titrimetric methods that are selective for Sn(II) are generally less sensitive. Disadvantages that may occur are related to so-called matrix effects, that is, the reaction of milieu, including the presence of ligands or stabilizers, low sensitivity, or slow formation of colored complex. Thus, iodometric titration as well as spectrophotometry by means of Dithiol-Reagent R are used to determine Sn(II) in, e.g., stock solution, but are not suitable to determine Sn(II) in kits. Methods that allow specification of Sn(II) and Sn(IV) are few. There is, as an example, a colorimetric determination of the Sn(II) concentration in Tc-mercaptoacetyltriglycine (MAG_3) kits based on a molybdate–thiocyanate complex formed in the presence of Sn(II) (Hoffmann et al. 1990). Pulse polarography has offered considerable advantages over the classical methods for the determination of the Sn(II) content, in particular an increase of sensitivity that allows suitable measurements of microgram amounts of tin(II) (Lejeune et al. 1996; Rakias and Zolle 1997).

References

Alberto R, Abram U (2003) ^{99m}Tc: labeling chemistry and labeled compounds. In: Vertés A, Nagy S, Klenscár Z (eds) Radiochemistry and radiopharmaceutical chemistry in life science, handbook of nuclear chemistry, vol. 4. Kluwer, Dordrecht, pp 211–256

Alvarez J (1975) Radiopharmaceuticals prepared with stannous chloride. J Radioanal Chem 27:475–482

Banerjee T, Singh AK, Sharma RK, Maitra AN (2005) Labelling efficiency and biodistribution of technetium-99m labeled nanoparticles: interference by colloidal tin oxide particles. Int J Pharm 289:189–195

Benjamin PP (1969) A rapid and efficient method of preparing ^{99m}Tc-human serum albumin: its clinical applications. Int J Appl Radiat Isot 20:187–194

Benjamin PP, Rejali A, Friedell H (1970) Electrolytic complexation of ^{99m}Tc at constant current: its application in nuclear medicine. J Nucl Med 11:147–154

Clarke MJ, Podbielski L (1987) Medical diagnostic imaging with complexes of ^{99m}Tc. Coord Chem Rev 78:253–331

Deutsch ME, Redmond ML (1972) Unitary freeze-dried kits for preparation of technetium-labeled human serum albumin. J Nucl Med 13(Abstr):426–427

Deutsch E, Elder RC, Lange BA, Vaal MJ, Lay DG (1976) Structural characterization of a bridged ^{99}Tc-Sn-dimethylglyoxime complex – implications for chemistry of 99mTc-radiopharmaceuticals prepared by Sn(II) reduction of pertechnetate. Proc Natl Acad Sci USA 73:4289

Deutsch E, Libson K, Becker CB (1980) Preparation and biological distribution of technetium diphosphonate radiotracers synthesized without stannous ion. J Nucl Med 21:859–866

Deutsch E, Glavan KA, Sodd VJ (1981) Cationic ^{99m}Tc complexes as potential myocardial imaging agents. J Nucl Med 22:897–907

Donaldson JD, Moser W (1960) Pure tin(II) sulfate. J Chem Soc 4000–4003

Dreyer R, Münze R (1969) Markierung von Serumalbumin mit ^{99m}Technetium [in German]. Wiss Z Karl Marx Univ Leipzig, Math-Naturwiss R 18:629–633

Dworkin HJ, Gutkowski RF (1971) Rapid closed system production of ^{99m}Tc-albumin using electrolysis. J Nucl Med 12:562–565

Eckelman WC, Richards P (1970) Instant ^{99m}Tc-DTPA. J Nucl Med 11:761–762

Eckelman WC, Meinken G, Richards P (1971a) Chemical state of ^{99m}Tc in biomedical products. J Nucl Med 12:596–600

Eckelman WC, Meinken G, Richards P (1971b) ^{99m}Tc-human serum albumin. J Nucl Med 12:707–710

Eckelman WC, Richards P (1972) Analytical pitfalls with ^{99m}Tc-labelled compounds. J Nucl Med 13:202–204

Eckelman WC, Steigman J (1991) Direct labelling with ^{99m}Tc. Nucl Med Biol 18:3–7

Fischer HW (1957) Colloidal stannic oxide – animal studies on a new hepatolienographic agent. Radiology 68:488–498

Gil MC, Palma T, Radicella R (1976) Electrolytical labelling of ^{99m}Tc-radiopharmaceuticals. Int J Appl Radiat Isot 27:69

Greenland WEP, Fogelman I, Blower PJ (2002) Direct labelling via simultaneous reduction of pertechnetate and a disulfide bond. In: Nicolini M, Mazzi U (eds) Tc, Re and other metals in chemistry and nuclear medicine 6. SG Editoriali, Padova, Italy, pp 495–498

Hoffmann I, May K, Noll B (1990) Determination of tin(II) concentration in ROTOP MAG-3 kit. Zentralinstitut Kernforschung Rossendorf, Dresden, Annual Report ZfK 711:46–47

Kalincak M, Machan V, Vilcek S (1982) ^{99m}Tc-labelled compounds prepared with Ti(III) as reducing agent. Isotopenpraxis 18:334–338

Kremer C, Leon A, Gambino D (1989) Solid-phase reduction of (TcO_4^-)-^{99m}Tc with zinc – a method for the preparation of difficult ^{99m}Tc compexes. J Labelled Comp Radiopharm 27:1331–1340

Lejeune R, Thunus J, Thunus L (1996) Polarographic determination of (Sn(II) in samples containing Sn(IV) such as in technetium-99m radiopharmaceutical kits. Anal Chim Acta 332:67–71

Lin SM, Winchell HS, Shipley BA (1971) Use of Fe(II) or Sn(II) alone for technetium labelling of albumin. J Nucl Med 12:204–211

Münze R (1980) Electrochemical investigation on the reduction of $^{99}TcO_4^-$ with tin(II) in the presence of citrate. Radiochem Radioanal Lett 43:219–224

Nakayama M, Terahara T, Wada M, Ginoza Y, Harada K, Sugii A, Nakayama H (1995) Insoluble macromolecular Sn(II) complex for the ^{99m}Tc labelling of protein-bearing mercapto groups. In: Nicolini M, Mazzi U (eds) Tc, Re and other metals in chemistry and nuclear medicine 4. SG Editoriali, Padova, Italy, pp 299–320

Noronha OPD (1978) Time-dependent characteristics on Sn-complexes for preparing ^{99m}Tc-labelled radiopharmaceuticals and their bioavailabilities – a review. Nuklearmedizin 17:110–125

Novotnik DP (1990) Physico-chemical concepts in the preparation of technetium radiopharmaceuticals. In: Sampson CB (ed) Textbook of radiopharmacy: theory and practice. Gordon and Breach, Philadelphia, pp 29–49

Persson RBR, Liden K (1969) ^{99m}Tc-labelled human serum albumin – a study of labelling procedure. Int J Appl Radiat Isot 20:241

Rakias F, Zolle I (1997) Stannous ion determination: importance and relevance for radiopharmaceutical kits. In: Bergmann H, Kroiss A, Sinzinger H (eds) Radioactive isotope in clinical medicine and research, vol. 22. Birkhauser, Basel, pp 401–409

Rhodes BA (1991) Direct labelling of proteins with ^{99m}Tc. Nucl Med Biol 18:667–676

Schwochau K (2000) Technetium – chemistry and radiopharmaceutical applications. Wiley, New York, p 44

Smith JE, Byrne EF, Cotton FA (1978) Thiol complex of technetium pertinent to radiopharmaceutical use of ^{99m}Tc. J Amer Chem Soc 100:5571–5572

Spies H, Johannsen B, Münze R, Unverferth K (1978) Die Reduktion von Pertechnetat mit Cystein [in German]. Z Anorg Allg Chem 447:215–220

Srivastava SC, Meinken G, Smith TD, Richards P (1977) Problems associated with stannous ^{99m}Tc-radiopharmaceuticals. Int J Appl Radiat Isot 28:83–95

Srivastava SC, Richards P (1983) Technetium-labelled compounds. In: Rayudu GVS (ed) Radiotracers for medical applications, CRC series in radiotracers in bioliology and medicine. CRC Press, Boca Raton, pp 107–185

Steigman J, Eckelman WC, Meinken G, Isaacs HS, Richards P (1974) The chemistry of technetium labeling of radiopharmaceuticals by electrolysis. J Nucl Med 15:75–80

Steigman J, Richards P (1974) The chemistry of technetium. Sem Nucl Med 4:269–279

Steigman J, Meinken G, Richards P (1975) The reduction of pertechnetate-99 by stannous chloride-I. The stoichiometry of the reaction in HCl, in a citrate buffer and in a DTPA buffer. Int J Appl Radiat Isot 26:601–609

Stern HS, Zolle I, McAfee JG (1965) Preparation of ^{99m}Tc-labelled serum albumin. Int J Appl Radiat Isot l6:283–288

Subramanian G, McAfee JG (1970) Stannous oxide colloid labeled with ^{99m}Tc or ^{113m}In for bone marrow imaging. J Nucl Med 11:365–366

Syhre R, Spies H, Johannsen B (1976) Die Organverteilung von ^{99m}Tc(Sn)-Hydroxidkolloid in der Ratte in Abhängigkeit vom Präparations-pH [in German]. Radiobiol Radiother 17:747–750

Thomas RW, Estes GW, Elder RC (1979) Technetium radiopharmaceutical development. I. Synthesis, characterization and structure of dichloro[hydrotris(1-pyrazolyl)borate]oxo-technetium(V). J Amer Chem Soc 101:4581

Vanderheyden JL, Ketring AR, Libson K, Heeg MJ, Roecker L, Motz P, Whittle R, Elder RC, Deutsch E (1984) Synthesis and characterization of cationic technetium complexes of 1,2-bis(dimethylphosphino)ethane (DMPE) – structure determinations of trans-[Tc-$(DMPE)_2(OH)(O)](F_3CSO_3)_2$, trans-[Tc-III$(DMPE)_2CL_2]F_3CSO_3$ and [Tc-I$(DMPE)_3]^+$ using x-ray-diffraction, EXAFS and ^{99}Tc NMR. Inorg Chem 23:3184–3191

Vilcek S, Kalincak M, Machan V (1982) Analysis of ^{99m}Tc-labelled compounds using Sorb-Gel method. Radiochem Radioanal Lett 52:55–64

Vilcek S, Machan V, Kalincak M (1984) ^{99m}Tc-labelled compounds prepared with molybdenum(III) as reducing agent. Int J Appl Radiat and Isot 35:228–230

Vilcek S, Kalincak M, Machan V (1985) 99mTc-labelled compounds prepared with tungsten(III) as the reducing agent. J Radioanal Nucl Chem 88:359–367

Wardell JL (1994) Tin: inorganic chemistry. In: King RB (ed) Encyclopedia of inorganic chemistry, vol. 8. Wiley, New York, pp 4159–4197

Yokoyama A, Kominami G, Harada S (1975) The role of ascorbic acid with the ferric ion in labelling human serum albumin with ^{99m}Tc. Int J Appl Radiat Isot 26:291–299

Zolle I, Oniciu L, Höfer R (1973) Contribution to the study of the mechanism of labelling human serum albumin (HSA) with Technetium-99m. Int J Appl Radiat Isot 24:621–626

Chapter 4

Quality Assurance of Radiopharmaceuticals

4

T. Bringhammar and I. Zolle

This chapter deals with neither the efficacy of radiopharmaceuticals nor the radiation protection of patients and workers in connection with the handling of radiopharmaceuticals.

4.1 Introduction

The quality of a radiopharmaceutical product may have an effect on the safety of the patient and the outcome of a diagnostic or therapeutic procedure. The purpose of setting standards for radiopharmaceuticals is to obtain radiopharmaceuticals with uniform properties that are effective with regard to their intended use and that entail a risk that is minimal compared with the benefit obtained.

4.2 Definitions

4.2.1 Quality Assurance

Quality Assurance (QA) is a wide ranging concept that covers all matters that individually or collectively influence the quality of a product.

The quality of radiopharmaceuticals must be assured and a system designed, documented, implemented, and controlled, to give a product appropriate for the intended use.

A QA system takes such measures as confirming that designated preparation and QA procedures are being followed by staff at all levels. Such evaluation should be carried out periodically and should result in a written report containing recommendations for action.

QA Incorporates Good Manufacturing Practice. A QA system ensures that radiopharmaceuticals are manufactured or prepared and controlled in such a way that they comply with set specifications throughout their shelf-lives. Specifications concerning radiopharmaceuticals may include parameters such as:

- Identity
- Radioactivity
- Specific radioactivity
- Assay
- Chemical purity
- Radiochemical purity
- Radionuclidic purity
- pH
- Particle size

- Sterility
- Amount of bacterial endotoxins

4.2.2 Good Manufacturing Practice

Good manufacturing practice (GMP) is that part of QA that is aimed at ensuring that radiopharmaceuticals are consistently manufactured or prepared and handled in hospitals to a quality appropriate to their intended use (Kristensen 1979; Nordic Council on Medicines 1989). It is concerned with both manufacture or preparation in hospitals, and quality control (QC). The principles and guidelines of GMP for medicinal products have been laid down in Council Directive 91/356/EEC, presented in volume IV of the European Regulations (European Economic Community 1997).

4.2.2.1 Quality Control

QC is that part of GMP that is concerned with analytical testing, documentation, and evaluation to ensure that the quality of the radiopharmaceutical complies with the relevant standard (European Pharmacopeia [Ph. Eur.], (Council of Europe 2005).

4.2.2.2 Surveillance

Surveillance is part of a QA program, designed to monitor quality of the radiopharmaceuticals as closely as possible. Drug defects and adverse reactions should be recorded and reported to national and/or international centers. All instances of unexpected biodistribution of radiopharmaceuticals in patients should be reviewed in order to detect whether any changes could have taken place in the quality of the radiopharmaceutical. If possible, the radiochemical purity of such a radiopharmaceutical should be determined.

4.2.2.3 Good Manufacturing Practice

GMP is not product specific. It concerns how a quality system on the manufacturing of a certain kind of pharmaceutical should be set up (constructed), and what then in practice should be done to reach and maintain the construction quality in order to produce a product of satisfactory quality.

The QC performed by the manufacturer of a ready-for-use product or a kit or the eventually performed QC in the hospital are not sufficient to ensure that the right quality is maintained.

To get a radiopharmaceutical of satisfactory quality each time, it is important that the manufacturing process as well as the process of preparation comply with current GMP. This is of utmost significance for radiopharmaceuticals that, due to their often-short shelf-lives (physical and chemical), have to be released before all QC measures are performed and evaluated. That is to say, radiopharmaceuticals in most cases are administrated to the patient before the results from all tests are available.

In most countries, the competent authorities have adopted GMP guidelines concerning the manufacturing of pharmaceuticals (European Economic Community 1997). These guidelines often contain supplements concerning radiopharmaceuticals. The most important deviations from the "normal" guidelines are arrangements in connection with radiation protection and that the release is permitted before all QC tests on the finished product have been performed and evaluated. Which control tests may be performed and evaluated after release is settled when an application for approval for marketing authorization is assessed by a competent authority. It is very important that methods used for manufacture are validated. The manufacturers are inspected by national and sometimes foreign competent authorities.

The same GMP guidelines are, where relevant, valid for the preparation of radiopharmaceuticals in hospitals. Some national competent authorities have issued detailed regulations about what has to be considered. Detailed requirements such as:

- Competence (education and training)
- Premises and equipment
- Documentation
- Distribution
- QC
- Self-inspection

The preparation of radiopharmaceuticals in hospitals will normally be inspected by the national inspectorates.

Within Europe GMP guidelines are governed by two international bodies:

1. *Pharmaceutical Inspection Convention* (PIC), which has been operating since 1970, was founded by the European Free Trade Association (EFTA) comprising initially the ten EFTA member states, i.e., Austria, Denmark, Finland, Iceland, Liechtenstein, Norway, Portugal, Sweden, Switzerland, and the United Kingdom; membership was subsequently expanded to include Hungary, Ireland, Romania, Germany, Italy, Belgium, France, and Australia.
2. *Pharmaceutical Inspection Cooperation Scheme* (PIC Scheme), which has been operating since 1995, was formed as an extension of PIC, offering a more flexible cooperation between PIC and health authorities outside Europe.
   a. PIC and PIC Scheme (jointly referred to as PIC/S) are two international instruments between countries and pharmaceutical inspection authorities, which together provide an active and constructive cooperation in the field of GMP (PIC guidelines) (European Free Trade Agreement 1992).
   b. PIC/S concern the following participating authorities: Australia, Austria, Belgium, Canada, Czech Republic, Denmark, Finland, France, Germany, Greece, Hungary, Iceland, Ireland, Italy, Latvia, Liechtenstein, Malaysia, Netherlands, Norway, Portugal, Romania, Singapore, Slovak Republic, Spain, Sweden, Switzerland, and the United Kingdom.

The main differences between the PIC Scheme and PIC are:

PIC Scheme	**PIC**
Scheme	Convention
Informal arrangement	Formal treaty
Between health authorities	Between countries
Has no legal status	Has legal status
Exchange of information	Mutual recognition of inspections

Other relevant GMP guidelines were elaborated by the United States and Japan. Furthermore, WHO has elaborated GMP guidelines to be used as basis for the elaboration of national or multinational guidelines.

4.2.3 Pharmacopeias

In contrast to GMP guidelines, monographs of a pharmacopeia concern a specific active ingredient or product.

A pharmacopeia contains standards concerning pharmaceuticals and comprises such subjects as:
- Analytical methods: identity reactions, chromatographic methods, methods of sterilization, etc.
- Packaging materials
- Ingredients
- General monographs on radiopharmaceuticals
- Ready-for-use pharmaceuticals

Monographs on radiopharmaceuticals are as a rule standards with which the ready-for-use radiopharmaceuticals must comply at the time of administration, regardless of whether the pharmaceutical is a ready-for-use product from a manufacturer or has been prepared in nuclear medicine.

The specifications set up in a pharmacopeia monograph are minimum standards only. At the time of an approval for marketing authorization, the manufacturer on its behalf may set tighter specifications.

4.2.3.1 Relevant Pharmacopeias

Ph. Eur. The *Ph. Eur.* is elaborated and published under the direction of the Council of Europe in accordance with the Convention on the Elaboration of a European Pharmacopeia (Council of Europe 2005). It is published in two official languages, English and French. In Great Britain, a separate pharmacopeia is published, *The British Pharmacopeia*, which contains the monographs of the *Ph. Eur.* together with national monographs. In the same way, separate pharmacopeias are published in France, Germany, and Spain. In the Nordic countries, national monographs are published in an annual publication.

United States Pharmacopeia (USP). The *USP* is elaborated by authority of the United States Pharmacopeial Convention (Unites States Pharmacopeia Convention 2005).

When a country has signed the European Pharmacopeial Convention, the *Ph. Eur.* is law in that country, and pharmaceuticals have to comply with relevant monographs.

For pharmaceuticals not in the *Ph. Eur.*, other pharmacopeias may be used.

These countries follow the *Ph. Eur.* in English, French, or equivalent translation: Albania (observer), Austria, Australia (observer), Belgium, Bosnia-Herzegovina, Bulgaria, Canada (observer), China (observer), Cyprus, Czech Republic, Denmark, Estonia, Finland, France, Germany, Greece, Hungary, Iceland, Ireland, Italy, Latvia, Lithuania, Luxembourg, Macedonia, Malta, The Netherlands, Norway, Poland (observer), Portugal,

Rumania, Serbia and Montenegro, Slovenia, Slovak Republic, Spain, Sweden, Switzerland, Syria (observer), Turkey, and the United Kingdom.

The appendix to this chapter comprises a list with English titles for *Ph. Eur.* monographs on radiopharmaceuticals together with the *USP* monographs. An existing *Ph. Eur.* monograph overrides a *USP* monograph.

4.2.4 Approval for Marketing Authorization of a Radiopharmaceutical

A manufacturer must have a marketing authorization before a product can be marketed. In the European Union, rules (directives and guidelines) are issued concerning application for approval for marketing authorization of a product regulating which documentation must be submitted with an application (Directive 75/318/EEC, Eudralex, vols. 2A and 2B [European Economic Community 1998 a, b]). The rules are not product specific. Each member state has the authority to grant a national marketing authorization.

Since 1993, there is an administrative authority responsible for medicines in the entire European Union, enacted by Council Regulation (EEC) No. 2309/93. This European Agency for the Evaluation of Medicinal Products (EMEA) is responsible for coordinating the scientific resources put at its disposal by the competent authorities of the member states for the evaluation and supervision of medicinal products for human use. A marketing authorization granted under the centralized procedure by EMEA is valid for all member states.

According to the guidelines the applicant must prove that:

- The manufacturing process complies with current GMP.
- The ingredients and packaging materials are tested according to validated methods.
- The finished product and its shelf-life (both before and after preparation) have been evaluated by validated methods.
- If a product must be prepared in nuclear medicine prior to administration, evaluation of the production process at another site has to be performed under the same circumstances.

If there is a monograph in the *Ph. Eur.* on a particular product, the manufacturer must prove that the product at least complies with the specifications of the monograph. However, normally the specifications used by the manufacturer of the finished product as well as the specifications used at in-process controls are tighter.

4.2.5 Quality Control

QC on Ready-for-Use Products from a Manufacturer. These radiopharmaceuticals are to be administrated to the patient without further preparation. As the manufacturing is inspected by competent authorities in order to ensure a high quality of the production process, the QC in the hospital in most cases can be reduced to control of transport documents, labels, and radioactivity. Tests on radionuclidic or radiochemical purity are normally not required.

QC on Products Intended to be Prepared in the Hospital. For these radiopharmaceuticals, the responsibility for the product is shared by the manufacturer(s) of starting materials and the hospital, where the radiopharmaceutical is prepared.

Before approval of marketing authorization, an applicant must prove that, when prepared in a hospital, the margins between an acceptable and an unacceptable radiopharmaceutical are broad enough to permit great variations between preparation routines in different hospitals and variations between different operators performing the work. If the margins are too narrow, the product is not likely to be approved, unless the medical benefit outweighs the efforts spent on QC of each preparation before its administration to the patient.

In most cases, QC on each prepared radiopharmaceutical is not necessary, except the control of radioactivity. On the other hand, QC of prepared radiopharmaceuticals should be a part of the hospital's validation of routines and working standards of personnel.

QC methods that are intended to be used in hospitals must be simple, robust, and validated. Furthermore, control methods on radiochemical purity, which is the most common parameter, should easily demonstrate the radiochemical purity. It is seldom necessary to know the precise percentage of every single impurity.

Manufacturers and competent national authorities have a great responsibility when evaluating the need for QC in hospitals. In other words, whether a validation of routines and working standards of personnel is sufficient, or whether it is necessary for each preparation to perform tests. At the same time, specifications on acceptable limits for clinical use must be given.

The methods used for preparation and QC of approved radiopharmaceuticals must be the same as those approved by the competent authority. QC methods recommended in a relevant pharmacopeia may also be used. In the European Union, the methods are part of the Summary of Product Characteristics (SPC), the purpose and scope of which is approved by the competent authority (European Economic Community 1983).

In an investigation of a patient in which something has gone wrong, the radiopharmaceutical is rarely to blame. There are many other parameters that may influence, e.g., the illness of the patient, medication, hydration, or weight. Likewise, when a defect radiopharmaceutical has been discovered, it is seldom a defect product from the manufacturer. In most cases, the fault lies with the personnel who prepared the radiopharmaceutical.

Radiopharmaceuticals Manufactured in the Hospital. For such radiopharmaceuticals, the hospital has the full responsibility, and the manufacturing must follow the same rules as those for the manufacture of other pharmaceuticals.

Appendix

European Pharmacopeia (*Ph. Eur.*) and *United States Pharmacopeia* (*USP*) monographs concerning radiopharmaceuticals

Ph. Eur. 5.0
Ammonia[^{13}N] injection
Carbon monoxide[^{15}O] injection
Chromium[^{51}Cr] edetate injection

USP 28
Ammonia N 13 injektion
Carbon monoxide [C11] injection
Chromium Cr 51 edetate injection
Chromic phosphate P 32 suspension

Ph. Eur. 5.0	**USP 28**
Cyanocobalamin[^{57}Co] capsules	Cyanocobalamin Co 57 capsules
Cyanocobalamin[^{58}Co] capsules	Cyanobalamin Co 58 capsules
Cyanocobalamin[^{57}Co] solution	Cyanobalamin Co 57 oral solution
Cyanocobalamin[^{58}Co] solution	
2-Fluoro-2-deoxy-D-glucose[^{18}F] injection	Fludeoxyglucose F 18 injection
	Fluorodopa F 18 injection
[5-methyl-^{11}C]flumazenil injection	Flumazenil C 11 injection
Gallium[^{67}Ga] citrate injection	Gallium citrate Ga 67 injection
Indium[^{111}In] chloride solution	Indium In 111 chloride solution
	Indium In 111 capromab pendetide injection
	Indium In 111 ibritumomab tiuxetan injection
Indium[^{111}In] oxinate solution	Indium In 111 oxyquinoline solution
Indium[^{111}In] pentetate injection	Indium In 111 pentetate injection
	Indium In 111 pentetreotide injection
	Indium In 111 satumomab pendetide injection
Iobenguane[^{123}I] injection	Iobenguane I 123 injection
Iobenguane[^{131}I] injection for diagnostic use	Iobenguane I 131 injection
Iobenguane[^{131}I] injection for therapeutic use	
Albumin[^{125}I] injection	Iodinated I 125 albumin injection
	Iodinated I 131 albumin injection
	Iodinated I 131 albumin aggregated injection
Iodinated[^{131}I] norcholesterol injection	
Krypton[^{81m}Kr] (gas)	Krypton Kr 81m (gas)
	Mespiperone C 11 injection
L-Methionine[methyl-^{11}C] injection	Methionine C 11 injection
Raclopride[methoxy-^{11}C] injection	Raclopride C 11 injection
	Rose Bengal sodium I 131 injection
	Rubidium chloride Rb 82 injection
Radiopharmaceuticals (general monograph)	Radioactivity (general monograph)
	Radiopharmaceuticals for positron emission tomography compounding
	Samarium Sm 153 lexidronam injection
Sodium[1^{11}C] acetate injection	Sodium acetate C 11 injection
Sodium fluoride[^{18}F] injection	Sodium fluoride F 18 injection
Sodium iodide[^{123}I] injection	Sodium iodide I 123 solution
Sodium iodide[^{131}I] injection	Sodium iodide I 131 solution
Sodium iodide[^{131}I] capsules	Sodium iodide I 131 capsules
	Sodium iodide I 123 capsules
	Sodium iothalamate I 125 injection
Sodium iodohippurate[^{123}I] injection	Iodohippurate sodium I 123 injection
Sodium iodohippurate[^{131}I] injection	Iodohippurate sodium I 131 injection
Sodium chromate[^{51}Cr] sterile solution	Sodium chromate Cr 51 injection
Sodium phosphate[^{32}P] injection	Sodium phosphate P 32 solution

Ph. Eur. 5.0	USP 28
Strontium[^{89}Sr] chloride injection	Strontium chloride Sr 89 injection
Sodium pertechnetate[^{99m}Tc] injection (fission)	Sodium pertechnetate Tc 99m injection
Sodium pertechnetate[^{99m}Tc] injection (nonfission)	
Technetium[^{99m}Tc] albumin (human) injection	Technetium Tc 99m albumin injection
	Technetium Tc 99m albumin colloid injection
	Technetium Tc 99m apcitide injection
	Technetium Tc 99m arcitumomab injection
	Technetium Tc 99m bicisate injection
	Technetium Tc 99m depreotide injection
	Technetium Tc 99m etidronate injection
Technetium[^{99m}Tc] exametazime injection	Technetium Tc 99m exametazime
Technetium[^{99m}Tc] gluconate injection	Technetium Tc 99m gluceptate injection
	Technetium Tc 99m disofenin injection
Technetium[^{99m}Tc] etifenin injection	Technetium Tc 99m lidofenin injection
	Technetium Tc 99m mebrofenin injection
Technetium[^{99m}Tc] macrosalb injection	Technetium Tc 99m albumin aggregated injection
Technetium[^{99m}Tc] medronate injection	Technetium Tc 99m medronate injection
Technetium[^{99m}Tc] mertiatide injection	Technetium Tc 99m mertiatide injection
Technetium[^{99m}Tc] microspheres injection	
	Technetium Tc 99m nofetumomab merpentan injection
	Technetium Tc 99m oxidronate injection
Technetium[^{99m}Tc] pentetate injection	Technetium Tc 99m pentetate injection
	Technetium Tc 99m (pyro- and trimeta-) phosphates injection
Technetium[^{99m}Tc] Tin pyrophosphate injection	Technetium Tc 99m pyrophosphate injection
Technetium[^{99m}Tc] colloidal rhenium sulfide injection	Technetium Tc 99m red blood cells injection
Technetium[^{99m}Tc] sestamibi injection	
Technetium[^{99m}Tc] succimer injection	
Technetium[^{99m}Tc] sulfur colloid Injection	Technetium Tc 99m sestamibi injection
	Technetium Tc 99m succimer injection
	Technetium Tc 99m sulfur colloid injection

Ph. Eur. 5.0	USP 28
	Technetium Tc 99m tetrofosmin injection
Technetium[^{99m}Tc] tin colloid injection	
Thallous[^{201}Tl] chloride injection	Thallous chloride Tl 201 injection
Tritiated[^{3}H] water injection	
	Urea C 14 capsules
Water[^{15}O] injection	Water O 15 injection
Xenon[^{133}Xe] injection	Xenon Xe 133 injection
	Xenon Xe 127 (gas)
	Xenon Xe 133 (gas)
	Yttrium Y 90 ibritumomab tiuxetan injection

References

Council of Europe (2005) European pharmacopeia, 5th edn. Maisonneuve, Sainte-Ruffine

European Economic Community (1983) Council Directive 83/570/EEC (26 October 1983), Summary of product characteristics. European Economic Community, Luxembourg

European Economic Community (1997) Pharmaceutical legislation: good manufacturing practices for medicinal products for human and veterinary use (Directive 91/356/EEC). Eudralex, vol. IV. European Economic Community, Luxembourg

European Economic Community (1998 a) Notice to applicants: procedures for marketing authorization. Medicinal products for human use. (Directive 75/318/EEC; Regulation (EEC) No 2309/93). Eudralex, vol. IIA. European Economic Community, Luxembourg

European Economic Community (1998 b) Notice to applicants: presentation and content of the dossier. Medicinal products for human use (Directive 75/319/EEC, resp. Regulation (EEC) No 2309/93; Directive 89/343/EEC). Eudralex, vol. IIB. European Economic Community, Luxembourg

European Free Trade Agreement (1992) Guide to good manufacturing practice for pharmaceutical products, the convention for the mutual recognition of inspections in respect of the manufacture of pharmaceutical products, Document 5/92 (PIC guidelines on GMP). European Free Trade Agreement, Geneva

United States Pharmacopeial Convention (2005) Official monographs: USP28. United States Pharmacopeia Convention, Rockville, Md.

Kristensen K (1979) Preparation and control of radiopharmaceuticals in hospitals. International Atomic Energy Agency, Vienna

Nordic Council on Medicines (1989) Radiopharmacy: preparation and control of radiopharmaceuticals in hospitals. National Library of Norway publication No. 26, Uppsala

Chapter 5

Performance and Quality Control of the $^{99}Mo/^{99m}Tc$ Generator

5

I. Zolle

Radionuclide generators serve as a convenient source of short-lived radionuclides for medical application. Daily supply with short-lived radionuclides is limited by the half-life and the cost of transport. Certain decay sequences, however, exist in an equilibrium state between the parent nuclide and the decay product, i.e., the daughter nuclide. In principle, the longer-lived parent radionuclide generates the shorter-lived daughter by radioactive decay. The type of equilibrium depends on the relative half-lives of the parent and daughter radionuclides. The growth of the daughter radionuclide continues until a maximum is reached; since parent and daughter radionuclides are different elements, the radioactive daughter can be separated by a simple chemical process, generally by liquid elution.

The lifetime of a generator system depends on the half-life of the parent nuclide. The longer the half-life of the parent, the longer can daughter radionuclide be eluted in adequate amounts. This condition is favorable for both transport and the use of short-lived radionuclides in diagnostic nuclear medicine (Eckelman and Coursey 1982; Richards 1966).

A number of generator systems have been developed for medical application, but only few have found widespread acceptance. Table 5.1 gives some examples of generators used in nuclear medicine (Boyd et al. 1985).

A radionuclide generator consists of a glass column filled with an adsorbent material such as aluminum oxide or an ion-exchange resin to which the parent nuclide is bound (Richards 1966). The column is fitted with a filter at the outlet to retain particulate matter. On top is the elution platform, where an evacuated sterile vial is connected with the outlet of the column through which saline or another suitable eluent is drawn from the eluent reservoir.

$^{99}Mo/^{99m}Tc$ generators produced for worldwide application have a sophisticated system for safe elution of the daughter radionuclide. The generator column is well shielded with lead, and the whole system must be adequately shielded to reduce radiation exposure of the operator to a permissible level (Fig. 5.1).

Table 5.1. Generator systems for medical application of short-lived radionuclides

Parent nuclide	Decay mode	$T_{1/2}$	Daughter nuclide	Energy (keV)	$T_{1/2}$
^{99}Mo	β^-	66.02 h	^{99m}Tc	141	6.02 h
^{113}Sn	EC	115.1 days	^{113m}In	392	1.66 h
^{81}Rb	EC, β^+	4.58 h	^{81m}Kr	190	13.3 s
^{82}Sr	EC	25 days	^{82}Rb	(β^+, EC)	1.25 min
^{68}Ge	EC	288 days	^{68}Ga	(β^+, EC)	68.3 min

$T_{1/2}$ half-life, *EC* electron capture

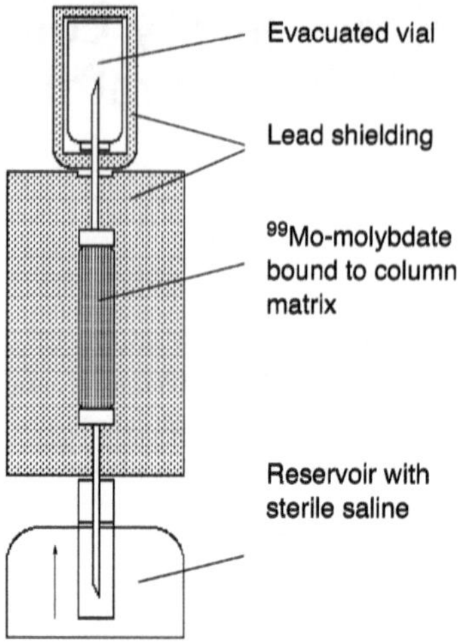

Fig. 5.1. Components of the ^{99}Mo/^{99m}Tc generator system

5.1 The Equilibrium State

Parent ^{99}Mo and the short-lived daughter ^{99m}Tc reach transient equilibrium, characterized by the decay of both parent and daughter radionuclides with one apparent half-life, namely that of the longer-lived ^{99}Mo (Richards 1966).

Transient equilibrium is established when the half-life of parent is long with respect to daughter radionuclide, but parent activity changes perceptibly during the period under consideration. Buildup of daughter ^{99m}Tc activity occurs until a maximum is reached, and then the "effective" half-life of the daughter activity will be essentially equal to the parent half-life, as long as parent activity continues to produce the daughter radionuclide. The ratio of daughter activity to parent activity is unchanging with respect to time, but the activity of each is declining with respect to time: A_2/A_1 = constant (Boyd 1982; Richards 1966).

The decay sequence of ^{99}Mo, with a half-life of 66.02 h, is shown in Fig. 5.2 (Boyd 1982).

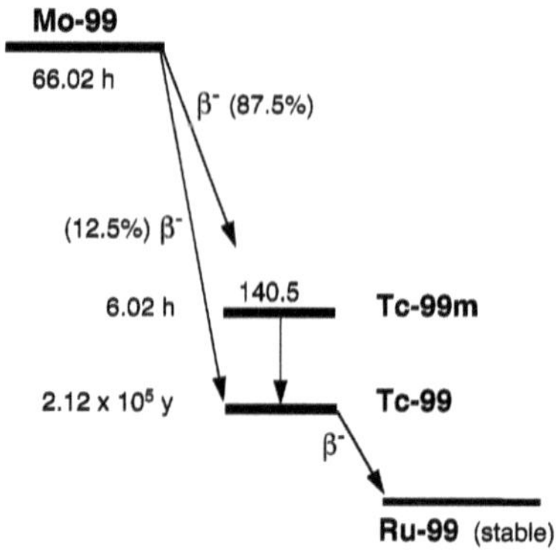

Fig. 5.2. Decay scheme of parent ^{99}Mo to stable ^{99}Ru

The ^{99}Mo parent radionuclide decays by several β-particle transitions (two shown), producing metastable ^{99m}Tc with 87.5% intensity, while 12.5% decay directly to long-lived ^{99}Tc. Subsequently, metastable ^{99m}Tc decays by isomeric transition to ^{99}Tc with a half-life of 6.02 h and emission of 140.5 keV gamma radiation. ^{99}Tc decays with a half-life of 212 000 years to stable ruthenium-99 (Boyd 1982).

5.1.1 Production of Molybdenum-99

Irradiation of metallic ^{98}Mo or molybdenum trioxide, Mo_2O_3 (natural or enriched in ^{98}Mo), with thermal neutrons in a nuclear reactor:

$$^{98}\mathrm{Mo}(n,\gamma)^{99}\mathrm{Mo}$$

Large quantities of natural target may be irradiated to give a high yield of ^{99}Mo (Robson 1972). High yields are also obtained by subjecting an enriched ^{98}Mo target to high-intensity neutron irradiation in excess of 10^{14} neutrons/cm^2·s. However, the price of this target material is very high (Boyd 1982). Consequently, for large-scale production, this method is rarely used. Radionuclidic impurities result from activation of metallic impurities present in the target material (Boyd 1973). Advantages of neutron activation are (1) post–irradiation processing is minimal, (2) radionuclidic contaminants are limited by target purity, and (3) only small quantities of radioactive waste are produced. A disadvantage is the low specific activity of ^{99}Mo (<10 Ci/g Mo).

Irradiation of ^{235}U with thermal neutrons and separation of ^{99}Mo from fission products:

$$^{235}\mathrm{U}(n,f)\ldots{}^{99}\mathrm{Mo}$$

An advantage of fission-product ^{99}Mo is the high specific activity of ^{99}Mo ($>10^4$ Ci/g Mo). Disadvantages are the (1) elaborate and expensive post–irradiation processing facilities, (2) chemical separation of the highly-toxic α-emitters (transuranic radionuclides) and pure β-emitters, (3) special problems of quality control, and (4) large quantities of long-lived radioactive waste.

^{99}Mo is separated from other radionuclides generated by this process (Barnes and Boyd 1982). Strict limits for radionuclidic impurities are stated in the monograph on sodium pertechnetate [^{99m}Tc] injection solution (Council of Europe 2005 a).

5.1.2 Separation Methods

Several methods have been used to separate the daughter nuclide ^{99m}Tc from parent ^{99}Mo, the three most common methods are column chromatography, solvent extraction, and sublimation (Boyd 1982; Richards 1982).

The $^{99}Mo/^{99m}Tc$ generator used in nuclear medicine is based on the chromatographic separation of ^{99m}Tc-pertechnetate.

5.1.3 Design of the Generator Column

^{99}Mo is bound strongly to a bed of chromatographic-grade alumina. The daughter radionuclide ^{99m}Tc is eluted by selective elution from the Al_2O_3 column. Both ^{99}Mo and the decay product ^{99m}Tc are bound as anions; however, parent ^{99}Mo shows the strongest fixation. Due to considerable differences in the binding affinities, a pertechnetate anion may be replaced by a nitrate or chloride anion, as is indicated by the order of affinities of anions:

$$OH^- > MoO_4^{2-} > Cl^- > NO_3^- > TcO_4^-$$

Sterile saline is used for elution of the ^{99m}Tc activity as Tc(VII)O_4-pertechnetate anion.

The sterile, isotonic eluate may be used directly in patients; however, most of the ^{99m}Tc activity is needed for the preparation of ^{99m}Tc radiopharmaceuticals with preformed kits.

Aluminum oxide is pretreated by activation at high temperature (250 °C) to resist structural changes in the high-radiation environment and to reduce acid solubility. Silver coating of the alumina particles has a similar effect and provides a higher capacity of the alumina matrix for fixation of ^{99}Mo-molybdate (Barnes and Boyd 1982).

Fission-produced molybdenum-99 offers considerable advantages for generator production:

- High specific activity ^{99}Mo-molybdate is applied to the column, permitting small amounts of alumina and small columns.
- Small chromatographic columns result in a small elution volume and thus, a higher concentration of activity in the eluate.
- Smaller columns can be shielded more easily, not increasing the weight of the generator. This is important for the transport and handling of the generator in the medical facility.

With fission-produced ^{99}Mo, the column for a 500-mCi generator requires approximately 1.2 g of aluminum oxide. The applied mass is by a factor of 1000 less than with previously used irradiation-produced ^{99}Mo; the elution volume is generally 5 ml, containing 80% of the theoretically available activity. To date, practically all commercially available generators are produced with fission-produced ^{99}Mo; problems with ^{99}Mo breakthrough do not exist. An acceptable contamination of the eluate with ^{99}Mo is highest in the first eluate, decreasing with daily elutions.

In comparison, former irradiation generators used larger columns suitable for approximately 20 g of alumina and an elution volume between 15 and 20 ml. The resulting activity concentration was much lower and the so-called ^{99}Mo breakthrough posed a considerable hazard (Richards and O'Brien 1969). Typically, contamination of the eluate with ^{99}Mo was increasing with daily elutions.

5.1.4 The Generator Eluate

The ^{99}Mo/^{99m}Tc generator is a closed system providing sterile ^{99m}Tc eluate on a daily basis. Generators are available from manufacturers with a wide range of ^{99}Mo activities (2–43 GBq; 54–1.162 mCi); sterile, evacuated vials are supplied with the generator, offering also variable elution volumes (5, 10, 15, and 20 ml).

Sterile saline is used for elution of the generator. The eluate is a clear, colorless, isotonic solution of ^{99m}Tc(VII)-pertechnetate.

The concentration of the ^{99m}Tc activity in the eluate (^{99m}Tc activity/volume) depends on the available ^{99}Mo activity and the elution volume. With daily elution of the generator, high specific activity is assured.

5.2 Performance of the ^{99}Mo/^{99m}Tc Generator System

The *European Pharmacopeia* (*Ph. Eur.*) contains two separate monographs for sodium pertechnetate [^{99m}Tc] injection solutions, depending on the source of ^{99}Mo used for production of the generator, natrii pertechnetatis [^{99m}Tc] fissione formati solutio iniectabilis (monograph 124) (Council of Europe 2005a), and natrii pertechnetatis [^{99m}Tc] sine fissione formati solutio iniectabilis (monograph 283) (Council of Europe 2005b).

The following quality criteria are stated in the *Ph. Eur.* and should be evaluated for each generator:

- Elution efficiency (percentage)
- Radionuclidic purity of the eluate
- Radiochemical purity of the eluate
- Chemical purity of the eluate
- pH of the eluate

5.2.1 Elution Efficiency

The performance of a generator system is expressed by the elution efficiency, which is defined as the fraction of eluted ^{99m}Tc activity of the theoretically available radioactivity at the time of elution, usually given as a percentage. The yield varies with different generator systems, and is generally between 80 and 100% of the theoretical value; however, daily elutions should represent a constant percentage, fluctuations as shown in the graph (Fig. 5.3) indicate poor performance.

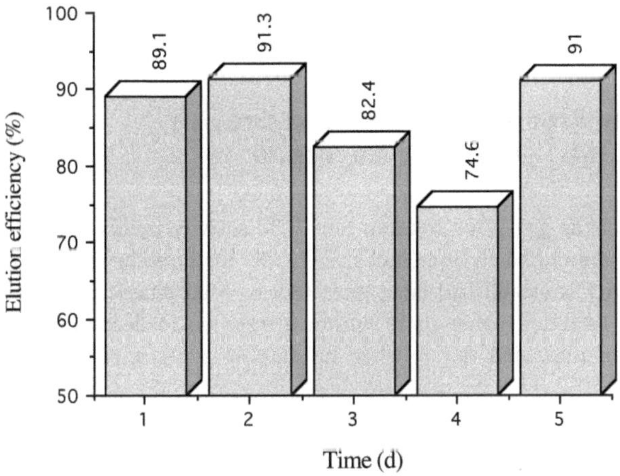

Fig. 5.3. Elution efficiency of a ^{99}Mo/^{99m}Tc generator (daily elution)

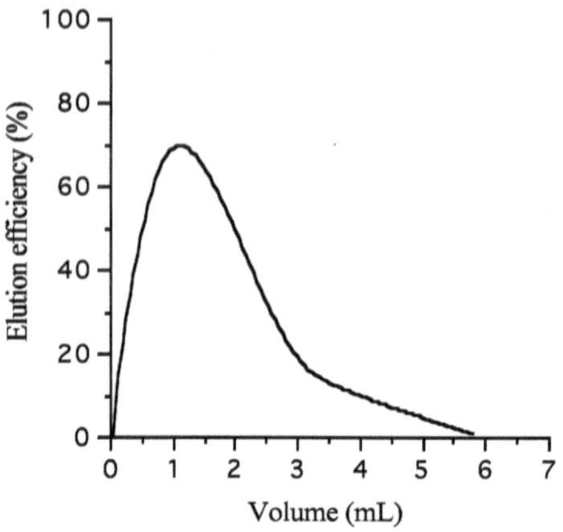

Fig. 5.4. Typical elution profile of a fission ^{99}Mo generator using three fractions (2.1, 2.4, and 2.6 ml), eluting 86% of the theoretically available radioactivity in the first 3 ml

Generators produced with high specific activity fission molybdenum-99 may be eluted with a small volume of saline, showing a narrow elution profile, as indicated by the graphic presentation (Fig. 5.4). Still, there are differences in the elution pattern between generator systems from different producers. Therefore, it may be beneficial to know the elution profile of each generator used at a medical facility.

The elution profile may be obtained from the manufacturer on request, or it may be determined by fractional elution using 1- to 2-ml fractions and measuring the eluted activity of each sample. Shown are elution efficiencies obtained with two generator systems manufactured by different producers. Generator I shows 61.6% of the eluted radioactivity in the first 3 ml shown in Fig. 5.5, whereas elution of generator II shows the major activity peak in the fractions between 3 and 5 ml (Fig. 5.6). Both generators are totally eluted with 5 mL of saline.

Fractional elution may be indicated when small volumes of high activity are needed for bolus injection or for labeling. In the case of generator I, it will suffice to elute a total volume of 3 ml, while two 3-ml fractions should be obtained in the case of generator II, since the main activity peak is collected in the second fraction.

5.2.2 The Kinetics of Decay and Growth of the $^{99}Mo/^{99m}Tc$ Generator

After elution of the generator column, the ^{99m}Tc activity builds up until a maximum is reached after approximately four half-lives (22.89 h). Equilibrium between mother and daughter activity is established later, after 43.9 h. A schematic presentation depicts the growth of ^{99m}Tc activity after daily elution in relation to decreasing parent ^{99}Mo, and the equilibrium state after day 5, when no elution is performed (Boyd 1982; Richards 1966; Fig. 5.7).

The theoretically available ^{99}Mo activity of a 100-mCi generator and buildup of ^{99m}Tc activity is shown in the decay–growth graph (Fig. 5.7).

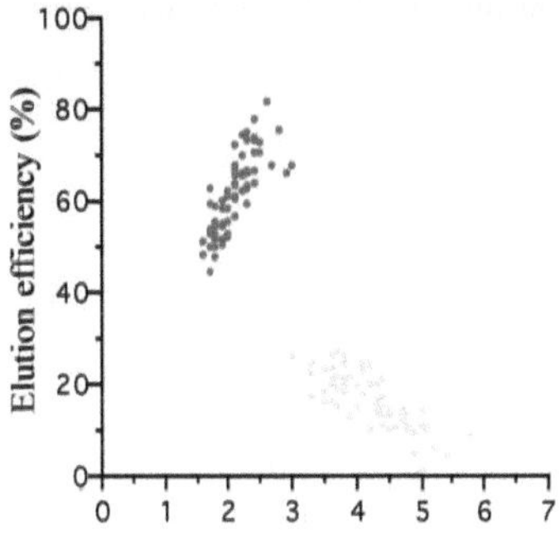

Fig. 5.5. Elution of generator I ($n=67$), using fractions of 1.6–3.0 ml, showing 61.6±8.3% in the first fractions and 16.2±5.9% in the second fractions, with an overall elution efficiency of 77.8±4.2%

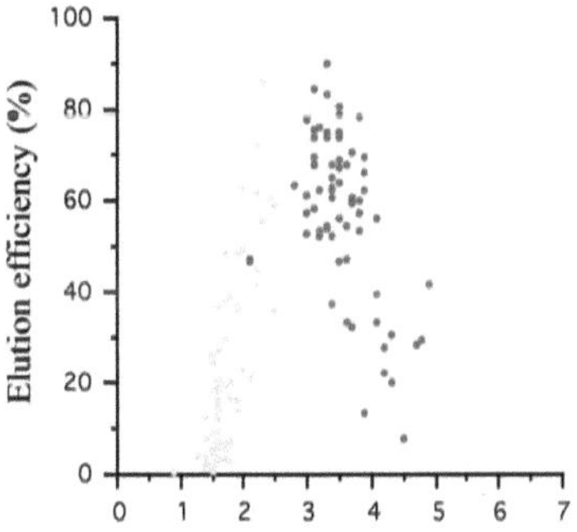

Fig. 5.6. Elution of generator II ($n=67$) using fractions of 1.0–2.5 ml, showing 21.8±20.4% in the first fractions and 57.2±18.3% in the second fractions, with an overall elution efficiency of 80.4±10.8%

Calculation of the available ^{99m}Tc activity:

$$A_1 = A_0 e^{-\lambda_1 t} \quad \text{indicates parent activity at any time} \tag{1}$$

$$A_2 = A_0 e^{-\lambda_1 t} - e^{-\lambda_2 t} \quad \text{indicates daughter activity at time, } t \tag{2}$$

$$A_{\text{Tc-99m}(t)} = A_{\text{Mo-99}(t=0)} \times 0.96(1 - e^{-0.105t}) \tag{3}$$

$A_0 =$ ^{99}Mo activity at time of calibration
$A_1 =$ ^{99}Mo activity at time, t (h)
$A_2 =$ ^{99m}Tc activity at time, t (h)
$\lambda_1 =$ decay constant of ^{99}Mo = 0.0105 (h^{-1})
$\lambda_2 =$ decay constant of ^{99m}Tc = 0.1155 (h^{-1})

Parent activity (A_1) is expressed by Eq. 1, considering the radioactive decay during the time (t). Daughter activity (A_2) is a function of parent decay and the decay of daughter during the time (t), expressed by Eq. 2. The available ^{99m}Tc activity at any time may be calculated using Eq. 3.

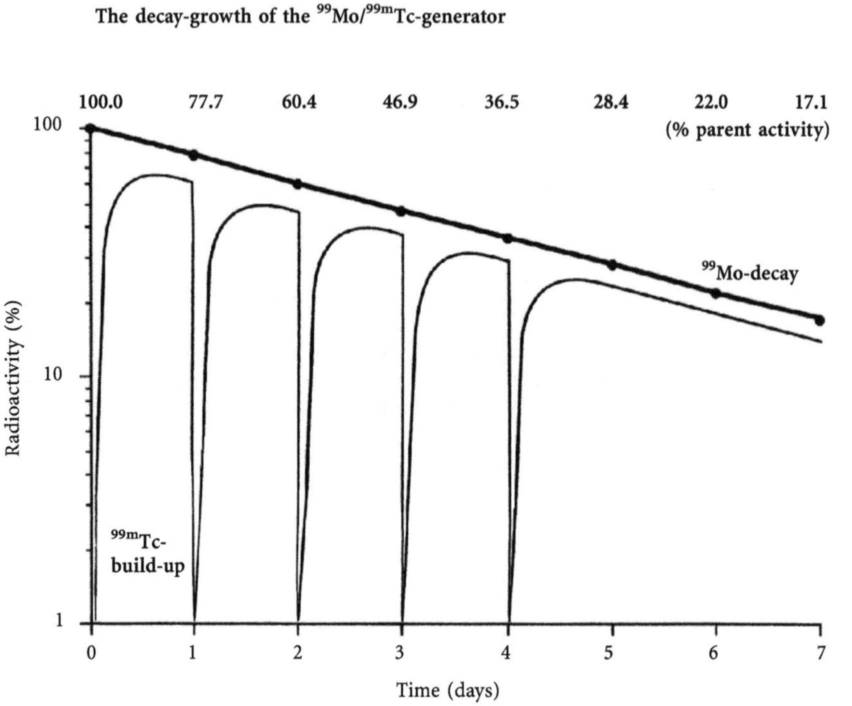

Fig. 5.7. Radioactive decay of parent ^{99}Mo and buildup of ^{99m}Tc activity within 24 h after elution, calculated for 7 days

A ^{99}Mo/^{99m}Tc generator (100 mCi) provides sufficient ^{99m}Tc activity for 5 days, as is indicated by the radioactive decay of parent ^{99}Mo (Table 5.2). It is apparent from Table 5.3 that the daily available ^{99m}Tc activity may be simply obtained by subtracting 6.25% from A_1 (decay of ^{99m}Tc during 24 h). After ^{99m}Tc activity has reached its maximum at 22.89 h, the available activity is effectively determined by the decay of parent activity.

For any time (t) that is short with respect to the half-life of parent ^{99}Mo, Eq. 4 may be used for calculating the available daughter activity after the last elution.

$A_2 = A_0(1 - e^{-\lambda_2 t})$ indicates build-up of daughter activity from zero to maximum activity with respect to time $(t \ll T_{1/2P})$ (4)

Table 5.2. Radioactive decay of parent ^{99}Mo (A_0 100 mCi)

Time (h)	$e^{-\lambda_1 t}$	A_1 (mCi)
24	0.7772	77.72
48	0.6041	60.41
72	0.4695	46.95
96	0.3649	36.49
120	0.2837	28.37
144	0.2205	22.05
168	0.1714	17.14

Table 5.3. ^{99}Mo decay and ^{99m}Tc buildup

Time (h)	$e^{-\lambda_1 t}$	$e^{-\lambda_2 t}$	A_2 (%)
1	0.9896	0.8909	9.87
3	0.9690	0.7072	26.18
6	0.9389	0.5001	43.88
9	0.9098	0.3536	55.62
12	0.8816	0.2501	63.15
18	0.8278	0.1251	70.27
24	0.7772	0.0625	71.47
48	0.6041	0.0039	60.02
72	0.4695	0.0002	46.93
96	0.3649	0.0000	36.49

5.2.3 Factors Affecting the Elution Yield

Actually, the elution yield of ^{99m}Tc activity is considerably less than calculated due to retention of certain ^{99m}Tc species and other factors. Small concentrations of dissolved organic impurities in the eluent have been shown to cause reduced efficiencies (Molinski 1982). Purification of the eluent by passage through activated charcoal maintained high elution efficiencies. If the generator is constructed of materials that release organic contaminants on autoclaving, the generator may be poisoned; this cannot be restored by the use of a charcoal filter in the inlet line. Extracts from plastic materials, residual glue (cyclohexanone), and disinfectant (propan-1-ol), all have been identified to reduce the elution yield (Boyd et al. 1985). The effect of radiolysis, on the other hand, has been discussed controversially.

5.2.4 Elution of Carrier ^{99}Tc

Long-lived ^{99}Tc atoms pose an inherent isotopic impurity that is eluted from the column and is also generated in the eluate by radioactive decay of ^{99m}Tc. Since the concentration of ^{99m}Tc in the eluate is very low, chemical effects may result when generators are not eluted at regular intervals. The total mass of technetium (^{99m}Tc + ^{99}Tc) in the 24-h eluate is calculated as 6.90×10^{-10} g/37 MBq (1 mCi). (For calculation of specific activity, see Appendix A2 of this chapter)

For a given generator, the amount of ^{99}Tc carrier atoms in the eluate is dependent on:

- Time elapsed since last elution
- Time interval between elution and use
- Elution efficiency

A considerable increase of ^{99}Tc carrier has been demonstrated when there is a gap between elutions, as is the case over weekends (Bonnyman 1983). Table 5.4 shows the calculated amounts of ^{99}Tc carrier, assuming 10 GBq (270 mCi) of eluted activity and different times of decay before elution of the generator.

The effect of low elution efficiency on subsequent elutions has also been investigated, resulting in an increase of ^{99}Tc carrier with each elution (Boyd et al. 1985).

Table 5.4. Carrier ^{99}Tc in eluates, depending on time elapsed between elutions

	Activity: 10 GBq (270 mCi)			
Time between elutions	24 h	27 h	3 days	4 days
Amount of ^{99}Tc carrier (μg) in eluate	0.186	0.208	0.679	1.036
Expressed as 10^{-10} g per 37 MBq (1 mCi)	6.9	7.7	25.15	38.37

Therefore, generators should be eluted regularly and completely; this is especially important for those generator types with reduced efficiencies. Insufficient understanding of these factors can lead to a significant loss of quality of the ^{99m}Tc eluate.

5.3 Purity of Generator Eluate (*Ph. Eur.*)

5.3.1 Radionuclidic Purity

Radionuclidic impurities in the eluate are directly related to the production mode of parent ^{99}Mo. In any case, ^{99}Mo is the most important impurity. The limit of contamination with ^{99}Mo stated in the *Ph. Eur.* is 0.1% of the total eluate activity. With irradiation-produced molybdenum-99, other radionuclidic impurities are limited to 0.01% of the total radioactivity. Identified impurities include Sn-113, Au-198, Au-199, Cs-134, and Nb-92.

On the other hand, fission-produced molybdenum-99 may contain a number of radionuclidic impurities (Briner and Harris 1974; Hammermaier et al. 1986). Limits of contamination are listed in Table 5.5 with reference to the radionuclide and type of radiation.

Table 5.5. Limits of radionuclidic impurities in the eluate of fission ^{99}Mo generators (*European Pharmacopeia*)

Radionuclide	Type of radiation	Limit (%)
^{99}Mo	(γ, β^-)	1×10^{-1}
^{131}I	(γ, β^-)	5×10^{-3}
^{103}Ru	(γ, β^-)	5×10^{-3}
^{89}Sr	(pure β^--emitter)	6×10^{-5}
^{90}Sr	(pure β^--emitter)	6×10^{-6}
α-emitters		1×10^{-7}
γ-emitters		1×10^{-2}

5.3.2 Radiochemical Purity

Technetium may exist in seven oxidation states. In the ^{99m}Tc eluate, the chemical species is ^{99m}Tc(VII)-pertechnetate. The *Ph. Eur.* states that not less than 95% of the radioactivity is identified as sodium pertechnetate [^{99m}Tc] by paper chromatography.

5.3.3 Chemical Purity

The eluate may contain certain chemical impurities, originating from either the generator column or the eluent. Aluminum cations are formed during absorption of ^{99}Mo, when the alumina bed is subjected to strong acid (1 M HNO_3), and may be detected in the eluate (Boyd 1973; Lin et al. 1971; Shukla et al. 1977). The limit stated in the *Ph. Eur.* has been 20 ppm, which was based on the formerly used large columns. The *Ph. Eur.* is recommending a colorimetric test using chromazurol-S for quantification (Feigl 1958).

Eluates of fission ^{99}Mo generators were analyzed by atomic absorption spectroscopy for identification of trace amounts of chemical impurities (Hammermaier et al. 1986).

Certain chemical impurities have been shown to reduce the labeling yield and alter the in vivo distribution of radioactivity (Boyd 1973; Lin et al. 1971; Ponto et al. 1987).

5.3.4 pH of Eluate

The *Ph. Eur.* permits a pH range between 4.0 and 8.0.

5.4 Methods and Results

Care should be taken that generators are eluted regularly and completely. The volume of the eluate and total activity are recorded with the date and time of elution.

5.4.1 Determination of the Elution Efficiency

It is accepted that the activity stated on the generator refers to the time of calibration; however, due to precalibration the actual amount of ^{99m}Tc activity at delivery might be considerably higher (up to ten times). The high activity load and long intervals between production and delivery of the generator result in a build up of ^{99}Tc carrier, which is coeluted with ^{99m}Tc and affects labeling. Therefore, radiopharmacies perform a test elution of the generator upon receipt from the manufacturer as part of the quality assurance program (Nordic Council on Medicines 1989). This "primary eluate" is not used for patient studies, but serves to remove ^{99}Tc carrier and to demonstrate the generator function:

- Calculate the elution efficiency
- Determine the ^{99}Mo content of the eluate

5.4.2 Determination of the ^{99}Mo Content of the Eluate

The primary eluate is allowed to decay for 5 days, so that the ^{99m}Tc activity does not interfere with the measurement of trace amounts of radionuclidic impurities.

Measurements of the primary eluate were performed over an energy range of 35 keV to 2.0 MeV, using a Ge(Li)-detector connected to a 4,096-channel pulse-height analyzer;

the acquired spectral data were analyzed by computer. Energy calibrations and efficiency determinations had been performed with standard solutions of several radionuclides (standard mixture). The activity peak at an energy of 740 keV (^{99}Mo) was expressed as nanocuries of ^{99}Mo contamination in the primary eluate at the time of elution (corrected for ^{99}Mo decay).

Example of calculation:	
Primary eluate obtained:	11 May, 9:00 hours
Total activity eluted:	402.2 mCi in 11.3 ml
Elution efficiency:	84.6%
Date of measurement of ^{99}Mo:	16 May, 13:00 hours
Time elapsed since elution:	124 h ($f=3.68$)
Measured ^{99}Mo activity:	293 nCi
Amount (corrected for decay):	1.078 nCi/402.2 mCi/11.3 ml
^{99}Mo impurity in primary eluate:	2.68 nCi/mCi (2.68×10^{-4}%)
	95.4 nCi/ml (3.53 kBq/ml)

Since the limit of ^{99}Mo contamination is 10^{-1}% (1 µCi/mCi) of the total ^{99m}Tc activity (*Ph. Eur.*), the measured ^{99}Mo contamination is acceptable.

Table 5.6 presents the results obtained with 140 generators from two different manufacturers; shown are the elution efficiencies and the ^{99}Mo activity in the first eluate, expressed as 10^{-4}% of the ^{99m}Tc activity and also as kilobecquerel per milliliter.

Other radionuclidic impurities stated in the *Ph. Eur.* were also determined. I-131 was detected in very small amounts; Ru-103 was not detectable in most eluates. Generators from one manufacturer were an exception, showing 0.27 nCi/mCi of I-131, and values between 0.012 and 0.069 nCi/mCi of Ru-103 (Hammermaier et al. 1986).

Table 5.6. Elution efficiency and ^{99}Mo impurity

	Primary eluate			
	Volume (mL)	Elution efficiency (%)	Mo-99 Activity	
			(10^{-4}%)	(kBq/mL)
Generator				
Number of samples	68	72	68	65
Range	11.0–13.2	64.2–86.4	1.52–14.7	2.04–27.64
Mean ± SD	12.76 ± 0.34	74.75 ± 4.15	6.38 ± 2.84	8.20 ± 4.25
Median	**12.8**	**74.6**	**5.58**	**6.95**
Generator II				
Number of samples	68	68	68	68
Range	8.7–10.8	72.1–97.5	0.01–6.93	0.01–11.69
Mean ± SD	9.63 ± 0.57	80.6 ± 6.99	2.35 ± 1.99	4.03 ± 3.42
Median	**9.55**	**79.4**	**2.17**	**3.58**

5.4.3 Determination of the Radiochemical Purity of the Eluate

In the *Ph. Eur.*, sodium pertechnetate [^{99m}Tc] and other ^{99m}Tc species are analyzed by descending paper chromatography (Council of Europe 1997a). The solvent used consists of eight parts of methanol and two parts of water. The chromatogram is developed for 2 h, dried, and the distribution of radioactivity is measured with a suitable detector. Not less than 95% of the recovered radioactivity is located in the spot corresponding to pertechnetate ion, which has an R_f value of approximately 0.6. Reduced, colloidal forms remain at the start of the chromatogram.

In the laboratory, ascending thin-layer chromatography on silica gel glass fiber sheets is used for analysis (Brandau et al. 1996). Acetone is recommended as solvent. Sodium pertechnetate [^{99m}Tc] migrates with the solvent front ($R_f = 1.0$). Reduced, hydrolized activity is analyzed in saline ($R_f = 0$).

Analysis of generators from seven manufacturers used in nuclear medicine showed high radiochemical purity of the eluates, between 99.5 and 100.0% (Hammermaier et al. 1985).

5.4.4 Determination of the Chemical Purity of the Eluate

Aluminum cations in the eluate may cause a low labeling yield and affect the biodistribution of ^{99m}Tc radiopharmaceuticals. Interference has been reported with aluminum levels below the permitted concentrations (Ponto et al. 1987). The limit specified in the *Ph. Eur.* was reduced to 10 µg/ml, in analogy to Al^{3+} concentrations stated in the *United States Pharmacopeia*, which must not exceed 10 µg/ml (United States Pharm. Convention 2005). A modified quinalizarin-based spot test is used for colorimetric evaluation of eluates against a known standard dilution.

Eluates of fission ^{99}Mo generators were analyzed by atomic absorption spectroscopy. Hammermaier et al. (1986) reported data below the detection limit of 1 µg of aluminum per milliliter of eluate. A more recent study presented a comparison between the two methods and showed that with the colorimetric test concentrations of Al^{3+} below 10 ppm were not detectable (Table 5.7; Marengo et al. 1999).

Table 5.7. Summary of quality control of ^{99m}Tc eluates (Marengo et al. 1999)

^{99}Mo (10^{-4}%)	^{103}Ru (10^{-8}%)	Al^{3+} (ppm)	pH	TcO_4^-	Volume (ml) ^{99m}Tc activity	
					50%	95%
2.37	9.19	<10	4.3–7.0	99.64	1.56	4.27

5.5 Conclusions

Generator function and the purity of ^{99m}Tc eluates of the available fission ^{99}Mo generators in Europe have been investigated by several European research institutions over the years. Much information on the benefits of improved technology has been documented, testifying to high standards of manufacturing and to the reliability of validation processes implemented by the regulatory agencies for safe application of the $^{99}Mo/^{99m}Tc$ generator in nuclear medicine.

Several European centers were engaged in implementing GMP standards for the preparation of ^{99m}Tc radiopharmaceuticals in nuclear medicine, and thus have contributed to establish a protocol for analytical procedures, which can validate the performance of generator systems from different manufacturers and assure the quality of short-lived generator eluates. The protocol is presented in Appendix A1 of this chapter.

Acknowledgments

The Preparation and Quality Control of Tc-99m pharmaceuticals was supported by the Austrian Ministry of Science and Research as a collaboration between the Dept. of Nuclear Medicine, AKH, in Vienna, Austria and the National Institute of Pharmacy in Budapest, Hungary. Part of this manuscript was translated from German by Dr. P.O. Bremer; his contribution is gratefully acknowledged.

Appendix

A1 Protocol for the Quality Control of $^{99}Mo/^{99m}Tc$ Generator Systems

Multicenter studies were performed in Italy between the Depts. of Nuclear Medicine and Medical Physics in Bologna, Cesena, Ferrara, Genova, Milano, Pavia, and Reggio Emilia; and in Austria and Hungary between the Dept. of Nuclear Medicine, AKH-Wien, Vienna, and the National Institute of Pharmacy, Budapest.

Quality assessment included approximately 250 generators from seven suppliers in Italy, and 150 generators from two commercial sources in Austria.

The protocol concerns the performance of the generator system and the analysis of the eluate according to the requirements of the *Ph. Eur.*

1. Performance of the generator system

First elution upon shipment of generator – analysis of primary eluate

Elution characteristics:

- Elution efficiency (standard elution, percent of nominal activity)
- Elution profile (fractional elution, each fraction as percent of sum of activities)
- Elution volume containing 50% resp. 95% of activity
- Radioactive concentration of eluate (MBq/ml)

2. Purity of generator eluate (*Ph. Eur.*)

Radionuclidic impurities in primary eluate:	<0.1% ^{99}Mo of total activity
Radiochemical purity:	>95% ^{99m}Tc-pertechnetate ion
Chemical purity:	<10 ppm ionic aluminum
pH value:	4.0–8.0

Elution of the generator is performed using aseptic techniques and special, evacuated, sterile vials. To facilitate determination of the volume of eluate, graduated vials are used. The vial is placed into the dose calibrator and the radioactivity is recorded.

Each vial containing ^{99m}Tc-pertechnetate injection solution must have a radioactive label with relevant information:

- Name of radionuclide and chemical name: ^{99m}Tc-pertechnetate i.v.
- Date and time of elution: 06/13/05, 8:10 hours
- Activity of eluate: 6.62 GBq (179 mCi)
- Volume of eluate: 8.5 ml
- Activity concentration: 778 MBq/ml (21.0 mCi/ml)
- Expiration: 6 h

A2 Calculation of the Specific Activity of [^{99m}Tc]Technetium

The specific activity of a radionuclide relates activity (A) (the number of radioactive atoms decaying) to the total number (N) or total mass of atoms present:

$$A/N = \lambda$$

Since there are 6.023×10^{23} atoms in one gram·atom (mole) of the radionuclide, specific radioactivity may be expressed according to the equation:

$$A\ (\text{dps/gram}) = \lambda N = \frac{6.023 \times 10^{23}\ \text{atoms} \times \lambda(\text{sec})}{M(\text{g})}$$

$$1\ \text{Ci/g} = \frac{6.023 \times 10^{23} \times \lambda\,(\text{sec})}{\text{Atomic weight} \times 3.7 \times 10^{10}}$$

$$\textbf{Curie/g} = \frac{\mathbf{1.628 \times 10^{13} \times \lambda\,(sec)}}{\textbf{Atomic weight}}$$

where

$$N(\text{g}) = \frac{6.023 \times 10^{23}\ \text{atoms}}{M(\text{g})}$$

$$1\ \text{gram} = \frac{6.023 \times 10^{23}}{\text{Atomic weight}}$$

$$1\ \text{Curie} = 3.7 \times 10^{10}\ \text{dps}$$

Example: Technetium-99m (M = 99)

$T_{1/2} = 6$ hr; $\lambda = \frac{0.693}{T_{1/2}}$, therefore: $\lambda(s) = 0.693/6 \times 60 \times 60$

1 Ci = 37 GBq; 1 GBq = 2.7×10^{-2} Ci

$$\text{Ci/g} = \frac{1.628 \times 10^{13} \times 0.693}{6 \times 60 \times 60 \times 99}$$

$$\text{Ci/g} = \frac{1.139 \times 10^{11}}{2.16 \times 10^{4}}$$

$$\mathbf{Ci/g} = \mathbf{5.27 \times 10^{6}}$$

$$\mathbf{GBq/g} = \mathbf{1.95 \times 10^{8}}$$

The specific activity of carrier-free technetium-99m is 5.27×10^6 Ci/g (mCi/mg, µCi/µg). Consequently, 1 Ci (mCi, µCi) of carrier-free technetium-99m has a weight (mass) of 1.90×10^{-7} g (mg, µg).

Upon daily elution of the generator (24-h interval), the eluate contains ^{99m}Tc + ^{99}Tc corresponding to a mass of 6.9×10^{-7} g/Ci (37 GBq).

However, longer intervals (> 2 days) result in a considerable increase in ^{99}Tc carrier; if a generator has not been eluted for 3 days, the eluate on day 4 contains 3.64 times more carrier corresponding to 25.15×10^{-7} g/Ci (37 GBq).

References

Barnes RK, Boyd RE (1982) The chromatographic extraction and purification of ^{99}Mo from uranium solutions by use of a silver impregnated alumina stationary phase. Int J Appl Radiat Isot 33:479–481

Bonnyman J (1983) Effect of milking efficiency on Tc-99 content of Tc-99m derived from Tc-99m generators. Int J Appl Radiat Isot 34:901–906

Boyd RE (1973) Recent developments in generators of ^{99m}Tc (IAEA-SM-171/94) In: Radiopharmaceuticals & labelled compounds 1973. IAEA Vienna, pp 3–26

Boyd RE (1982) Technetium-99m generators – the available options. Int J Appl Radiat Isot 33:801–810

Boyd RE, Hetherington ELR, Moore PW (1985) Radionuclide generator technology. In: Radiopharmaceuticals & labelled compounds 1984. IAEA Vienna, 79–94

Brandau W, Hotze L-A, Meyer G-J (1996) Radiochemie. In: Büll U, Schicha H, Biersack H-J, Knapp WH, Reiners Chr, Schober O (eds) Nuklearmedizin [in German]. Georg Thieme, Stuttgart, pp 79–113

Briner WH, Harris CC (1974) Radionuclide contamination of eluates from fission-product molybdenum-technetium generators. J Nucl Med 15:466–467

Council of Europe (2005 a) Sodium pertechnetate[^{99m}Tc] injection (fission). Monograph No. 124. European Pharmacopoeia, 5th edn., Maisonneuve, Sainte-Ruffine

Council of Europe (2005 b) Sodium pertechnetate[^{99m}Tc] injection (nonfission). Monograph No. 283. European Pharmacopoeia, 5th edn. Maisonneuve, Sainte-Ruffine

Eckelman WC, Coursey BM (1982) Technetium-99m: generators, chemistry and preparation of radiopharmaceuticals. Int J Appl Radiat Isot 33:793–890

Feigl F (1958) Spot tests in inorganic analysis, 5th edn. Elsevier, Amsterdam, p. 145

Hammermaier A, Reich E, Endrulat HJ, Bögl W (1985) Investigations on the quality of eluates from seven fission Mo-99/Tc-99m-generators. Institute for Radiation Hygiene of the Federal Health Office, Neuherberg, Germany. ISH Report 64.

Hammermaier A, Reich E, Bögl W (1986) Chemical, radiochemical, and radionuclidic purity of eluates from different commercial fission ^{99}Mo/^{99m}Tc generators. Eur J Nucl Med 12:41–46

Lin MS, McGregor RD, Yano Y (1971) Ionic aluminium (III) in generator eluate as an erythrocyte-agglutinating agent. J Nucl Med 12:297–299

Marengo M, Aprile C, Bagnara C, Bolzati C, Bonada C, Candini G, Casati R, Civollani S, Colombo FR, Compagnone G, Del Dottore F, Di Guglielmo E, Ferretti PP, Lazzari S, Minoia C, Pancaldi D, Ronchi A, Sanita di Toppi G, Saponaro R, Torregiani T, Uccelli L, Vecchi F, Piffanelli A (1999) Quality control of ^{99}Mo/^{99m}Tc-generators: results of a survey of the radiopharmacy working group of the Italian Association of Nuclear Medicine (AIMN) Nucl Med Commun 20:1077–1084

Molinski VJ (1982) A review of ^{99m}Tc-generator technology. Int J Appl Radiat Isot 33:811–820

Nordic Council on Medicines (1989) Nordic guidelines. Radiopharmacy – preparation and control of radiopharmaceuticals in hospitals. NLN Publication No. 26, Uppsala

Ponto JA, Swanson DP, Freitas JE (1987) Clinical manifestations of radio-pharmaceutical formulation problems. In: Hladik WB III, Saha GB, Study KT (eds) Essentials of nuclear medicine science. Williams & Wilkins, Baltimore, pp 270–274

Richards P (1966) Nuclide generators. In: Radioactive pharmaceuticals. USAEC Symposium Series, No. 6, (CONF-651111). Oak Ridge, Tennessee, pp 155–163

Richards P, O'Brien MJ (1969) Rapid determination of ^{99}Mo in separated ^{99m}Tc. J Nucl Med 10:517

Richards P, Tucker WD, Srivastava SC (1982) Technetium-99m: an historical perspective. Int J Appl Radiat Isot 33:793–800

Robson J (1972) Process for the production of technetium-99m from neutron irradiated molybdenum trioxide – assigned to the AAEC patents: UK (40273/72), USA (284,868), Sweden (10868/72), Japan (85901/1972)

Shukla SK, Manni GB, Cipriani C (1977) Effect of aluminium impurities in the generator-produced pertechnetate-99m ion on thyroid scintigrams. Eur J Nucl Med 2:137–141

United States Pharmacopeial Convention (2005) Official monographs: USP 28, sodium pertechnetate Tc^{99m} injection. United States pharmacopeia (USP) 28 – national formulary (NF) 23, p 1861

Chapter 6

Preparation of Technetium ^{99m}Tc Pharmaceuticals

6

J. Mallol and I. Zolle

6.1 Introduction

Technetium-99m (^{99m}Tc) is widely used in radiopharmaceutical preparations due to its excellent physical and chemical properties. In fact, more than 80% of all radiopharmaceuticals used in diagnostic nuclear medicine are based on this short-lived radionuclide, which is obtained by elution of a ^{99}Mo/^{99m}Tc generator system that is available in any radiopharmacy and nuclear medicine facility.

6.1.1 Physical Characteristics

^{99m}Tc decays with a half-life of 6 h by isomeric transition and emission of 140.5-keV gamma radiation. Large amounts of radioactivity may be used with the single-photon emission computed tomography (SPECT) technology, producing high-contrast images with the gamma camera. In fact, the energy window of the gamma camera is optimized to 140.5 keV (110–220 keV).

6.1.2 Chemical Characteristics

^{99m}Tc is eluted from the generator as a pertechnetate anion. It has been demonstrated that heptavalent technetium must be reduced to a lower valency state in order to be chemically reactive for labeling. Most ^{99m}Tc pharmaceuticals comprise complexes of ^{99m}Tc at various oxidation states (I–V).

The preparation of ^{99m}Tc pharmaceuticals is greatly facilitated by the availability of commercial cold kits containing the chemical ingredients as a lyophilized formulation ready for labeling with ^{99m}Tc-pertechnetate.

6.2 Kit Preparations

The preparation of any ^{99m}Tc pharmaceutical is performed by using a commercial cold kit and adding the required ^{99m}Tc activity in a certain volume of ^{99m}Tc eluate (pertech netate). Kits offer convenience and ease of preparation for ad hoc labeling.

Cold kits are considered as semimanufactured products, which need marketing authorization as medicinal products (European Economic Community 1998). Cold kits are commonly produced at the industrial level. The preparation of a ^{99m}Tc pharmaceutical by a nuclear medicine facility or a nuclear pharmacy by adding ^{99m}Tc eluate to an approved cold kit according to the instructions given by the producer does not need a specific authorization (Nordic Council on Medicines 1998). The labeling procedure is termed as "reconstitution of the kit", which may be interpreted as preparation of an official formula as described in the European regulations (Cox 1993). Similarly, ^{99m}Tc pharmaceuticals described in the *European Pharmacopeia* in specific monographs (Council of Europe 2005), as well as the preparation of any medicinal product according to pharmacopeia specifications, is conceptually an official formula.

6.2.1 General Considerations

The concept reconstitution of a kit is erroneous, because radiolabeling is an active process by which a new chemical compound (a ^{99m}Tc pharmaceutical) is formed involving chemical reactions, in any case the reduction of pertechnetate, and in most cases complex formation, which might need heating in a boiling water bath. Moreover, a kit may be perfectly reconstituted (dissolved) but labeling might be inadequate.

Preparation of ^{99m}Tc radiopharmaceuticals by using aseptic techniques has to comply with the regulations issued in the Good Manufacturing Practice Guidelines ([GMP] European Economic Community 1997; European Free Trade Agreement 1992). Labeling by a closed procedure is defined as a procedure whereby a sterile radiopharmaceutical is prepared by the addition of sterile ingredients to a presterilized closed container via a system closed to the atmosphere (Lazarus 1994).

^{99m}Tc eluate is added to the sterile vial (kit) with a syringe through the rubber stopper. An excess of pressure in the vial is avoided by withdrawing an equal volume of gas with the same syringe. A breather needle should not be used because oxygen may affect the stability of the radiopharmaceutical, and may cause microbial contamination.

The instructions given by the manufacturer of the kit should be strictly followed, particularly with respect to the maximum activity and volume of ^{99m}Tc eluate that is transferred to the kit for labeling. The labeled product is a sterile, pyrogen-free solution suitable for intravenous injection. Any abnormality observed by visual inspection of the injection solution is a cause to reject the preparation. ^{99m}Tc radiopharmaceuticals have a short shelf-life (generally 6 h); they are used right after preparation.

6.2.2 Cold Kits

Cold kits are prepacked sets of sterile ingredients designed for the preparation of a specific radiopharmaceutical. Kits are fully tested to verify the specific characteristics, which are guaranteed by the producer.

A kit contains the active ingredient, a reducing agent, and may contain authorized excipients and additives, such as antimicrobial agents, antioxidants, buffer, a nitrogen atmosphere, etc. The active ingredient is the compound to be labeled with the radionuclide. The reducing agent is responsible for the reduction of pertechnetate to a lower valency state; it is considered an essential material. Without reduction, there is no labeling reaction.

In fact, the stability of kits had been of considerable concern until freeze-drying was applied to kit production. Exact filling of the sterilized vials is performed automatically by a sterile dispensing/stoppering device, whereby a certain volume (1 ml) of the kit formulation is delivered and subjected to lyophilization. When closing the vial, nitrogen gas is introduced through a sterile filter. By this computer-assisted automatic filling device with subsequent lyophilization, kit contents are stable for long periods; additionally, oxidative processes no longer affect the labeling yield. The shelf-life of kits is usually in excess of 1 year. It is important that kits are stored according to the specific conditions (temperature, humidity) indicated on the package, since radiolabeling depends on the integrity of the reducing agent.

Cold kits may also be prepared in a radiopharmacy similar to other medicinal products prepared in a pharmacy. The radiopharmacist then functions as the qualified person responsible for preparation of an official formula (formulation, pharmaceutical requirements, stability, etc.)

6.2.3 ^{99m}Tc-Pertechnetate

The ^{99m}Tc eluate used for radiolabeling must comply with the specifications stated in the pharmacopeia. Additionally, the specific activity (activity/Tc carrier) and the activity concentration (activity/ml) should be known.

Since the specific activity of the ^{99m}Tc eluate is related to the time elapsed between elutions, daily elution of the generator at an interval of 24 h will produce eluates with the best quality. Kits are compatible with eluates from different commercial generators; however, high specific activity is mandatory with some kits, especially when labeling biomolecules.

The total ^{99m}Tc activity and the volume injected into the vial (kit) should comply with the recommendations by the manufacturer. Dilutions should be performed with isotonic saline.

6.2.4 Incubation

After dissolving the lyophilisate in the added volume, incubation is an essential step to obtain the radiolabeled medicinal product. In this phase, the chemical reactions take place, resulting in ^{99m}Tc labeling. If incubation is inadequate, the labeling reaction may not be completed, and the radiopharmaceutical may not be suitable for administration.

Each kit requires specific incubation conditions, but in general, this process is carried out at room temperature in a clean area. In certain cases, the incubation must be performed in a boiling water bath; in such cases it is necessary to operate carefully.

Incubation is normally performed with occasional agitation of the shielded reaction vial.

6.2.5 Quality Control

Quality is directly related to the labeling yield, which is measured by the amount of unbound ^{99m}Tc activity. Limits of radiochemical impurities are stated in the official monographs (Council of Europe 2005).

The determination of the radiochemical purity lies in the responsibility of the user (Theobald 1994). To assure safety and efficacy of a ^{99m}Tc radiopharmaceutical, the product should be tested regularly, in certain cases before application to the patient.

Radiochemical purity is analyzed by thin-layer chromatography, described in this book. Poor quality of a radiopharmaceutical would affect the clinical information and cause unnecessary radiation exposure of a patient.

6.2.6 Dispensing

Dispensing is the compliance of a prescribed medicinal product with the required quality standards (Lazarus 1994). For dispensing a prescribed amount of radioactivity, it is necessary to determine the total radioactivity and the radioactivity concentration of the radiopharmaceutical.

A single dose may be withdrawn aseptically from the multidose vial by using a suitable syringe; each syringe must be measured in the dose calibrator to verify the prescribed amount of radioactivity for a patient.

Syringes with individual doses of the radiopharmaceutical may be prepared in advance. In this case, the correct identification of each syringe is mandatory, stating on the label:

- Identification of patient (name and/or number)
- Name of the radiopharmaceutical
- Amount of radioactivity and time of preparation
- Relevant comments (storage, cautions, etc.)
- Name of dispenser, if necessary

References

Cox PH (1993) EEC Directives and guidelines applicable to radiopharmaceuticals. Eur J Nucl Med 20:712–715

Council of Europe (2005) European pharmacopeia, 5th edn. Maisonneuve, Paris

European Economic Community (1997) Pharmaceutical legislation for medicinal products for human and veterinary use in the European Union. Good manufacturing practices. (Directive 91/356/EEC). Eudralex, vol. IV. European Economic Community, Luxembourg

European Economic Community (1998) Notice to applicants for marketing authorization for medicinal products for human use in the European Union, Volume IIA (III/3567/92)

European Free Trade Agreement (1992) Guide to good manufacturing practice for pharmaceutical products, the convention for the mutual recognition of inspections in respect of the manufacture of pharmaceutical products, Document 5/92 (PIC guidelines on GMP).

Lazarus CR (1994) Techniques for dispensing of radiopharmaceuticals. In: Sampson CB (ed) Textbook of radiopharmacy, theory and practice, 2nd enlarged edn. Gordon and Breach, Philadelphia, pp 59–68

Nordic Council on Medicines (1989) Radiopharmacy: preparation and control of radiopharmaceuticals in hospitals. National Library of Norway (NLN) publication No. 26, Uppsala

Theobald AE (1994) Quality control of radiopharmaceuticals. In: Sampson CB (ed) Textbook of radiopharmacy: theory and practice, 2nd enlarged edn. Gordon and Breach, Philadelphia, pp 103–125

Chapter 7

Lyophilization Technique for Preparing Radiopharmaceutical Kits

7

E. Chiotellis

Freeze-drying, or lyophilization, is the drying of frozen materials by sublimation, i.e., the direct transition from the frozen state into vapor, without any intermediate liquid phase (Neuman et al 1962).

The lyophilization technique is the most common method for preparing radiopharmaceutical kits with a prolonged shelf-life.

Most radiopharmaceutical kits available today are technetium-99m compounds, while a limited number involves indium-113m complexes. Few kits are also used for the preparation of iodine-123 radiopharmaceuticals.

7.1 History of the Lyophilization Technique

The freeze-drying process was applied for first time in 1890 by the Leipzig anatomist Altmann. He froze samples of tissue required for microscopic investigations at very low temperatures. Then, the tissues were placed in a desiccator filled with sulfuric acid and cooled on the outside. The air was evacuated while the tissues remained frozen and the water yielded to the sulfuric acid by sublimation. In this way, Altmann obtained anhydrous tissue preparations, without any shrinkage of the cells.

Twenty years later, Shackell demonstrated that the freeze-drying method could be used with very good results for the investigation of easily destructible sensitive compounds in animal and human organs. At the same time, Hammer and Rogers (in 1911 and 1914) demonstrated that suspensions of bacteria may be dehydrated by the freeze-drying technique, without being destroyed. In the years that followed until 1930, many experiments were performed on sensitive proteins that lose their biological properties with other drying methods. It was found that they could be retained in their original native form if subjected to freeze-drying. Numerous products such as blood plasma, enzyme preparations, and sensitive drugs were examined and found unchanged by this particular dehydration method.

In 1939 the term "lyophile" was used for the first time by Reichel, Masucci, and Boye for characterization of products obtained by this particular process of dehydration.

Up to World War II, the lyophilization technique was used only on a laboratory scale. The sudden increase of demands in freeze-dried blood plasma resulted in its application on an industrial scale. In the first post-war years, the freeze-drying method was applied to various antibiotics, and very quickly became a valuable method of preserving medicinal products.

In the years that have passed, a remarkable progress on both technical and physical aspects of freeze-drying was performed, and safe and reproducible methods required for industrial application have been established.

7.2 Principles of Lyophilization

Lyophilization is a high-technology method of preserving chemicals or sensitive biological materials from humidity: enzymes, hormones, vitamins, blood products, antibiotics, drugs, radiopharmaceutical kits, etc.

Lyophilized products are characterized by prolonged shelf-life, and chemical bacteria or enzymatic changes do not easily occur. The sterility is more guaranteed and solubility assured. In addition, transportation is easier. Finally, certain compounds or radiopharmaceutical kits exist only as lyophilized products. However, freeze-dried products suffer from certain disadvantages. The reentry of moisture may destroy the products. Direct optical control of lyophilized products cannot be performed. Therefore, the risk of particle contamination of the final product is high. Bacterial contamination can only be avoided by using the proper installations (clean rooms) for manufacturing injectable lyophilized products.

The lyophilization method is similar to ordinary vacuum distillation with one essential difference: the material to be dried must first be solidly frozen and then subjected to a very low absolute pressure (high vacuum) and controlled heat input. Under these conditions, the water content, in the form of an ice matrix, is selectively removed via sublimation, completely bypassing the intermediate liquid phase. The solid particles are locked into the matrix during drying and cannot interact. The following conditions are necessary for freeze-drying:

1. The product must be solidly frozen, usually below its eutectic point (–10 to –50 °C).
2. A condensing surface must be provided (–40 to –50 °C) for trapping ice vapors.
3. A powerful evacuating system must be provided, capable of evacuating to an absolute pressure between 5 and 25 mHg.
4. A thermostatically controlled heating source is necessary for heat input to the product. The heat drives the water from the solid to the vapor state (heat of sublimation).

Time of drying depends on the nature and volume of the product. Drying temperatures vary and depend on each individual product. Compounds sensitive to heat are dried at low temperatures.

From experience, most radiopharmaceutical kits dried at 4–10 °C result in final products of good quality. For example, a batch of 1000 vials (volume = 2 ml) requires a 24-h primary drying cycle (4–10 °C), followed by another 24-h secondary drying cycle.

In some kits sensitive to heat, like dimercaptosuccinic acid (DMS) (V), low drying temperatures are required (0 °C).

The whole principle of the freeze-drying process depends on the fact that water at temperatures below 0 °C immediately transforms from the solid state of aggregation into the vaporous state. In other words, the technique is based on the sublimation properties of water.

Beyond 0 °C, ice melts and another phenomenon, evaporation, takes place. Melting during the freeze-drying process may destroy the product and should be avoided (back melting).

About 700 kcal (2,777 BTU) of energy is required for the sublimation of 1 kg of ice. This amount of heat must be brought into contact with the frozen material from the outside in such a way that there is absolutely no danger of the material having time to thaw.

The same amount of heat that is required in the drying chamber for the sublimation of the water is subsequently liberated as the water vapor condenses at the ice condenser, and must be removed by means of refrigeration units. Therefore, three basic conditions have to be fulfilled in order to carry out the freeze-drying process:

1. The temperature of the material must be regulated in such a way that on one hand, thawing is avoided, and on the other hand, vapor pressure above the material does not sink due to overcooling.
2. The water vapor molecules escaping from the frozen material should be removed in such way that saturation of vapor pressure above the substance to be dried is avoided.
3. The process should take place in a vacuum so that removal of water molecules may not be impeded by the presence of residual gases.

7.3 Apparatus for Freeze-Drying

In constructing any freeze-drying apparatus, care should be taken to obtain optimum values of all the above-mentioned conditions. A freeze-drying plant should contain a drying chamber, an ice condenser, a refrigeration unit, and a vacuum pump.

7.3.1 The Drying Chamber

This chamber is a vacuum-tight container in which the material to be dried is placed. There are practically no limits to the dimensions and shape of the drying chamber. Rack surfaces either heated or cooled are provided for the reception of material to be dried (ampoules, vials, dishes, liquid, or solid drying substances).

7.3.2 The Ice Condenser

The ice condenser is a pipe coil system or in the case of smaller apparatus, a cold trap, fitted in a vacuum-tight container, which is connected to the drying chamber.

In smaller laboratory devices, the ice condenser is normally cooled by means of carbon dioxide or liquid nitrogen, while in larger plants, by refrigeration units.

7.3.3 The Refrigeration Unit

During the freeze-drying process, the water contained in the material passes three states of aggregation. The product should be cooled, frozen, and then subjected to the sublimation stage. During this stage, heat must be applied to the material to compensate for the sublimation cold developed. This performance is regulated by the refrigeration unit.

7.3.4 The Vacuum Pump

The vacuum pump keeps both chamber and ice condenser sufficiently free from residual gases so that the water vapor is able to flow unimpeded from the drying material process.

In practice, the freeze-drying process of solutions proceeds as following:

1. The bulk solution is dispensed into the vial from a 0.5 to 2.5 cm height. The amount of solution in the vial is very important for efficient drying. The vials are loosely capped with the special rubber stoppers, and then are put into the vacuum chamber.
2. The vials with the product are gradually frozen below the eutectic point of the solution (usually –30 to –40 °C).
3. The condenser is turned on, and the condenser indicator is observed until it reaches its maximum value (usually –50 °C).
4. At this point, the vacuum pump is turned on, and vacuum is applied to the chamber until the lowest value of vacuum is obtained (usually 10–20 mHg).
5. Heat is then applied to the product through a controlled heating–cooling system. The heat facilitates the sublimation process (heat of sublimation). This step of freeze-drying is the primary drying cycle, and the temperature, the drying temperature. The point at which the temperature indicator reaches the preset temperature is theoretically considered the end of the freeze-drying process.
6. A secondary drying cycle, ranging in time, is usually applied so that the complete absence of humidity from the product is assured.
7. After completion of drying cycles, the product is closed with a stopper in the vacuum chamber under either a low-vacuum or nitrogen atmosphere. The stoppering device could be pneumatic or mechanical, depending on the specifications of the freeze-drying plant.

Apart from the high cost, the freeze-drying technique remains an excellent and almost unique method for preserving a variety of compounds, among which are valuable medicinal products sensitive to humidity.

References

Neumann K, Koln P (1962) Freeze drying. A modern application of refrigeration. Rep Science Technol, pp 45–53

Nail SN, Catlin LA (1985) Advances in control of production freeze-dryers. J Parenter Sci Technol 39:16–27

Pyle HAA, Eilenberg H (1971) Continuous freeze drying. Vacuum 21:103–114#

Chapter 8

Cellular Labeling with ^{99m}Tc Chelates: Relevance of In Vitro and In Vivo Viability Testing

H. Sinzinger and M. Rodrigues

8.1 Introduction

Labeling of blood cells has gained importance as routine procedure in nuclear medicine. A variety of different blood cells such as red blood cells (RBC), platelets, and white blood cells ([WBC] neutrophils, lymphocytes and monocytes and recently stem cells) can be radiolabeled and applied for diagnosis and therapeutic monitoring in specific disease states.

In order to label specific cell types, the separation of that particular type of cell must be carefully performed.

Two main categories of cell labeling have been used, cohort and random population labeling. In cohort labeling, the label binds to marrow precursors, which appear after a few days as a labeled cell population of uniform age in the circulation, enabling the study of the cells during their life span. Cohort labeling is useful for the study of cell production rates and survival. In random labeling, the circulating cell population (of all ages) is uniformly labeled, allowing mean life span determination. Random labeling is carried out usually in vitro on a small sample of venous blood. Random labeling methods have been more successfully used than cohort labeling techniques.

Accuracy in diagnosis depends largely on the labeling efficiency (LE) of blood cells, which can be affected by many factors. The specificity of the labeling, maintenance of viability, and normal physiological function of the labeled cells, together with sterility, apyrogenicity and radiopharmaceutical purity, and stability have to be carefully controlled.

The methods involved in the labeling of blood cells with ^{99m}Tc are reviewed and discussed in this overview.

8.2 Red Blood Cells

Labeling of RBC is easily facilitated as they are relatively abundant in the blood, easily separated and handled in vitro, not very susceptible to damage from physical or chemical manipulations, not as dependent on energy and nutritional requirements as the other cellular elements in vitro, and have a variety of cellular transport mechanisms and hemoglobin within that is rich in active metal-binding sites (Srivastava and Rao Chervu 1984).

Random labeling methods of RBC have been more widely utilized than have cohort labeling techniques.

Besides ^{99m}Tc, several radionuclides such as ^{3}H, ^{11}C, ^{14}C, ^{15}N, ^{32}P, ^{51}Cr, ^{55}Fe, ^{59}Fe, ^{67}Ga, ^{68}Ga, and ^{111}In have been evaluated for labeling of RBC. In vivo, in vitro, and in vivo/in vitro techniques have been developed and are presently available for routine use in nuclear medicine (for imaging as well as kinetic studies).

Technetium as pertechnetate is in the +7 valency state and as such, is not bound firmly to RBC, and moves in and out of the RBC rather freely. In contrast, reduced technetium cannot readily cross the cell membrane (Srivastava and Rao Chervu 1984) and leave the cell, and binds rapidly and irreversibly mainly to the β-chain of the globin part of hemoglobin (Eckleman et al. 1979). Reducing agents thus have to be used for labeling RBC in vivo and in vitro.

For in vivo labeling, stannous salts are injected intravenously approximately 15–20 min before injection of ^{99m}Tc-pertechnetate. Stannous ions diffuse into the cell, become bound to a cellular component, have a relatively slow clearance (Srivastava and Rao Chervu 1984), and reduce ^{99m}Tc when it enters the RBC. The labeling of RBC thus occurs within the intravenous space. Using the in vitro technique, the labeling of RBC is performed in a sterile vial. Most methodologies contact RBC, generally in whole blood, with stannous ions, using a suitable tin (II) preparation (Srivastava and Rao Chervu 1984). In the in vitro technique, stannous ions are injected into the patient, and labeling with ^{99m}Tc-pertechnetate of a smaller number of RBC is performed in a closed system (e.g., butterfly needle placed in a peripheral vein). Thereafter, the labeled RBC (in plasma) are reinjected into the patient (Berger and Zaret 1984; Winzelberg et al. 1982). In vivo labeling is more widely used. However, RBC compete for intravenous ^{99m}Tc-pertechnetate with the stomach, thyroid, and kidney. In vivo methods thus have frequently variable and irreproducible LE, which is generally lower than that of the in vitro technique. In the in vitro techniques, quality control in order to avoid free ^{99m}Tc-pertechnetate and thus, unnecessary radiation exposure to the patient, is performed prior to the injection of ^{99m}Tc-RBC. In vitro techniques are therefore the methods of choice when high LE is required as there is almost no free pertechnetate in the patient (Berger and Zaret 1984; Winzelberg et al. 1982).

Several parameters are influencing RBC labeling and thus, LE (which is determined by measuring the cell-bound radioactivity as compared with the total radioactivity in percentage) (Table 8.1). Carrying out an in vitro technique the type of anticoagulant used can modify the ^{99m}Tc labeling of RBC as well as the ^{99m}Tc-RBC distribution. Controversy exists about the effect of the type of anticoagulants that have been used (ethylenediaminetetraacetic acid [EDTA], heparin, acid-citrate-dextrose [ACD], sodium citrate, sodium oxalate and others) for RBC labeling in vitro. LE, in vitro and in vivo stability of the label with EDTA are higher as compared with saline. EDTA reduces the stannous tin content of plasma to a minimum level and assures good labeling yields, which is particularly important with samples of blood with very low hematocrit (Srivastava and Rao Chervu 1984). The use of ACD results in superior RBC labeling relative to heparin (the most utilized methods) (Porter et al. 1983) and ^{99m}Tc-RBC/ACD images are superior to those of ^{99m}Tc-RBC/heparin (Wilson and Hung 1992). Heparin in the catheter was shown to affect RBC labeling (Hegge et al. 1978). In addition, ACD preserves the cellular function much better than does heparin (Srivastava and Straub 1990). ACD is thus the anticoagulant preferred for preparing in vivo/in vitro ^{99m}Tc labeling of RBC. The optimal conditions for RBC labeling, both in vitro and in vivo, evaluated with ^{99m}Tc-oxine (1 mCi/ml) and ACD or sodium citrate as anticoagulants, showed that the LE is also dependent on the density of RBC (Fig. 8.1), time (Fig. 8.2), pH (Fig. 8.3), and temperature of incubation (Figs. 8.1–8.4), as well as on the amount of the tracer (Fig. 8.4) (optimal conditions, LE >95%: 1×10^9 RBC/ml, 1 min, pH 5.0–8.0, 37 °C incubation, and 5 μg oxine/ml) (Reiter et al. 1984).

The LE, as well as the localization, of RBC can be markedly influenced by drugs circulating in the plasma and/or diseases, which can affect RBC (the membrane potential, among others) (Table 5.1). In patients who have recently received iodinated contrast

Table 8.1. Variables influencing red blood cell (*RBC*) labeling with ^{99m}Tc-pertechnetate

Collection injury
Density of RBC
Disease (e.g., leukemia, high fibrinogen level, sickle cell disease, other abnormalities of hemoglobin (Riordon and Nelp 1982)
RBC antibodies (Leitl et al. 1980)
Intravenous canulae
Anticoagulant
Quantity of stannous ions (II) (Rao et al. 1986)
Time delay between "tinning" of RBC and pertechnetate administration
Incubation
Medium (e.g., dextrose in water) (Berger and Zaret 1984)
pH
Temperature
Time
Speed of centrifugation
Drugs or pharmacological interventions (heparin (Berger and Zaret 1984), antimicrobials or antibiotics, anticonvulsants, tranquilizers, anti-inflammatory agents (Chervu et al. 1981; Hladik et al. 1982), nifedipine (Sampson 1993), prazosin, digoxin (Lee 1983), propanolol, hydralazine, methyldopa, verapamil, chlorothiazide, furosemide, ranitidine, iron therapy, etoposide+ doxontbicin (Sampson 1993, 1995), others)
Aluminum (Sampson 1995)
Iodinated contrast media (Tatum et al. 1983)

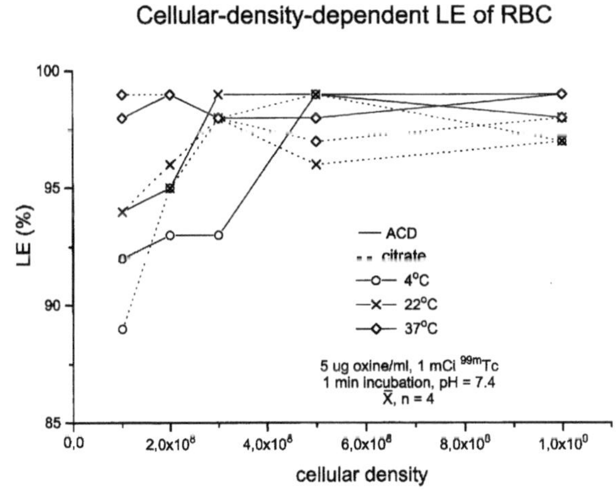

Fig. 8.1. Cellular density–dependent labeling efficiency (*LE*) of red blood cells (*RBC*)

media, a great reduction of LE (from the normal value of 90% to as low as 30%) can occur, probably due to a change in either stannous ion distribution or redox potential (Tatum et al. 1983). Stannous ions should be injected directly and not through an intravenous canulae, as synthetic material also diminishes LE (Sampson 1995).

For evaluation the function of labeled RBC, in vitro and in vivo viability testing have been used. No evidence of morphological alteration of RBC (histological and/or electron microscopic examinations (Fig. 8.5) after ^{99m}Tc-labeling was seen.

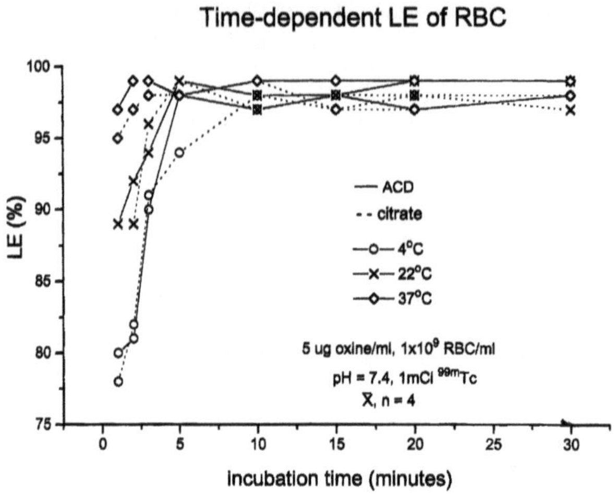

Fig. 8.2. Time-dependent LE of RBC

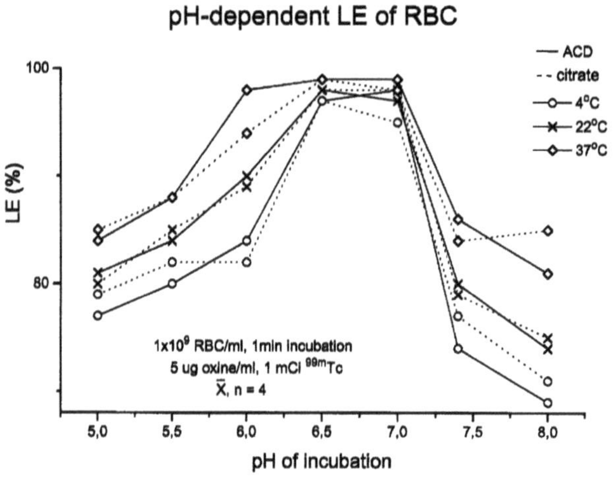

Fig. 8.3. pH-dependent LE of RBC

Recovery ([REC] i.e., the percentage of injected dose remaining cell bound in the circulation for 60 min) and in vivo biodistribution provide key information about cell integrity and function. Even after several isolation and incubation steps and being exposed to radioactive material, RBC have been shown not to alter their in vivo behavior significantly. Elution of ^{99m}Tc from RBC and/or its poor LE is reflected by gastric secretion of ^{99m}Tc-pertechnetate and accumulation of free pertechnetate in the thyroid, as well as increased urinary radioactivity.

^{99m}Tc labeling of RBC allows imaging with low-dose radiation to the patient. Applications of ^{99m}Tc-RBC are in widespread use clinically (Table 8.2). In particular, ^{99m}Tc-RBC have greatly contributed to the rapid growth of nuclear cardiology. However, due to the short half-life of ^{99m}Tc, serial imaging should be performed up to 12 h after the

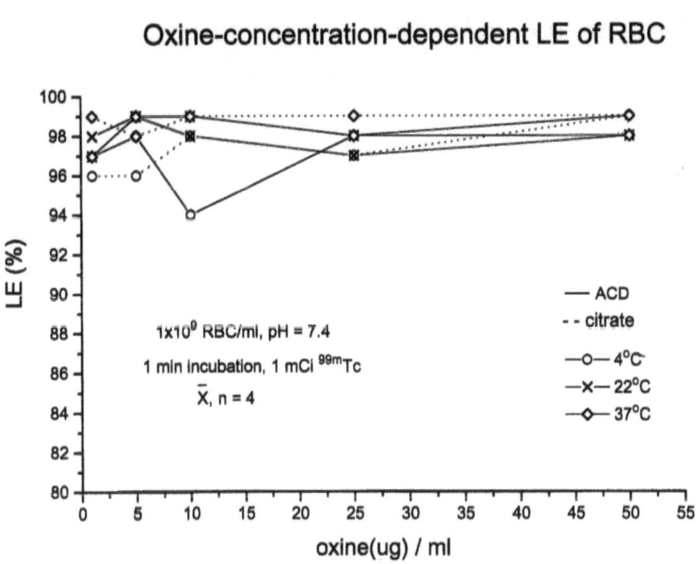

Fig. 8.4. Oxine concentration-dependent LE of RBC

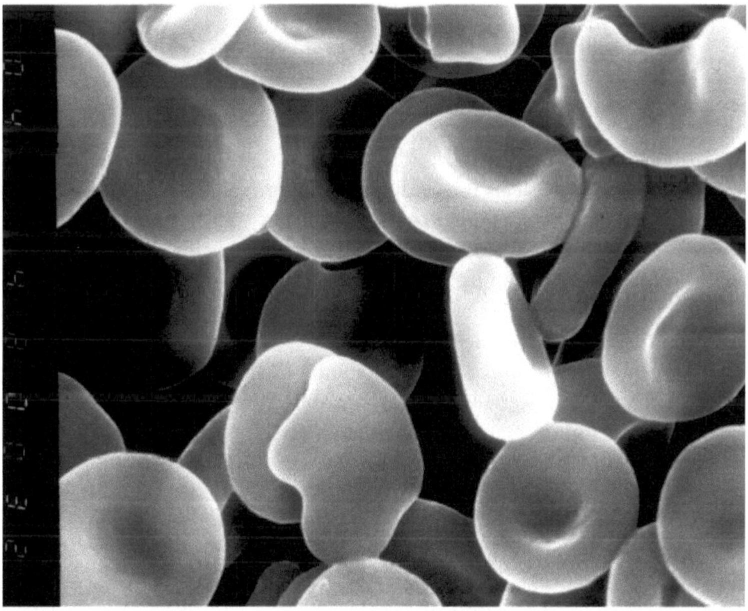

Fig. 8.5. Human RBC after radiolabeling, examined by scanning electron microscopy. No sign of morphological alteration is visible. Perfusion fixation, critical point drying, ×1500

Table 8.2. Clinical applications of ^{99m}Tc-RBC

Normal RBC	Cardiovascular procedures hemangiomas Blood pool imaging Vascular malformations Bleeding (GI, cerebral, others), cerebral blood flow stroke Deep vein thrombosis
Denaturated RBC	Spleen imaging Splenic infarcts Accessory spleen(s) localization Assessment of splenic function and size Space-occupying disease GI bleeding

GI gastrointestinal

radiolabeling. In addition, a spontaneous elution of the label in vivo of 4–10% per hour occurs (Ferrant et al. 1974). ^{99m}Tc-RBC is thus not applicable for studying RBC mass determination and lifespan measurement. For these and other clinical procedures, longer-lived labels (such as ^{51}Cr) are required.

For splenic studies (Table 8.2), heat-damaged ^{99m}Tc-RBC have to be used, the in vitro technique being the method of choice. High splenic uptake (~70%) allows the administration of low activities of the radiotracer (~1 mCi), thus reducing the radiation dose to patients considerably (Srivastava and Rao Chervu 1984).

8.3 Platelets

Random labeling of platelets has been more successfully employed than cohort labeling. Only autologous platelets are used nowadays. Platelets have to be isolated prior to labeling, as the radiopharmaceuticals presently available for platelet-labeling are nonspecific, labeling other blood cells and some plasma proteins as well (Najean 1986). The centrifugation steps and washing procedures are potentially damaging to platelets, and platelet isolation must thus be carefully performed in order to avoid cellular activation and in vitro damage. Great varieties of methods have been reported for platelet isolation (Bunting et al. 1982; Hawker et al. 1987; Sinzinger et al. 1984). In practice, however, the platelets are isolated from anticoagulated blood by simple differential centrifugation (Najean 1986; Sinzinger et al. 1984). This method allows obtaining platelets in high concentration, maintaining their functional qualities (in terms of viability, cell kinetics, survival, and sequestration sites). Complete separation of platelets from the other blood cells is, however, difficult to obtain. Recently (Sweeney et al. 1995), relatively rapid (average filtration time of 6.6 min) preparation of platelet concentrates with good platelet yields (total platelet concentrates platelet counts averaging $7.8 \pm 1.8 \times 10^{10}$) that are WBC-reduced (WBC count to levels below 7×10^5) before storage has been made possible by the introduction of filtration (using a large-capacity filter) of the platelet-rich-plasma (PRP). Prestorage filtration of PRP and the preparation of filtered platelets did not result in any significant beneficial or adverse effect on subsequent platelet quality. This approach seems thus to be a new and easy means of producing a PRP-derived WBC-reduced platelet concentrate (Sweeney et al. 1995).

Various factors influence platelet labeling (Rodrigues and Sinzinger 1994; Table 8.3). Platelet function depends on glycogen as an energy source, glucose being a key sub-

Table 8.3. Variables influencing platelet labeling with ^{99m}Tc-pertechnetate

Collection injury
Density of platelets
Anticoagulant
Amount of the complex
Incubation
Medium
pH
Temperature
Time
Speed of centrifugation
Type of plastic used for bags and tubes
Aging (Baldini and Myers 1980)
Calcium ions (Becker et al. 1988)
Lipids
Cholesterol (Granegger et al. 1988; Sinzinger 1987 a)
Low-density lipoprotein (LDL) (Granegger et al. 1988; Sinzinger 1987 a)
Additives
Prostaglandins (Sinzinger 1987 a)
Nitric oxide (Wagner et al. 1989)
Others

strate for platelet function (Schneider and Gear 1994). The medium used for resuspending the platelet pellet is thus particularly important for the maintenance of the functional properties of platelets. The choice of the anticoagulant is also crucial. Several anticoagulants have been used: heparin (may induce platelet activation and shortens the platelet survival), citrate (lacks dextrose for platelet nutrition), EDTA (may damage platelets), and ACD (pH of 6.5, contains dextrose, increases thus cellular viability). LE of platelets is by far higher with ACD and sodium citrate as compared with EDTA and heparin (Fig. 8.6).

Due to the absence of proteins and calcium, incubation of platelets in ACD-saline increases the membrane permeability and further the incorporation of tracers and LE. ACD is thus nowadays the anticoagulant of choice for platelet labeling. During pelleting, platelets may become activated and may become difficult to resuspend, unless activation is inhibited. The simplest way of performing such inhibition is to acidify (to pH 6.2–6.5) the platelet-rich plasma (Lötter et al. 1986). In addition, antiaggregatory prostaglandins (PG), either PGE_1 (Hawker et al. 1987) or prostacyclin ([PGI_2] 3.33 μg) (Sinzinger et al. 1981), which are protecting platelets from the activating effects of centrifugation and other damage by elevation of cAMP, among others, can be added and are commonly used.

The addition of PG affects labeling parameters and in vitro viability testing favorably (Sinzinger et al. 1987a), being especially useful in thrombocytopenia or other rare clinical conditions, to avoid damage of particularly sensitive platelets. PG stabilizes the platelet membrane (Gorman et al. 1977), minimizes artificial damage (Sinzinger et al. 1981, 1987 b), and reduces platelet trapping on the reinjection site (Sinzinger et al. 1981), thereby increasing REC significantly (Sinzinger et al. 1987 a). The duration of the effect of PG on platelets is prolonged, and PG can thus be added to the whole blood once, only at the start of the labeling procedure, without the need to further add the compounds at subsequent steps (of the procedure), even though these may involve platelet washing (Peters et al. 1986). The platelet pellet can then be easily resuspended

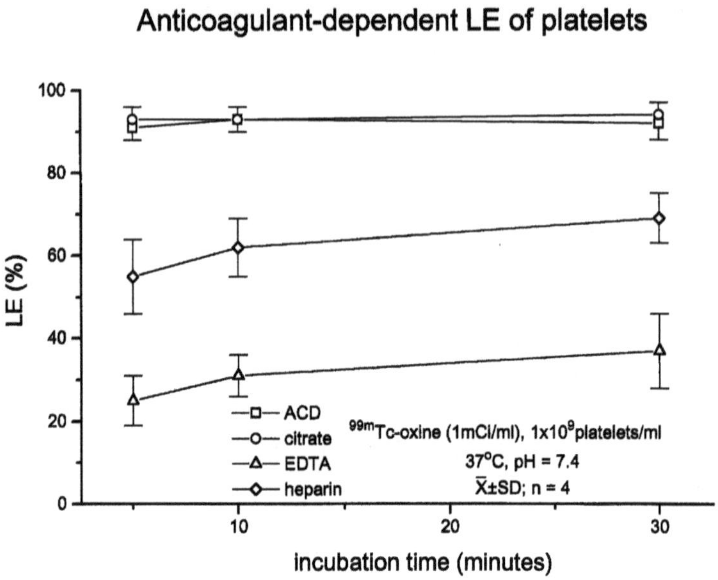

Fig. 8.6. Anticoagulant-dependent LE of platelets

in a small volume of buffer or plasma for labeling (Peters 1988). Nitric oxide (NO), the active compound of the endothelium derived relaxing factor (EDRF), improves the platelet metabolic activity by increasing cGMP, and may increase the REC and facilitate platelet handling (Wagner et al. 1989). It has to be added at a quite late stage after the preparation of the platelet pellet, because hemoglobin inactivates NO. Therefore, extremely high amounts of NO or NO donors (e.g., doses of more than 5–10 mg of molsidomine) per preparation added immediately and repeatedly would be necessary, which are rather expensive and thus unrealistic. Platelet function is better preserved in plasmatic environment and thus viability is superior, and a higher REC is seen when platelets are labeled in plasma as compared with buffer (Thakur et al. 1981). Use of plasma, however, requires longer incubation times to achieve acceptable LE as compared with buffered salt solutions (Scheffel et al. 1979). Keeping the time of incubation of platelets as short as possible is fundamental, as platelet viability due to damaging decreases with time (Mathias and Welch 1979). A closed Monovette technique (Sinzinger et al. 1984) made the techniques of labeling platelets accessible for wide clinical use, even in smaller units, at low costs. This method combines the requirements currently considered as optimal for platelet labeling (i.e., lowest amount of blood [16 ml], shortest ex vivo period of platelets (less than 60 min), and optimal labeling conditions (37 °C, 5 min).

Several techniques, radionuclides and tracers (such as ^{51}Cr, ^{111}In [-oxine, -oxine-sulfate, -tropolone, -mercaptopyridine-*N*-oxide, -acetylacetone, and -chlorotetraphenylporphyrin], ^{99m}Tc [-oxine, -hexamethylpropyleneamine oxime {HMPAO} and -phytate] and ^{123}I-metaiodobenzylguanidine [MIBG]) have been communicated in an attempt to achieve the requirements for optimal in vitro radiolabeling and subsequent in vivo studies. The labeling of platelets with ^{99m}Tc is still in its infancy. ^{99m}Tc has the advantage of easy storage, low cost, and it shows almost no reutilization. However, it has no specificity of labeling, and the elution rate (for platelet labeling) is high (8% per hour) (Becker et al. 1988 a). Various tracers have been evaluated.

^{99m}Tc-oxine has proved to be a good platelet label. The mean LE (91%) and REC (71%) data of ^{99m}Tc-oxine were found to be comparable to ^{99m}Tc-oxine-sulfate and ^{111}In-oxine (Angelberger et al. 1981). An LE of >90% was found after an incubation time of only 1 min at 37 °C (Fig. 8.2). ^{99m}Tc-oxine and ^{99m}Tc-oxine-sulfate, however, are not commercially available.

^{99m}Tc-HMPAO is an interesting platelet-labeling compound, mainly due to its lipophilicity and the good quality images of platelet accumulation over the lifetime of ^{99m}Tc (Peters 1988). However, ^{99m}Tc-HMPAO has an LE by far lower, and requires a much longer incubation time than do ^{99m}Tc- and ^{111}In-oxine, which favors platelet functional damage. Due to the relatively high elution rate from platelets, a significant gastrointestinal and renal excretion of free ^{99m}Tc-pertechnetate occurs (Becker et al. 1988a), which significantly impairs abdominal and pelvic imaging.

^{99m}Tc-phytate again requires a long incubation time (30 min), shows a low LE (20–50%) and a high elution rate. In addition, divalent cations, such as Ca^{2+} and Mg^{2+}, in the anticoagulant and the incubation buffer interfere negatively with the labeling (H. Sinzinger, unpublished data). Thus, ^{99m}Tc-phytate did not succeed as a platelet label, nor is it commercially available. The viability of platelets following radiolabeling has been evaluated by in vitro tests and by their in vivo distribution. The most commonly used in vitro methods have been testing the response of radiolabeled platelets to aggregating agents (e.g., adenosine diphosphate [ADP] (Sinzinger et al. 1987a) [Fig. 8.7], collagen, ristocetin, arachidonic acid, PAF, epinephrine (Mathias and Welch 1984; and others), their adhesion to foreign (e.g., vascular) surfaces (Peters 1988), their migration

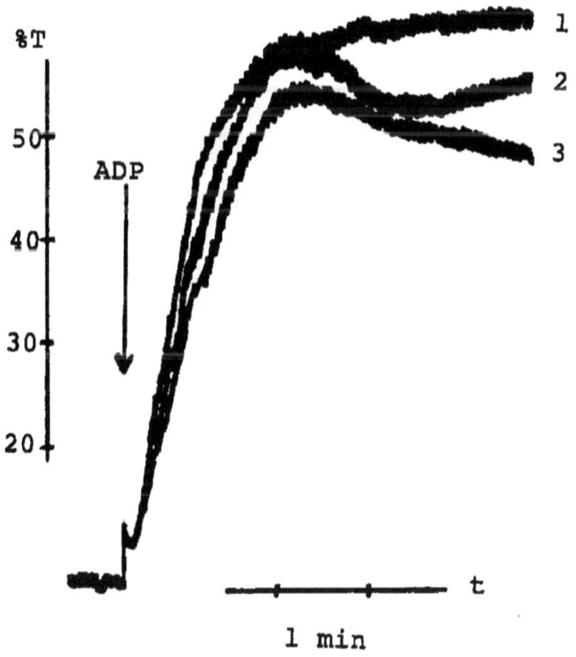

Fig. 8.7. Aggregation response before (*1*), during (*2*) and after (*3*) the handling of platelets, showing that the platelet function was not impaired as measured by the adenosine diphosphate (*ADP*)-induced aggregation. Inducing agent: ADP (10 μ*M*, 100 μl); Born-type aggregometer; 0.6-ml, platelet-rich plasma samples; platelet concentration: 250×10^3/μl. *T* light transmission (percentage)

(Sinzinger et al. 1987 a) or histological and/or electron microscopic examination, before the reinjection of the radiolabeled platelets.

Other parameters include the release of markers of platelet activation, such as β-thromboglobulin (Towler 1985). Electron microscopy is a sensitive indicator of platelet activation, manifested as pseudopodia formation (Badenhorst et al. 1982; Figs. 8.8 and 8.9). However, all the available in vitro tests failed to predict correctly collection and labeling injury, showing their limited practical value, and take a long time, which results in functional platelet deterioration and excludes their routine application (Sinzinger et al. 1987 a).

Labeling-induced injury is also reflected by the in vivo behavior of platelets. Platelets activated and/or damaged during the labeling procedure show an abnormal biodistribution, such as sequestration of radiolabeled platelets in the liver, spleen, and lungs, which results in lower REC in blood and shortened survival. REC is therefore the only reliable viability parameter, as morphologically damaged cells may recover completely after reinjection, while apparently intact ones, in contrast, may not. The normal values for platelet REC range from 55–72% of the injected dose (Baldini and Myers 1980; International Committee for Standardization in Hematology 1988; Mortelmans et al. 1986) due to pooling of about one third of the platelets in the liver and spleen (Baldini and Myers 1980). ^{99m}Tc in none of the tracers used until now caused by its own any sign of morphological or functional alteration due to radiation injury. Conclusive microdosimetry data are not yet available.

The too-short half-life of ^{99m}Tc for platelet survival studies and the high elution rate (instability of cell binding) with the tracers so far examined have limited considerably the clinical use, although with various applications (Table 8.4), of ^{99m}Tc-labeled platelets. ^{111}In-oxine is still nowadays the tracer of choice for the studies with radiolabeled platelets, which are long established and have attracted considerable interest.

Fig. 8.8. Human platelets after radiolabeling, examined by scanning electron microscopy. No sign of morphological alteration is visible. Perfusion fixation, critical point drying, ×8000

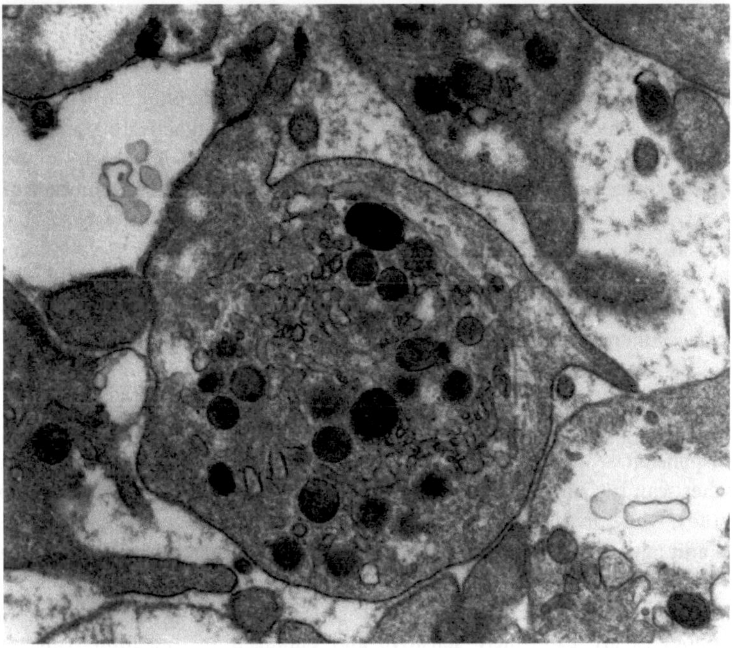

Fig. 8.9. Human platelets after the radiolabeling procedure, examined by transmission electron microscopy. Almost no signs of platelet activation, manifested as pseudopodia formation. Perfusion fixation, critical point drying, ×40 500

Table 8.4. Indications for studies with radiolabeled platelets

Kinetic studies
Evaluation of the normal platelet kinetics and biodistribution
Understanding of the role of platelets, platelet activation and consumption in diseases characterized by thromboembolic phenomena or thrombocytopenia (Baldini and Myers 1980; Lötter et al. 1986; International Committee for Standardization in Hematology 1988; Vallabhajosula et al. 1986) as well as in vascular disorders or endothelial (surface) abnormalities (Baldini and Myers 1980; Vallabhajosula et al. 1986; Harker et al. 1977)
Differentiation of the cause of thrombocytopenia
Evaluation of the effects of platelet-inhibitor drugs
Detection of prethrombotic states (Lötter et al. 1986)
Evaluation of the effects of risk factors (Rodrigues and Sinzinger 1884)
Evaluation of different cell separation techniques, storage or transfusion of platelets procedures, in transfusion medicine (International Committee for Standardization in Hematology 1988; De Vries et al. 1993; Mollison et al. 1987)
Imaging (Rodrigues and Sinzinger 1994)
Atherosclerosis
Surgical interventions
Synthetic material
Thrombosis
Transplantation

8.4 White Blood Cells

Several methods are available for labeling WBC. Their underlying principles are, however, similar and include isolation of WBC, prevention of cell activation and/or damage, and choice of ligand (Peters et al. 1986). Diverse factors influence the LE of WBC (Table 8.5). Apart from maintenance in plasma (a major factor in the promotion of granulocyte activation is their isolation from plasma), no further additives have been used (Peters et al. 1986) to inhibit granulocytes activation during their isolation and/or labeling procedure.

As an anticoagulant, ACD solution is preferred to heparin, because WBC show less tendency to adhere to plasticware with ACD (McAfee et al. 1984).

Antibiotics seem to be very unlikely to have a negative influence on LE (Sinzinger and Granegger 1988). WBC cannot be isolated from platelets and RBC by differential centrifugation alone. Random labels are the ones in common use for WBC. Either a mixed WBC suspension or a specific WBC type population, which requires more-sophisticated approaches, can be isolated, depending on the circulating WBC count, and the purpose of the study, while preserving cell viability. Homologous WBC can be used in leukopenic patients and have proven successful (Sinzinger and Granegger 1988). Monocytes are particularly difficult to separate from whole blood because they are normally very few in number, and their viability is affected by the dose of radiation (Fig. 8.10).

Lymphocytes are exquisitely sensitive and damaged by radiation (Chisholm and Peters 1980; Chisholm et al. 1979), and very few clinical studies using radiolabeled lymphocytes thus exist. After labeling WBC with 740 MBq of ^{99m}Tc-HMPAO, the radiation damage of the lymphocytes due to self-irradiation was estimated to be equivalent to 26 Gy of x-rays. Due to almost complete inhibition of the proliferative capacity at this high dose level, the increased risk for a lymphoid malignancy after administration of isolated lymphocytes or mixed WBC labeled with ^{99m}Tc-HMPAO activities sufficient for scintigraphy can, however, be regarded as small (Thierens et al. 1982). An injection of 100 MBq ^{99m}Tc-labeled autologous human granulocytes was found to give them an absorbed radiation dose of 1.8 Gy after 25 min and 8.4 Gy after 4 h. In vitro tests revealed no signs of radiation damage to the cells (Skretting et al. 1988).

Table 8.5. Variables influencing white blood cells (*WBC*) labeling with 99mTc- hexamethylpropyleneamine oxime (*HMPAO*)

Collection injury
Density of WBC
Anticoagulant
Sedimenting agents (type and volume) (Webber et al. 1994)
Amount of the complex
Time delay between labeling of WBC and administration (HMPAO)
Incubation
Medium
pH
Temperature
Time
Speed of centrifugation
Drugs: corticosteroids (Sampson 1995; McAfee et al. 1984), ethanol (McAfee et al. 1984), cyclosporin, azathioprine, ranitidine, nifedipine, procainamide (Sampson 1995), others

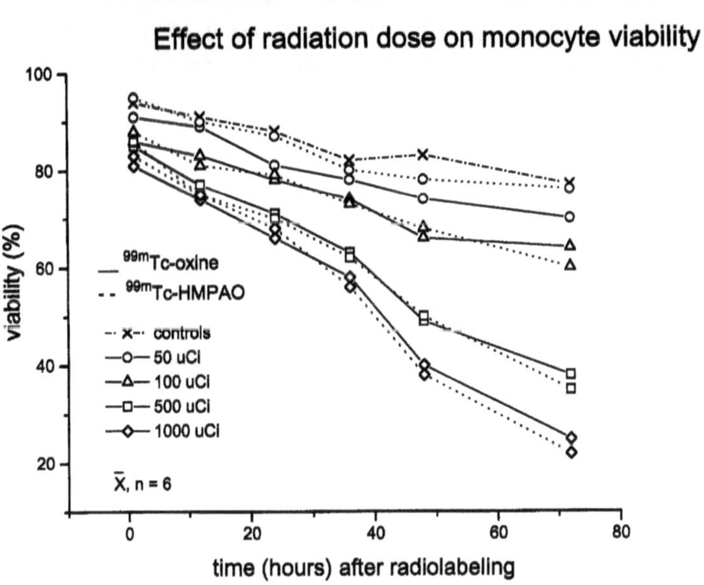

Fig. 8.10. Effect of radiation dose on monocyte viability

Separation of WBC involves the addition of a sedimenting agent (e.g., dextran, methylcellulose, and hydroxyethylstarch) to whole blood (Peters et al. 1986). Laborious cell harvesting techniques include newer density gradients for isopycnic centrifugation (nonionic contrast media), centrifugal elutriation, and flow cytometry, among others (McAfee et al. 1984). Relatively pure populations of granulocytes, lymphocytes, or monocytes can be harvested by elutriation (Berger and Ederson 1979; McAfee et al. 1984), which is, however, rather expensive. Granulocytes or lymphocyte–monocyte isolation can also be performed, achieving a high yield or high purity by isopycnic density centrifugation, such as Percoll-plasma and metrizamide-plasma density gradients, because of slight differences in cell density (Peters et al. 1986). With the surface adherence method, pure lymphocyte or monocyte populations can also be obtained (McAfee et al. 1984). However, some cell functions are altered by this manipulation (Fisher and Koren 1981). Monocytes and granulocytes may also be isolated by phagocytosis of radioactive particles, followed by density gradient centrifugation or flow through a magnetic field for in vitro experiments, with an LE generally only of 30–40%. However, this approach activates the cells metabolically, resulting in poor in vivo survival (McAfee et al. 1984), and thus is not acceptable for in vivo studies.

Several radiotracers such as ^{32}P-, ^{51}Cr-, ^{125}I-, ^{67}Ga-, ^{68}Ga-, and ^{111}In-lipophilic chelates, among others, have been developed for the purpose of labeling WBC. ^{111}In-oxine has become the standard for radiolabeling WBC. However, ^{99m}Tc has more favorable dosimetry and energy and many attempts have been made to replace ^{111}In with ^{99m}Tc, the use of ^{99m}Tc-compounds for WBC labeling having attracted considerable interest becoming an area of research undergoing much experimentation. Several ^{99m}Tc radiopharmaceuticals, such as ^{99m}Tc-HMPAO, -albumin colloid, -sulfur colloid, -sestamibi, -teboroxime, -mebrofenin, -disofenin, -gluceptate, -dimercaptosuccinic acid (DMSA) (Segall et al. 1994), -oxine, -phytate (H. Sinzinger, unpublished data) and -pyrophosphate (Uchida et al. 1979) have been tested. HMPAO showed the highest mean LE, 41–56% (Arndt et al.

1993; Becker et al. 1988b; Danpure et al. 1988; Lantto et al. 1991; Mountford et al. 1990; Roddie et al. 1988) and 79.4% (Segall et al. 1994), as compared with albumin colloid (12.1%, Segall et al. 1994), sulfur colloid (11.7%, McAfee et al. 1984; Segall et al. 1994), sestamibi (11.1%), and teboroxime (5.3–14.2%, Segall et al. 1994). Mebrofenin, disofenin, gluceptate, and DMSA do not label WBC (Segall et al. 1994). Pyrophosphate had a LE of 40–60% and showed poor in vivo recovery of the labeled WBC and poor abscess localization (Uchida et al. 1979). All the other agents have high labeling stability (ranging from 80 to >95% at 4 h and >80% at 24 h), except sestamibi, which washed out very rapidly (remaining 52% at 4 h and 5% at 24 h, Segall et al. 1994). ^{99m}Tc-oxine had an LE ranging from 50 to >80%, with 5 min of incubation, while ^{99m}Tc-phytate required a long incubation time (30 min), had low LE (20–50 %), and a high elution rate (H. Sinzinger, unpublished data). Attempts were recently made to use ^{99m}Tc-ethylcysteinate dimer (ECD) to label WBC (Kao et al. 1994). ECD is a neutral, lipophilic agent, introduced as a new brain perfusion agent. It has a higher radiochemical stability than HMPAO, being stable in vitro for at least 24 h, but it has lower WBC LE, worse stability of labeled WBC, and more rapid elimination than ^{99m}Tc- HMPAO from most tissues, thus being not a good choice as a WBC-labeling agent to replace ^{99m}Tc-HMPAO (Kao et al. 1994).

For assessing radiolabeled WBC viability, in vitro testing before administration to the patient and in vivo distribution of WBC can be performed. The adherence characteristics of WBC are measured in vitro and performed easily, but suffer from a lack of specificity and sensitivity (McAfee et al. 1984). Adherence of WBC is not altered by labeling with HMPAO, while it is decreased by albumin colloid, sestamibi, and teboroxime and increased by sulfur colloid (probably as a result of activation) (Segall et al. 1994). Routine tests of granulocyte function, namely random migration, chemotaxis, *Candida* killing, phagocytosis, and trypan blue exclusion test, are not sensitive enough to detect abnormalities on the status of the cell function before administration, which profoundly change further in vivo behavior (Peters et al. 1986) and are of limited value, only showing a difference in rather extreme situations. Besides, results of these tests generally are not available until several hours later. Labeled WBC smeared and observed with light microscopy provide a reliable, inexpensive, and quick insight into their purity, structural integrity, and their dispersal (McAfee et al. 1984), but do not replace the other functional tests. Transmission electron microscopy (Fig. 8.11) is expensive and rather time consuming.

The most valuable parameter remains the in vivo function of reinjecting the labeled WBC and measuring recovery and in vivo distribution, although too late for consequences if negative. Following injection, normal distribution of labeled WBC includes an initial transitory uptake in the lungs and, later, in the spleen, liver, and bone marrow. The relative amounts of the distribution of each of these sites vary with the labeling procedure used. A disadvantage of ^{99m}Tc complexes in general is nonspecific bowel accumulation, which makes their late abdominal imaging difficult. Damage to WBC or cell clumping during and/or because of the labeling procedure produces an abnormal distribution of the labeled WBC. A prolonged and delayed lung transit time and/or an abnormal high liver uptake are ominous signs that some damage of the labeled WBC occurred (Bowring 1986). Damage of lymphocytes by ^{111}In was detected as a failure of recirculation between blood and lymph (Peters et al. 1986). To minimize damage of the specific lymphocyte, activity must be restricted, requiring harvesting of a high number of cells for imaging.

HMPAO thus has become the best agent for labeling WBC. It was a radiopharmaceutical introduced for evaluating regional cerebral blood flow and is currently the only ^{99m}Tc-labeling agent for WBC commercially available in most countries. HMPAO is a li-

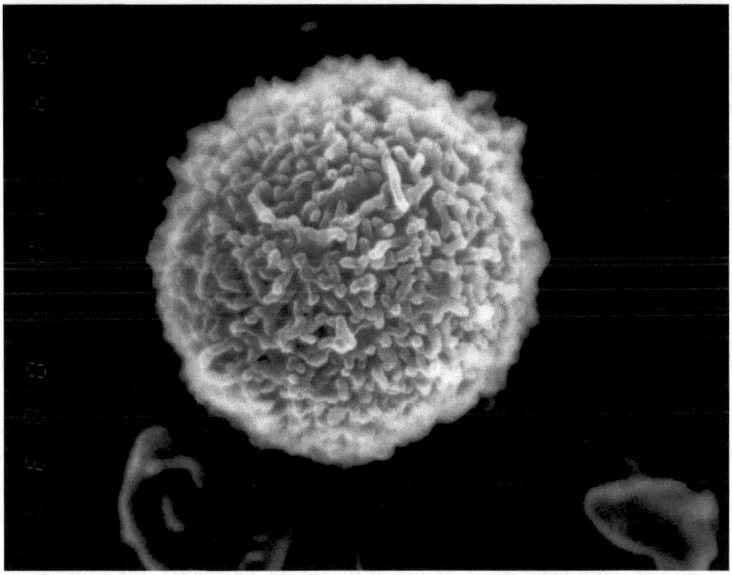

Fig. 8.11. Human lymphocyte after radiolabeling, examined by scanning electron microscopy. No sign of morphological alteration is visible. Perfusion fixation, critical point drying, ×2000

Table 8.6. Clinical Applications of ^{99m}Tc-WBC

Inflammatory bowel disease
Ulcerative colitis
Crohn's disease
Localization of sepsis
Localization of abscesses
Infections after surgery (e.g., grafts, prostheses)
Osteomyelitis (Mortelmans et al. 1989)
Pneumonia (Raptopoulos et al. 1982)
Endocarditis (Sfakianakis et al. 1982 a)
WBC accumulation around tumor infiltration (Sfakianakis et al. 1982 b)
Response to splenectomy (ratio of splenic to hepatic uptake) (Birnie et al. 1982)
WBC kinetics (e.g., quantification of fecal WBC excretion) (Saverymuttu et al. 1983), others

pophilic compound that diffuses through the cell membrane and converts with time to a secondary complex that is less lipophilic. The conversion to a less lipophilic form inhibits back diffusion across the cell membrane (Segall et al. 1994). However, HMPAO is limited in that its rapid decomposition in vitro requires its usage within 30 min of preparation (Neirinckx 1987). In addition, elution of ^{99m}Tc is a major problem, 20% (Mortelmans et al. 1989) to 25% (Segall et al. 1994) of the label eluting from WBC by 24 h. The feasibility of using other chelating agents, such as the neutral lipophilic complex teboroxime, still needs further investigation.

The labeling of WBC with ^{99m}Tc chelates, in particular HMPAO dramatically increased the widespread availability and use of these studies in nuclear medicine. Several clinical applications of radiolabeled WBC exist (Table 8.6), but the main clinical indication of imaging WBC distribution has been the detection of inflammatory bowel disease.

Peptides, proteins, and monoclonal antibodies or its fragments, with the potential for labeling specific subpopulations of WBC by binding to specific surface receptors or antigens of WBC, may be challenging and exciting in the future for selective cell labeling and in vivo studies.

8.5 Stem Cells

Stem cell therapy is gaining considerable interest, mainly in cardiovascular medicine and oncology (Barbash et al. 2003, Bengel et al. 2005). Visualization and monitoring of the therapeutically administered cells could provide a better understanding of the underlying mechanisms and open a new field of research for clinical nuclear medicine. Preliminary attempts to label stem cells in vitro have been successful (Brenner et al. 2004, Chin et al. 2003). However, while labeling with In-111-oxine is above 80%, not affecting cell viability, a rather high elution rate was observed when labeling was performed with either ^{99m}Tc-HMPAO or ^{99m}Tc-oxine. Together with its shorter half-life, Tc-99m therefore does not allow longterm in vivo monitoring. Since the clinical use of labeled stem cells in patients is in a rather early stage, not enough information on the in vitro and in vivo viability is as yet available, in particular, the number of labeled cells required to correctly target the few percent of accumulating cells is not known. Other tracers, such as Cu-64 and ^{18}F-labeled deoxyglucose (FDG) do not offer any advantage over In-111-oxine. Like with platelets, the labeling efficiency of stem cells is negatively affected by risk factors of cardiovascular disease (Sinzinger et al. 2006), such as cigarette smoking, diabetes mellitus, and hyperlipidemia.

8.6 Conclusions

Developments in both instrumentation and cell labeling have enabled a great expansion of their use in both clinical practice and research.

Studies with ^{99m}Tc-RBC are in widespread use in clinical practice. In contrast, ^{99m}Tc has not been successfully applied to platelet labeling. In contrast to ^{99m}Tc-RBC-research, there have been more methodological reports with ^{99m}Tc-WBC than clinical applications at the present. Much progress has been achieved in the techniques of harvesting and labeling WBC, but many challenges remain. HMPAO is currently the best ^{99m}Tc agent for labeling WBC, and played a particularly important role for the successful expansion of diagnostic imaging with radiolabeled WBC in nuclear medicine.

New approaches, such as the use of peptides, proteins, antibodies, and molecular recognition unit technologies, may result in substantial improvements in the labeling methodology and could yield labeled cells with the least damage and high in vivo stability in the future.

Acknowledgments

The authors gratefully acknowledge the collaboration and assistance by Judith Bednar, Susanne Granegger, Ulli Horvarth, Brigitte Jankovic, J. Kaliman, G. Karanikas, H.

Kolbe, H. Kritz, W. Linkesch, H. Ludwig, Ch. Müller, Irmgard Neumann, J. O'Grady, Adelaide Osei-Bonsu, H. Pesl, Cornelia Pones, Sonja Reiter, Eva Strobl-Jäger, H. Vinazzer, K. Widhalm, and Ch. Zielinski.

References

Angelberger P, Sinzinger H, Kolbe H, Leithner C (1981) Labelling of human platelets with ^{111}In-oxine and ^{99m}Tc-oxine. Comparison of in vitro and clinical use for monitoring of kidney transplants. Progr Radiopharmacol 2:211–220

Arndt JW, van der Sluys Veer A, Blok K et al (1993) Prospective comparative study of technetium-99m-WBCs and indium-111-granulocytes for the examination of patients with inflammatory bowel disease. J Nucl Med 34:1052–1057

Badenhorst PN, Lotter MG, Heyns A et al (1982) The influence of the "collection injury" on the survival and distribution of indium-111-labelled canine platelets. Br J Haematol 52:233–240

Baldini MG, Myers TJ (1980) Utilitá e limitazioni degli studi di sopravvivenza delle piastrine. Una valutazione critica [in Italian]. Haematologica 65:689–716

Barbash IM, Chouraqui P, Baron J, Feinberg MS, Etzion S, Tessone A, Miller L, Guetta E, Zipori D, Kedes LH, Kloner RA, Leor J (2003) Systemic delivery of bone marrow-derived mesenchymal stem cells to the infarcted myocardium: feasibility, cell migration, and body distribution. Circulation 108:863–868

Becker W, Borner W, Borst U (1988 a) ^{99m}Tc-hexamethylpropyleneamine-oxime (HMPAO) as a platelet label: evaluation of labelling parameters and first in vivo results. Nucl Med Commun 9:831–842

Becker W, Schomann E, Fischbach W et al (1988 b) Comparison of ^{99m}Tc-HM-PAO and ^{111}In-oxine labelled granulocytes in man: first clinical results. Nucl Med Commun 9:435–437

Bengel FM, Schachinger V, Dimmeler S (2005) Cell-based therapies and imaging in cardiology. Eur J Nucl Med Mol Imaging 32:404–416

Berger CL, Ederson RL (1979) Comparison of lymphocyte function after isolation by Ficoll-Hypaque flotation or elutriation. J Invest Dermatol 73:231–235

Berger HJ, Zaret BL (1984) Radionuclide assessment of cardiovascular performance: radiopharmaceuticals. In: Freeman LM (ed) Freeman and Johnson's clinical radionuclide imaging. Grune & Stratton, Orlando, Fla., pp 366–367

Birnie, GG, Eadie AS, Hosie CJ et al (1982) ^{111}In-labelled white blood cells in the diagnosis of Felty's syndrome. J Clin Pathol 35:74–76

Bowring CS (1986) Imaging and quantitative scanning. In: Radionuclides in hematology. Lewis SM, Baily RJ (eds) Churchill Livingstone, Edinburgh, pp 151–172

Brenner W, Aicher A, Eckey T, Massoudi S, Zuhayra M, Koehl U, Heeschen C, Kampen WU, Zeiher AM, Dimmeler S, Henze E (2004) In-111-labeled CD34+ hematopoietic progenitor cells in a rat myocardial infarction model. J Nucl Med 45:512–518

Bunting RW, Callahan RJ, Finkelstein S, Lees RS, William MS (1982) A modified method for labelling human platelets with indium-111 oxine using albumin density-gradient separation. Radiology 145:219–221

Chervu LR, Huq SS, Joseph JA et al (1981) Medication-induced changes in biodistribution of radiopharmaceuticals. J Nucl Med 22:72

Chin BB, Nakamoto Y, Bulte JW, Pittenger MF, Wahl R, Kraitchman DL (2003) In-111-oxine labelled mesenchymal stem cell SPECT after intravenous administration in myocardial infarction. Nucl Med Commun 24:1149–1154

Chisholm PM, Peters AM (1980) The effect of 111-indium labelling on the recirculation of rat lymphocytes. In: Thakur ML, Gottschalk A (eds) ^{111}In-labelled neutrophils, platelets and lymphocytes. Trivirum, New York, pp 205–211

Chisholm PM, Danpure HJ, Healey G, Osman S (1979) Cell damage resulting from the labelling of rat lymphocytes and HeLa S3 cells with In-111-oxine. J Nucl Med 20:1308–1311

Danpure HJ, Osman S, Carroll MJ (1988) The development of a clinical protocol for the radiolabelling of mixed leukocytes with ^{99m}Tc-hexamethylpropyleneamine oxime. Nucl Med Commun 9:465–475

De Vries RA, De Bruin M, Marx JJM, Van De Wiel A (1993) Radioisotopic labels for blood cell survival studies: a review. Nucl Med Biol 20:809–817

Eckleman W, Richards P, Hauser W, Atkins H (1979) Technetium-labelled red blood cells. J Nucl Med 20:22–24

Ferrant A, Lewis SM, Szur L (1974) The elution of ^{99m}Tc from red cells and its effect on red cell volume measurement. J Clin Pathol 27:983–985

Fisher DG, Koren HS (1981) Isolation of human monocytes. Adams DO, Edelson PJ, Koren HS (eds) In: Methods for studying mononuclear phagocytes. Academic, New York, 43–47

Gorman, RR, Bunting S, Miller OV (1977) Modulation of human platelet adenylate cyclase by prostacyclin (PGX). Prostaglandins 13:377–388.

Granegger S, Flores J, Widhalm K, Sinzinger H (1988) Increased low-density lipoproteins (LDL) negatively affect human labelling. Folia Haematol 115:451–453

Harker LA, Slichter SJ, Sauvage LR (1977) Platelet consumption by arterial prostheses: the effects of endothelialization and pharmacologic inhibition of platelet function. Ann Surg 186:594–601

Hawker RJ, Hawker LM, Wilkinson AR (1987) Indium-labelled human platelets. Optimal method. Clin Sci 58:243–248

Hegge FN, Hamilton GW, Larson SM et al (1978) Cardiac chamber imaging: a comparison of red blood cells labelled with Tc-99m in vitro and in vivo. J Nucl Med 19:129

Hladik WB, Nigg KK, Rhodes BA (1982) Drug-induced changes in the biologic distribution of radiopharmaceuticals. Semin Nucl Med 12:184

International Committee For Standardization In Hematology (1988) Panel on diagnostic applications of radionuclides. Recommended method for indium-111 platelet survival studies. J Nucl Med 29:564–566

Kao CH, Hunang WT, Wang YL et al (1994) A comparative study of ^{99m}Tc-HM-PAO and ^{99m}Tc-ECD as a leukocyte labelling agent. Nucl Med Comm 15:294–297

Lantto EH, Lantto TJ, Vorne M (1991) Fast diagnosis of abdominal infections and inflammations with technetium-99m-HM-PAO labeled leukocytes. J. Nucl Med 32:2029–2034

Lee HB, Wexler JP, Scharf SC et al (1983) Pharmacologic alterations in Tc-99m binding by red blood cells: concise communication. J Nucl Med 24:397–401

Leitl GP, Drew HM, Kelly ME et al (1980) Interference with Tc^{99m}-labelling of red blood cells (RBCs) by RBC antibodies. J Nucl Med 21:44

Lötter MG, Heyns AD, Badenhorst PN et al (1987) Evaluation of mathematic models to assess platelet kinetics. J Nucl Med 27:1192–1201

Mathias CJ, Welch MJ (1979) Labelling mechanism and localization of indium-111 in human platelets. J Nucl Med 20:659

Mathias CJ, Welch MJ (1984) Radiolabelling of platelets. Semin Nucl Med 14:118–127

McAfee JG, Subramanian G, Gagne G (1984) Technique of leukocyte harvesting and labeling: problems and perspectives. Semin Nucl Med 14:83–106

Mollison PL, Engelfriet CP, Contreras M (1987) The transfusion of platelets, leukocytes, haemopoietic cells and plasma components. In: Blood transfusions in clinical medicine. Blackwell, Oxford, pp 7–27

Mortelmans L, Verbruggen A, De Roo M, Vermylen J (1986) Evaluation of three methods of platelet labeling. Nucl Med Comm 7: 519–529

Mortelmans L, Malbrain S, Stuyck J et al (1989) In vitro and in vivo evaluation of granulocyte labeling with (^{99m}Tc)D,L-HM-PAO. J Nucl Med 30:2022–2028

Mountford PJ, Kettle AG, O'Doherty MH, Coakley AJ (1990) Comparison of technetium-99m-HM-PAO leukocytes with indium-111-oxine leukocytes for localizing intraabdominal sepsis. J Nucl Med 31:311–315

Najean Y (1986) The choice of tracer for platelet kinetics and scintigraphic studies. Nucl Med Biol 13:159–164

Neirinckx R, Canning L, Piper I et al (1987) Technetium-99m-D,L-HM-PAO: a new radiopharmaceutical for SPECT imaging of regional cerebral blood perfusion. J Nucl Med 28:191–202

Peters AM (1988) Review of platelet labelling and kinetics. Nucl Med Commun 9:803–808

Peters AM, Saverymuttu SH, Danpure HJ, Osman S (1986) Cell labeling. In: Bailey RJ, Lewis SM (eds) Radionuclides in haematology. Churchill Livingstone, Edinburgh, pp 79–109

Porter WC, Dees SM, Freitas JE, Dworkin HJ (1983) Acid-citrate-dextrose compared with heparin in the preparation of in vivo/in vitro technetium-99m red blood cells. J Nucl Med 24:383–387

Rao SA, Knobel J, Collier D et al (1986) Effect of Sn (II) ion concentration and heparin on technetium-99m red blood cell labelling. J Nucl Med 27:1202–1206.

Raptopoulos V, Doherty PW, Goss TP et al (1982) Acute osteomyelitis: advantage of white cell scans in early detection. Am J Roentgenol 139:1077–1082

Reiter S, Angelberger P, Kolbe H, Sinzinger H (1984) In vitro- und In vivo-Untersuchungen mit ^{111}Indium-Oxin, ^{111}Indium-Oxin-Sulfat und ^{99m}Tc-Oxin zur Erythrozytenmarkierung [in German]. Wien Klin Wschr 3:120–123
Riordon F, Nelp WB (1982) Binding capacity of normal and sickle cell hemoglobin for technetium atoms. J Nucl Med 23:91
Roddie ME, Peters AM, Danpure HJ et al (1988) Inflammation: imaging with Tc-99m HM-PAO-labelled leukocytes. Radiology 166:767–772
Rodrigues M, Sinzinger H (1994) Platelet labeling – methodology and clinical applications. Thromb Res 76:399–432
Sampson CB (1993) Adverse reactions and drug interactions with radiopharmaceuticals. Drug Saf 8:280–294
Sampson CB (1995) Influence of drugs on cell labelling. 6th European Symposium on Radiopharmacy and Radiopharmaceuticals. Graz, March 1995
Saverymuttu SH, Peters AM, Lavender JP et al (1983) Quantitative fecal indium-111 labelled leukocyte excretion in the assessment of activity in Crohn's disease. Gastroenterology 85:1333.
Scheffel U, Tsan MF, McIntyre M (1979) Labeling of human platelets with (^{111}In)8-hydroxyquinoline. J Nucl Med 20:524–531
Schneider W, Gear ARL (1994) Significance of glucose and glycogen metabolism for platelet function. Am J Physiol 3:225–232
Segall GM, Lang HV, Chaovapong W (1994) In vitro evaluation of white blood cell labelling with ^{99m}Tc radiopharmaceuticals. Nucl Med Commun 15:845–849
Sfakianakis GN, Al-Sheikh W, Heal A et al (1982) Comparison of scintigraphy with In-111 leukocytes and Ga-67 in the diagnosis of occult sepsis. J Nucl Med 23:618
Sfakianakis GN, Mnaymneh W, Ghandus-Mnaymneh L et al (1982) Positive In-111 leukocyte scintigraphy in a skeletal metastasis. Am J Roentgenol 139: 601–603
Sinzinger H, Granegger S (1988) The effect of various antibiotics on the labelling efficiency of human white blood cells with ^{111}In-oxine. Nucl Med Comm 9:597–601
Sinzinger H, Angelberger P, Höfer R (1981) Platelet labelling with ^{111}In-oxine: benefit of prostacyclin (PGI_2) – addition for preparation and injection. J Nucl Med 22:292
Sinzinger H, Flores J, Widhalm K, Granegger S (1987a) Platelet viability (aggregation, migration, recovery) after radiolabelling of platelets from hypercholesterolemics using various tracers (oxine, oxine-sulphate, tropolone, MPO). Eur J Nucl Med 14:358–361
Sinzinger H, Fitscha P, Kaliman J (1987b) Prostaglandin I_2 improves recovery, but does not change platelet half-life and platelet uptake over active human lesion sites. Prostaglandins 33:787–790
Sinzinger H, Granegger S (2006) Radionuclide vascular imaging and characterization of human metabolic endothelial surface lining. Q J Nucl Med Mol Imaging 50:226–235
Sinzinger H, Kolbe H, Strobl-Jäger E, Höfer R (1984) A simple and safe technique for sterile autologous platelet labelling using "Monovette" vials. Eur J Nucl Med 9:320–322
Skretting A, Benestad HB, Sundrehagen H (1988) Whole body distribution of ^{99m}Tc labelled autologous human granulocytes and radiation dose to cells and organs. Eur J Nucl Med 14:1–7
Srivastava S, Rao Chervu L (1984) Radionuclide-labelled red blood cells: current status and future prospects. Semin Nucl Med 14:68–82
Srivastava S, Straub RF (1990) Blood cell labelling with ^{99m}Tc: progress and perspectives. Semin Nucl Med 20:41–51
Sweeney JD, Holme S, Heaton WAL, Nelson E (1995) White cell-reduced platelet concentrates prepared by in-line filtration of platelet-rich plasma. Transfusion 35:131–136
Tatum JL, Burke TS, Hirsh JI et al (1983) Pitfall to modified in vivo method of technetium-99m red blood cell labelling. Iodinated contrast media. Clin Nucl Med 8:585–587
Thakur ML, Welch MJ, Malech HL (1981) Indium-111 labelled human platelets: improved method, efficacy and evaluation. J Nucl Med 22:381–385
Thierens HM, Vral AM, Van-Haelst JP et al (1982) Lymphocyte labelling with technetium-99m-HM-PAO: a radiotoxicity study using the micronucleus assay. J Nucl Med 33:1167–1171
Towler HMA (1984) Effect of ^{111}In-oxinate and tropolonate labelling on human platelet aggregation and β-thromboglobulin release. In: Thakur ML, Ezekowitz MD, Hardeman MR (eds) Radiolabeled blood cellular elements. Plenum, New York
Uchida T, Nemoto T, Yui T et al (1979) Use of technetium-99m as a radioactive label to study migratory patterns of leukocytes. J Nucl Med 20:1197–2000
Vallabhajosula S, Machac J, Goldsmith SJ et al (1986) ^{111}Indium platelet kinetics in normal human subjects: tropolone versus oxine methods. J Nucl Med 27:1669–1674

Wagner X, Granegger S, Dembinska-Kiec A, Sinzinger H (1998) Nitric oxide (NO) for radiolabelling of human platelets. In: Sinzinger H, Thakur ML (eds) Nuclear medicine research. Facultas, Vienna, p 58

Webber D, Nunan TO, Doherty JO (1994) The effect of varying type and volume of sedimenting agents on leukocyte harvesting and labelling in sickle cell patients. Nucl Med Comm 15:735–741

Wilson ME, Hung JC (1992) Evaluation of heparin and anticoagulant citrate dextrose in the preparation of technetium-99m-red blood cells with Ultra Tag® RBC kit. J Nucl Med 33:306–307

Winzelberg GG, McKusick KA, Froelich JW et al (1982) Detection of gastrointestinal bleeding with ^{99m}Tc-labelled red blood cells. Semin Nucl Med 12:139–146

Chapter 9

Quality Control Methods of ^{99m}Tc Pharmaceuticals

9

9.1 Determination of Radiochemical Purity

9.1.1 Thin-Layer Chromatography Methods

C. Decristoforo and I. Zolle

Thin-layer chromatography (TLC) is commonly used for the determination of radiochemical purity in nuclear medicine. TLC was described as early as 1967 for testing radiopharmaceuticals (Hoye 1967). Since the introduction of TLC (Izmailov and Shraiber 1938; Stahl 1956), a variety of modifications and new applications have been reported (Fairbrother 1984).

The principle of this analytical method is that a mobile phase (solvent) moves along a layer of adsorbent (stationary phase) due to capillary forces. Depending on the distribution of components between the stationary and the mobile phase, a radioactive sample spotted onto the adsorbent will migrate with different velocities, and thus, impurities are separated. The distance each component of a sample migrates is expressed as the R_f value. The R_f is the relative migration of a component in relation to the solvent front (SF):

$$R_f = \frac{\text{Distance from origin of the component}}{\text{Distance of the SF}}$$

The R_f values range from 0–1. If a component migrates with the SF, the R_f is 1. If a component remains at the point of application (origin), the R_f is 0. For a given TLC system, which is defined by the mobile and the stationary phases, the R_f value of a pure chemical compound is specific and reproducible.

The main principles of separation are adsorption (electrostatic forces), partition (solubility), and ion exchange (charge). Information on the theoretical background of TLC is presented elsewhere (Miller 2004). Depending on the movement of the mobile phase, TLC may be ascending or descending; in the nuclear medicine laboratory, ascending TLC is the method of choice (Robbins 1983).

For the analysis of radiopharmaceuticals, techniques should be fast and safe. TLC offers reliable separation properties with easy and rapid performance. The applied sample remains quantitatively on the plate, and therefore, no losses of radioactivity during analysis occur. The commercially available ready-to-use stationary phases combine adsorbent with ionic or hydrophobic properties and are suited for separation of a variety of molecules using polar or nonpolar solvents. This chapter emphasizes methods for routine use in nuclear medicine and describes materials, techniques, and methods for quantification (Carpenter 1986; Hammermaier et al. 1986; Thoebald 1984).

9.1.1.1 ^{99m}Tc Species Separated by TLC

The main impurities in ^{99m}Tc pharmaceutical preparations are free pertechnetate ($^{99m}TcO_4^-$) and reduced, hydrolized technetium (colloidal ^{99m}Tc). These two ^{99m}Tc species may be separated from ^{99m}Tc pharmaceuticals by simple TLC procedures.

The migration properties of free pertechnetate may be influenced by the choice of different mobile and stationary phases. When silica gel or paper is used as stationary phase, the migration of free pertechnetate depends on the solubility of this anion in the solvent. In a polar solvent like saline, 80% methanol, acetone or 2-butanone (methyl ethyl ketone, or MEK) pertechnetate migrates with the SF (R_f=0.6–1.0). If a nonpolar, lipophilic solvent (e.g., ethylacetate, chloroform) is used and the sample is dried (no water content), free pertechnetate remains at the origin. In addition, when using an anion-exchange material in for stationary phase (e.g., aluminum oxide), free pertechnetate will be retained at the start.

Colloids do not migrate in most TLC systems since insoluble material will stay at the origin. Changing the mobile or the stationary phase will not affect the migration properties of colloidal ^{99m}Tc. This is the reason why hydrolized ^{99m}Tc species are not determined in colloidal and particulate preparations (e.g., macroaggregated albumin [MAA]), microspheres, or high-molecular ^{99m}Tc species such as monoclonal antibodies. The major impurity recognized by the pharmacopeia in the case of these radiopharmaceuticals is free pertechnetate, moving with the SF (Robbins 1983; Zimmer and Pavel 1977).

9.1.1.2 Stationary Phases

Standard TLC Materials. Standard TLC plates are available as glass plates, and as plastic or aluminum foils covered with the stationary phase. Aluminum or plastic foils have the advantage that they are easily cut into pieces for measurements of radioactivity. A broad range of stationary phases is commercially available including silica gel, reversed-phase silica, aluminum oxide, synthetic resins (ion-exchange chromatography), and cellulose (partition chromatography). The length of plates (foils) may vary between 10 and 20 cm, although miniaturized systems have been introduced (<5 cm). Generally, the developing distance will depend on the number of components in a sample and the relative retention properties in a system as well as the time used for migration.

For determination of the radiochemical purity of ^{99m}Tc pharmaceuticals, methods using standard TLC materials have been described (Zimmer and Pavel 1977). Reversed-phase materials offer advantages with respect to more polar solvents that are miscible with the aqueous medium of the sample (Carpenter 1986). Alumina plates are used to separate anionic ^{99m}Tc-pertechnetate from neutral or positively charged complexes.

The main limitation of standard TLC techniques is the time required for analysis. Due to the particle size (20 μm) of adsorbent materials, the developing time is usually >30 min. This is too long considering additional time for measurements and quantification.

The main advantage of standard TLC materials lies in the comparably high resolution, exemplified by the separation of two ^{99m}Tc-dimercaptosuccinic acid (DMSA) complexes in one TLC system: The trivalent ^{99m}Tc-DMSA complex is separated as an impurity in a preparation of pentavalent ^{99m}Tc-DMSA (Westera et al. 1985).

High-Performance TLC (HPTLC). HPTLC materials have a smaller average particle size (5 vs 20 μm) and a narrower particle size distribution when compared with conventional materials. Actually, the development of the chromatogram is faster and the time for analysis shorter. However, HPTLC materials are preferable for more complex separations; their use for determinations of the radiochemical purity in nuclear medicine is limited.

Instant TLC (ITLC). ITLC materials are the most frequently used stationary phases in nuclear medicine. ITLC methods fulfill the need for rapid and accurate analysis of the radiochemical purity of radiopharmaceuticals and have thus been accepted by the European Pharmacopeia.

ITLC plates are made of fiberglass sheets, impregnated with an adsorbent, usually silica gel (e.g., SG). Due to the fine mesh material, the migration properties are increased many-fold by the TLC materials. The time for the development of any chromatogram may be reduced to <5 min, without affecting the separation of radiochemical impurities.

Although ITLC materials are more expensive, fiberglass sheets offer high economy, since the flexible material may be cut to any size. Chromatographic systems utilizing these materials have been described (Frier and Hesslewood 1980; Theobald 1994; Zimmer and Pavel 1977).

The separation properties of silica gel 60 depend to a certain extent on the "activated" state of the adsorbent, which is related to the water content. However, when using organic solvents, it is recommended to dry the ITLC plate before use by heating for 10–20 min at 110 °C (Frier M and Hesslewood 1980).

The most common impurities in radiopharmaceuticals remain at the origin ($R_f=0$) or migrate with the SF ($R_f=0.8$–1.0). In order to separate and quantify two (or more) impurities, two (or more) analytical systems are generally used. A typical procedure, using different solvents, is shown in Fig. 9.1.1.1. Using an organic solvent (MEK) for separation, colloidal forms and the Tc complex remain at the origin, and free pertechnetate migrates with the SF. This system is suited to quantify free ^{99m}Tc-pertechnetate. In saline, free pertechnetate and the ^{99m}Tc complex migrate with the SF; while reduced, hydrolyzed ^{99m}Tc remains at the start. Using saline, the colloidal ^{99m}Tc species is quantified. With some experience, the whole procedure may be performed in <15 min, before application of the radiopharmaceutical to patients. Attention must be paid when ITLC strips are marked to indicate the spotting area or SF, since the material is fragile and easily damaged, which may affect results.

Silica Gel. The chemical properties are based on siloxane (Si–O–Si) and silanol (Si–OH). Polar groups are responsible for the interaction of the adsorbent with water and with the sample to be analyzed. Silica stationary phases (3–8 μm) have been produced for ITLC as silica gel (ITLC-SG) and silicic acid (ITLC-SA). ITLC-SG is the most frequently used adsorbent for routine radiochemical purity determinations (Table 9.1.1.1).

RP Phases. The surface properties of silanol may be modified by reaction with alkylating agents that bind to the surface of porous silica gel 60.

Hydrophobic modification includes binding of nonpolar groups such as dimethyl (RP-2), octyl (RP-8), octadecyl (RP-18), and phenyl. The hydrophobic properties increase within the different groups from RP-2 to RP-18, and with the degree of saturation of the hydroxyl groups that has been attained.

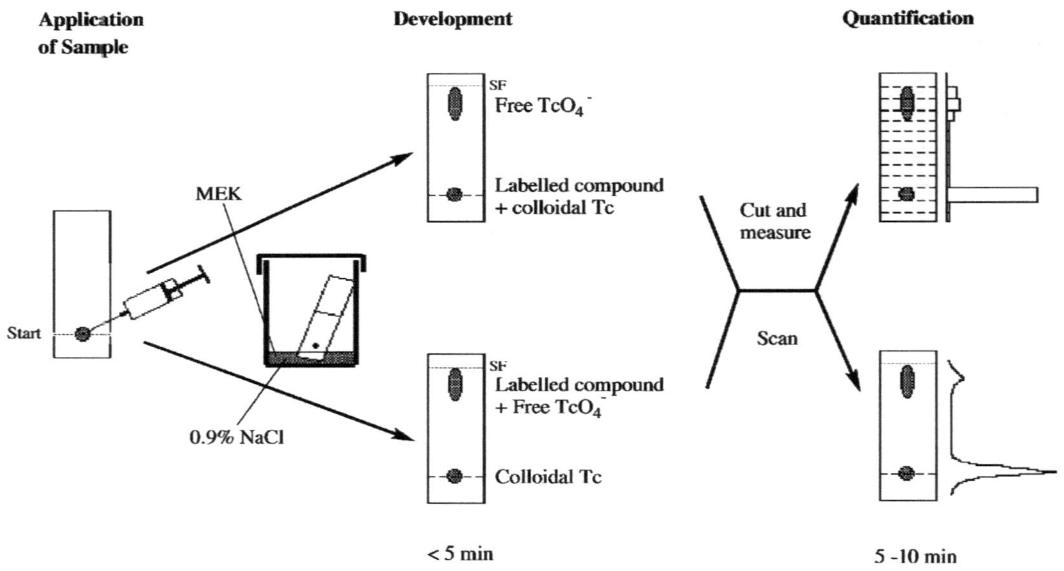

Fig. 9.1.1.1. Determination of the radiochemical purity of a radiopharmaceutical by instant thin-layer chromatography (ITLC) using two different solvents, methyl ethyl ketone (MEK) and saline

Table 9.1.1.1. Materials for thin-layer chromatography (*TLC*)

Layer type	Format (cm)	Ordering no.	Quantity
TLC glass plates (layer thickness 0.2 and 0.25 mm)			
Silica gel 60 F_{254}	20×20	Merck 1.05715.000	25 plates
Silica gel 60 F_{254}	5×10	Merck 1.05789.000	25 plates
Silica gel 60 F_{254}	2.5×7.5	Merck 1.15327.000	100 plates
RP-18 F_{254}	20×20	Merck 1.15389.000	25 plates
RP-18 F_{254}	5×20	Merck 1.15683.000	50 plates
RP-18 F_{254}	5×10	Merck 1.15685.000	25 plates
HPTLC Si 60 F_{254}	5×5	Merck 1.05635.000	100 plates
HPTLC LiChrospher Si 60 F_{254}	20×10	Merck 1.15445.000	25 plates
TLC aluminum sheets (layer thickness 0.2 mm)			
Silica gel 60	20×20	Merck 1.05553.0001	25 sheets
Silica gel 60 F_{254}	5×10	Merck 1.16834.0001	50 sheets
Silica gel 60 F_{254}	5×7.5	Merck 1.05549.0001	20 sheets
Aluminum oxide 60 F_{254} neutral	20×20	Merck 1.05550.0001	25 sheets
Aluminum oxide 150 F_{254} neutral	20×20	Merck 1.05551.0001	25 sheets
HPTLC LiChrospher Si 60 F_{254}	20×20	Merck 1.05586.0001	25 sheets
TLC plastic sheets (layer thickness 0.2 mm)			
Silica gel 60 F_{254}	20×20	Merck 1.05735.0001	25 sheets
Aluminum oxide 60 F_{254}	20×20	Merck 1.05581.0001	25 sheets
Baker-flex silica gel 1B-F	20×20	J.T. Baker 4463-04	25 sheets
Baker-flex silica gel 1B-F	2.5×7.5	J.T. Baker 4463-02	200 sheets
Baker-flex aluminum oxide 1B-F	20×20	J.T. Baker 4467-04	25 sheets
Baker-flex aluminum oxide 1B-F	2.5×7.5	J.T. Baker 4467-02	200 sheets
ITLC fiberglass sheets (layer thickness 0.2 mm)			
Silica gel (ITLC-SG)	20×20	Pall Gelman 61886	25 sheets
Silica gel (ITLC-SG)	5×20	Pall Gelman 61885	50 sheets
Silicic acid gel (ITLC-SA)	20×20	Pall Gelman 51432	25 sheets

HPTLC high-performance thin-layer chromatography, *ITLC* instant thin-layer chromatography

RP phases offer higher selectivity for separation of nonpolar compounds or molecules with nonpolar groups. In certain cases, highly polar, ionic compounds may also be separated due to selective retention on these modified RP materials. Partially modified silica gel is also available as RP-18 alumina sheets.

Silica gel may also be modified with hydrophilic groups that are attached using short-chain nonpolar spacers. Hydrophilic modifications with polar groups include amino, cyano, and diol functionalities. Amino-SG offers weakly basic ion-exchange properties; diol-SG considerably less affinity for water when compared with unmodified SG-60. These HPTLC materials offer increased selectivity for complex separations of biomolecules and drugs.

Aluminum Oxide. Some separations of radiopharmaceuticals are based on adsorption chromatography with aluminum-coated plates. Aluminum oxide (Al_2O_3) has polar properties; it is also a weak anion exchange material. The pH adjusted adsorbent is available in three pH ranges: neutral (pH 7.0–8.0), basic (pH 9.0–10.0), and acidic (pH 4.0–4.5). Standardized products include aluminum oxide 60, 90, and 150. Plates are also available for UV detection.

Table 9.1.1.2. Materials for paper chromatography

Specification	Format (mm)	Ordering no.	Quantity
Whatman 1	100×300	3001845	100 sheets
Whatman 1	200×200	3001861	100 sheets
Whatman 3MM	50×250	30306122	50 sheets
Whatman 3MM	100×130	30306123	50 sheets
Whatman 3MM	150×150	3030286	100 sheets
Whatman 3MM	200×200	3030861	100 sheets
Whatman 31 ET	460×570	3031915	25 sheets

Cellulose. This organic material consists of polymerized glucose fibers (400–500 molecules) in nature and also as a synthetic product (40–200 glucose molecules). Cellulose interacts with water and serves as a stationary phase for the separation of polar substances by paper chromatography. As a powder, it is used as an adsorbent for TLC. Separation of polar substances by paper chromatography is described in the *European Pharmacopeia* for identification of ^{99m}Tc-pertechnetate (Council of Europe 1982).

Paper. Paper was the first material used for chromatography of radioactive compounds (Dickey 1953). Paper materials show low-resolution properties; however, since paper is robust and easy to cut, paper chromatography is still used and recommended for many applications. The mechanism of separation is probably different; nevertheless, paper chromatography is used "ascending" or "descending", like ITLC. Likewise, a developing distance of 8–10 cm is usually sufficient for the separation of free pertechnetate and colloidal impurities. The developing time might be slightly increased, but usually finished in <10 min, if small sized paper strips are used.

Whatman 3MM is the material of choice for determination of the radiochemical purity by partition chromatography. (For information on materials, Tables 9.1.1.1 and 9.1.1.2.)

9.1.1.3 Mobile Phases

The saline/MEK system is applied for the analysis of most radiopharmaceuticals that contain free pertechnetate and/or colloidal Tc. Acetone has been replaced by MEK because artificially high values of pertechnetate have been obtained, caused by its higher water content (Carpenter 1986).

More recently, developed ^{99m}Tc complexes (e.g. mercaptoacetyltriglycine [MAG_3], monodentate methoxyisobutyl isocyanide [MIBI], hexamethylpropylene amine oxime [HMPAO], etc.) require more sophisticated analytical methods to detect additional impurities in the injection solution. This is inherent in the chemical properties of certain ^{99m}Tc pharmaceuticals that need more complex kit formulations and contain several labeled impurities. In addition, special composed solvent systems have to be used.

9.1.1.4 Spotting of the Sample

The sample size has a considerable effect on the separation characteristics of a certain system. Therefore, the sample diameter on the plate should be kept as small as possible (<3 mm). Inefficient separations and artificial results are caused by too-large spots. The standard TLC technique for applying a sample onto the plate is to use micropipets or glass capillaries for single use. The volume is typically 5 µl. A certain drawback of this technique is the fact that the sample is withdrawn aseptically with a syringe from the vial and needs to be transferred for spotting.

Therefore, to reduce handling, the sample is withdrawn using a 1-ml syringe with a fine needle (>25 gauge) and is spotted directly onto the plate using a single drop from the needle. The volume corresponding to a drop from a vertically held needle (25 gauge) is approximately 6 µl. If the syringe is held in the horizontal position, the volume might double (Robbins 1983). With experience, reproducibly small spots are applied to the TLC plate, suitable for analysis. The technique also avoids contamination and minimizes sample exposure to air, which may affect certain ^{99m}Tc radiopharmaceuticals.

9.1.1.5 Development of the Chromatogram

For the development of chromatograms, a small beaker might be used, which is closed with a glass plate or covered with a foil to maintain a saturated atmosphere inside the vessel. Small tanks for chromatography are recommended, especially when organic solvents are used. If an open beaker is used, evaporation of the solvent may affect the separation properties of a system (Levit 1980; Manger 1986).

The solvent should cover the bottom not more than 5 mm high; the solvent level has to be below the start line on the plate/strip.

After applying the sample, the plate must be placed into the tank and developed immediately, without drying the spot. Dried samples may lead to artificial results due to oxidation of ^{99m}Tc complexes and formation of free pertechnetate (Mallol 1990; Manger 1986).

An exception to this rule is made for mobile phases that do not dissolve the aqueous sample (e.g., ethylacetate, chloroform). In this case, a wet sample will produce a poor resolution and high background activity along the track. The sample spot should be dried in a gentle stream of nitrogen, and heating the plate must be avoided.

The plate/sheet is placed vertically into the chamber, carefully avoiding any damage of the surface. Materials with limited mechanical resistance like certain chromatography paper should be supported (e.g., clipped to the lid of the chamber), otherwise they will slip into the solution or touch the chamber wall during development.

The usual developing time is between 2 and 15 min, depending on the stationary phase and (to a lesser extent) on the solvent. The SF is not always visible, especially when using ITLC materials. Therefore, SF (end of chromatogram) is marked with a color pen (water soluble for aqueous solutions; water resistant for organic solvents) (Fig. 9.1.1.2). When the solvent gets in contact with the marker, the color migrates with the solvent, indicating the end of the development. The strip should be marked in such a way that the color does not interfere with the sample track to avoid artificial results (Levitt 1980). When the solvent has reached SF, the strip is removed from the chamber and then dried.

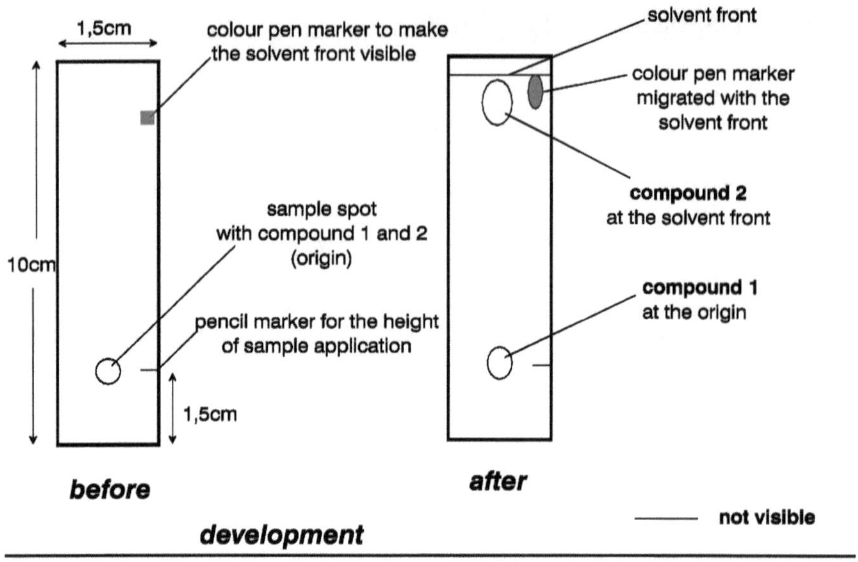

Fig. 9.1.1.2. ITLC plates before and after development (typical dimensions). The application point is marked with a line; a color pen marker may be used to mark the solvent front. When the solvent reaches the marker, the color migrates with the solvent. Markers are positioned outside the track to avoid interference with the sample

Procedure for the determination of radiochemical purity by ITLC or paper chromatography:

- Fill a 100-ml beaker with the solvent (about 10 ml, solvent 3–5 mm high); close the beaker with a tight lid or parafilm.
- Prepare the strip: Mark the solvent front with a color pen and the start with a pencil.
- Take a small sample of the preparation ready for injection (< 100 μl).
- Apply one small drop of the sample with a thin needle onto the strip; the drop must not dry.
- Immediately put the strip into the beaker, the spot must remain above the solvent level.
- When the solvent has reached the front, take the strip out and let it dry.
- Quantify the regional distribution of radioactivity on the strip.

9.1.1.6 Measurement of Radioactivity

There are several methods for the quantification of radiochromatograms. Depending on the available instrumentation, the resolution of analysis will vary, and the amount of radioactivity used for analysis will differ considerably. In any case, the results of measurements are used to calculate the radiochemical purity of a radiopharmaceutical (percentage) as the ratio between the radioactivity corresponding to the main component, divided by the total recovered radioactivity of the chromatogram:

$$\text{Radiochemical purity (\%)} = \frac{\text{Radiopharmaceutical (dpm)}}{\text{Total recovered activity (dpm)}} \times 100$$

The relative merits of quantification methods are:

- Resolution
- Sensitivity
- Linearity
- Time needed to perform
- Practicability
- Costs

Autoradiography. Autoradiography is one of the oldest methods used for semiquantitative measurement of radioactivity. The chromatogram is placed on an x-ray film and exposed in the dark for usually less than 1 h (a detailed procedure is given elsewhere) (Theobald 1994). This method is no longer in use since it is time-consuming and inaccurate. Visualization of the radioactivity distribution on a film may be useful for documentation purposes.

Gamma Camera. Measurement of a radiochromatogram with a gamma scintillation camera might be first choice when no specific instrumentation is available. The dried strip is placed at close distance to the head of the gamma camera, and images are acquired. Using the regions of interest (ROI) technique (drawing regions over distinct areas of activity), the radiochemical purity is expressed as a fraction of the total recovered activity. A certain advantage is the use of undiluted sample for chromatography (measurement of high count rates). The main disadvantage of this technique is that it is a rather time-consuming procedure (measurement and analysis) at a time when the camera is needed for patients.

In cases when no other device is available, visualization of the radioactivity distribution by an activity profile might be a good indicator of the amount of impurities present in the injection solution.

Ionization Chamber. This method is frequently used and is recommended in many official procedures (Robbins 1983). It might be used for simple separation techniques (compounds with an R_f of 0 or 1). The strip is cut into two segments (one corresponding to the main compound and the other to the impurity) and measured in the ionization chamber. This facilitates quick analysis before the radiopharmaceutical is injected.

Limitations are low resolution of measurements, an overestimation of the amount of impurities, and if not cut properly, results may be wrong. This problem can be avoided by cutting the strip into more segments, which are measured separately (Bish et al. 1980).

Second, the sensitivity is limited by the sensitivity of the ionization chamber. In practicee, 3 MBq or more should be applied on the plate, which would correspond to an injection solution of 600 MBq/ml if 5 μl are applied (Robbins 1983). Results with lower concentrations of radioactivity should be judged with caution.

Cut and Count. Chromatographic plates may be analyzed using a NaI(Tl) scintillation counter. The plates are cut in segments (up to ten) and transferred into tubes to be counted. The radioactivity on the plate can be plotted as a histogram. The sensitivity is much higher compared with the ionization chamber (even dilution may be necessary), and the resolution is dependent on the number of segments. It is important to keep the same counting geometry for all segments at the bottom of the counting tube. Cutting the plate in many segments permits assessment of the quality of separation. The background radioactivity is subtracted from each count leading to a more accurate analysis. The main limitations of this method are that the high activities of the main component may paralyze the counter (Robbins 1983), and the procedure is time-consuming (cutting samples and filling into test tubes, time of measurements, and calculation of the relative percentages).

Chromatogram Scanning. The advantage of this method is the fact that a slit-collimated detector is moved along the thin-layer plate, and the radioactivity distributed between the start and SF of a chromatogram is recorded. The detector is coupled to a scaler ratemeter, with the ratemeter output signal passed on to a chart recorder producing a radioactivity profile (Janshold et al. 1980). A useful addition is a chromatography integrator to measure the area of radioactive peaks, which is proportional to the detected radioactivity. Quantification of single peaks as a percentage of the total measured radioactivity will produce a purity report for each radiopharmaceutical.

For measuring ^{99m}Tc and other gamma emitters, a NaI(Tl) scintillation detector is used. The resolution of the scanner is dependent on the width of the slit-collimator, the distance between chromatogram and detector, and the window settings on the scaler. Artificial results may be obtained if the peaks are not symmetrical and comparable.

Linear Analyzer. The linear analyzer was developed for the measurement of beta-emitting radionuclides and was introduced in 1980 by Berthold Analytical Instruments (Filthuth 1986). It operates as a position-sensing proportional counter, measuring a fixed number of channels along the length of the chromatographic plate. The system is equipped with analytical software for quantification.

Further developments have increased the potential of proportional counting by placing a grid of anode wires between two large plates, 1.5 mm above and below, serving as cathodes. Ions formed in the gas-filled volume drift toward the nearest anode wire where they undergo multiplication; each event is automatically localized. Thus, two-dimensional localization may be accomplished by these position-sensing detectors.

The *digital autoradiograph* is a multiwire proportional counter; each electrode is composed of 100 wires (gold-plated tungsten) arranged at a 2-mm distance. The gas-filled proportional chamber has a window of 24×24 cm for the measurement of chromatograms at a fixed distance (0.5 mm). Each channel is connected with a time analyzer for registration of position specific pulses (nanoseconds). Finally, the analog signals are digitalized for computer analysis and quantification.

Gas flow proportional counters exhibit higher sensitivity for beta-emitters, but also pure gamma-emitters like ^{99m}Tc are measured with high resolution. When placing a chromatogram into the counting chamber the developed strip must be completely dry.

The main advantages over a conventional TLC scanner are:

- Increased sensitivity because of the high specific ionization produced in the gas-filled proportional chamber using argon/methane (89:10)
- Two-dimensional positioning of radioactivity
- A "photographic" image of the radioactivity distribution is generated
- Several chromatograms (strips) are analyzed during one exposure
- The linear analyzer and the digital autoradiograph may be used with any radionuclide
- Quantification of chromatograms, storage, and display of data
- Resolution may decrease with high activities.

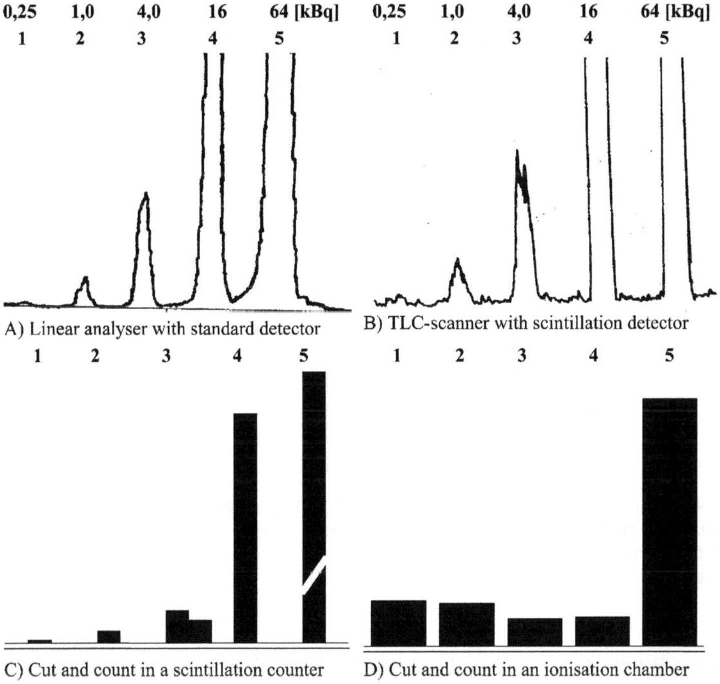

Fig. 9.1.1.3. "Phantom" chromatographic plate analyzed by four different methods of measurement. The highest sensitivity is achieved with A and C, the best resolution with B and C. Using the ionization chamber (D), activities less than 20 kBq (peak 1, 2, 3, 4) are not resolved accurately.

A comparison of four different methods of quantification is shown in Fig. 9.1.1.3. A "phantom" chromatographic plate with increasing amounts of radioactivity (0.25, 1.0, 4.0, 16.0, and 64.0 kBq) spotted at exact intervals was measured by: A. linear analyzer, B. conventional scanner, C. cut/measure, and D. ionization chamber (Capintec). The highest sensitivity was achieved with A. and C., the best resolution with B. and C. Using the ionization chamber indicated low detection efficiency of radioactivity below 20 kBq (peaks 1, 2, 3, and 4).

Linear analyzers and conventional TLC scanners are standard equipment in nuclear medicine. These instruments offer high detection efficiency and resolution characteristics combined with speed of analysis, which is required for the analysis of short-lived radiopharmaceuticals.

Electronic Autoradiography. New methods have been developed with the aim of replacing the time-consuming film autoradiography. Two different systems, the so-called phosphor imagers and the microchannel plate analyzers, are currently available for quantifying TLC plates, ITLC, or paper strips. While phosphor imagers require two steps for quantification of an autoradiogram (exposure and "development"), microchannel plate analyzers provide a direct measurement of radioactivity. The main advantages are a short analysis time, high sensitivity, a broad dynamic range, system stability at high radioactive concentrations, online two-dimensional imaging, direct quantification of radioactivity, and detectability of all radionuclides used in nuclear medicine. A de-

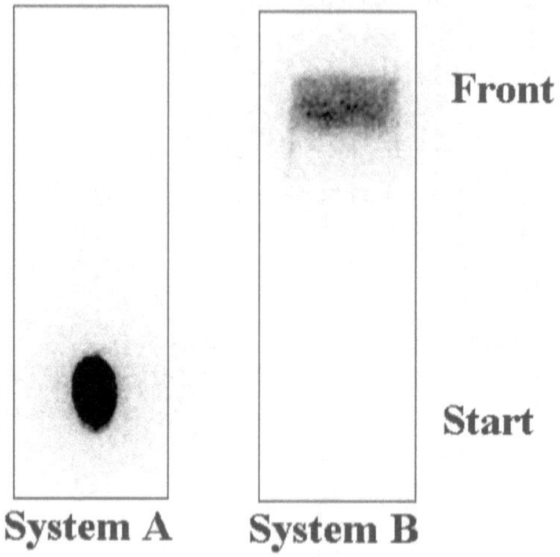

Fig. 9.1.1.4. Electronic autoradiogram of two ITLC-SG strips showing the radiochemical purity of ^{99m}Tc-diethylene triamine pentaacetate (DTPA) in two solvent systems: *A* MEK: ^{99m}Tc-DTPA remains at the start, *B* MeOH (80%): ^{99m}Tc-DTPA moves with SF. Strip dimensions: 1.5×10 cm. (Instant Imager, Canberra Packard, acquisition time: 60 s)

tailed description of such a system (Instant Imager, Canberra Packard) for radiopharmaceutical applications is described in (Decristoforo 1977). Figure 9.1.1.4 shows an electronic autoradiogram of the radiochemical purity of ^{99m}Tc-diethylene triamine pentaacetate (DTPA). These instruments offer a wide range of radioanalytical applications besides radiochemical purity determinations, yet the high cost may limit their use in nuclear medicine.

9.1.1.7 Summary

The determination of the radiochemical purity in nuclear medicine may be performed with little expenditure of material.

Minimal equipment for routine determination of the radiochemical purity with TLC:

- Two chromatographic chambers for small volumes or laboratory beakers (about 100 ml size, 10 cm high), foil or lid for covering the beakers
- 1 ml syringes, fine needles (25 gauge), gloves
- Small flasks for solvents (MEK, saline, etc.)
- Chromatographic plates:
  - ITLC-SG plates (Gelman No. 61886, 20×20 cm, cut into strips [1.5×10 cm])
  - Paper (Whatman 3MM No. 3030861, 20×20 cm, cut into strips [1.5×10 cm])
- Pencil for marking the strips, color pens
- Scissors for cutting
- Plastic tubes for the measurement of the segments
- Ionization chamber

For handling of radioactive samples, only single-use materials should be used in order to minimize the risk of contamination of the sample and the staff. The TLC analysis should be performed in an area of the hot lab with a low probability of contamination from other working procedures. If possible, a separate working area should be used. There should be a place with adequate ventilation (hood) for drying the strips after development in organic solvents (MEK, ethylacetate, etc.).

The determination of the radiochemical purity of radiopharmaceuticals by ITLC requires an experienced operator to produce accurate results. Quality control should be performed daily and not on demand, when abnormal distribution or poor image quality is reported.

References

Bish RE, Silverstein D, Bede JI (1980) Radiopharmaceutical-radiochromatography quality assurance with a radioisotope calibrator. J Radioanal Chem 57:565–573

Carpenter AP (1986) Radioanalytical techniques: ITLC, TLC, mini columns and electrophoresis. In: Wieland DM, Tobes MC, Mangner TJ (eds) Analytical and chromatographic techniques in radiopharmaceutical chemistry. Springer, Berlin Heidelberg New York

Council of Europe (1982) Sodium pertechnetate [^{99m}Tc] injection (fission), Monograph (124). European pharmacopeia. Strasbourg

Decristoforo C, Zaknun J, Kohler B, Oberladstaetter M, Riccabona G (1997) The use of electronic autoradiography in radiopharmacy. Nucl Med Biol 24:361–365

Dickey EE (1953) Separation of radium D, E and F by paper chromatography. J Chem Ed 30:525

Fairbrother JE (1984) Advances in TLC in two parts: Part 1 – Techniques. Pharm J 10 March:293–297; Part 2 – Applications. Pharm J 7 July:23–28

Filthuth H (1986) Detection of radiochromatograms and electropherograms with position-sensitive wire chambers. In: Wieland DM, Tobes MC, Mangner TJ (eds) Analytical and chromatographic techniques in radiopharmaceutical chemistry. Springer, Berlin Heidelberg New York

Frier M, Hesslewood S (1980) Quality assurance of radiopharmaceuticals – a guide to hospital practice. Nucl Med Comm, Special Issue 3

Hammermaier A, Reich E, Bögl W (1986) Qualitätskontrolle von in-vivo Radiopharmaka, ein Handbuch für das nuklearmedizinische Labor zur Analytik radiochemischer Verunreinigungen [in German]. ISH-Heft 94

Höye, A (1967) Determination of radiochemical purity of some radiochemicals and pharmaceuticals by paper chromatography, thin-layer chromatography and high voltage electrophoresis. J Chromatograpy 28:379

Izmailov, N.A. Shraiber MS (1938) The application of analysis by drop chromatography to pharmacy. Farmezia 3:1

Janshold, AL, Krohn KA, Vera DR, Hines HH (1980) Design and validation of a radiochromatogram scanner with analog and digital output. J Nucl Med Tech 8:222

Levit N (1980) Radiopharmacy laboratory manual for nuclear medicine technologists. University of New Mexico College of Pharmacy Albuquerque, pp 69–80

Mallol J (1990) Analytical artifacts in quality control of Tc-99m radiopharmaceuticals. Nucl Med Biol 17:609–611

Mangner TJ (1986) Potential artifacts in the chromatography of radiopharmaceuticals. In: Wieland DM, Tobes MC, Mangner TJ (eds) Analytical and chromatographic techniques in radiopharmaceutical chemistry. Springer, Berlin Heidelberg New York

Miller JM (2004) Chromatography: concepts and contrasts, 2nd edn. Wiley, New York

Robbins PJ (1983) Chromatography of technetium-99m radiopharmaceuticals – a practical guide. The Society of Nuclear Medicine, New York

Stahl E (1956) Dünnschicht-Chromatgraphie [in German]. Pharmazie 11:633

Theobald AE (1994) Quality control of radiopharmaceuticals. In: Sampson CB (ed) Textbook of radiopharmacy: theory and practice, 2nd enlarged edn. Gordon and Breach, Philadelphia

Westera G, Gadze A, Horst W (1985) A convenient method for the preparation of 99m Tc(V)dimercaptosuccinic acid (^{99m}Tc(V)-DMSA). Int J Appl Radiat Isot 36:349–355

Zimmer AM, Pavel DG (1977) Rapid miniaturized chromatographic quality-control procedures for Tc-99m radiopharmaceuticals. J Nucl Med 18:1230–1233

9.1.2 Column Chromatography

F. Rakiás and J. Imre

Column chromatography, or liquid chromatography, is a separation method by which a ^{99m}Tc pharmaceutical is resolved into its components when passed through a chromatographic column with a mobile phase (solvent). Interaction of the sample with the solvent and the column matrix is affected by solubility, electrostatic forces, and charge, resulting in the retention of components on the chromatographic column and separation of impurities. Physical and chemical forces include van der Waal's, dipole, hydrogen bonding, dielectric, and electrostatic interactions, which cause separation of components with different physicochemical properties.

Several separation methods were derived from liquid chromatography, depending on the choice of the stationary phase and the solvent used for a particular compound, i.e., adsorption chromatography, partition chromatography (normal or reversed phase), ion-exchange chromatography, ion-pair chromatography, gel-permeation chromatography, and affinity chromatography.

For ^{99m}Tc-labeled pharmaceuticals, the most frequently used separation techniques are gel permeation chromatography at low pressure, and high-performance reversed-phase, ion-pair, ion-exchange, and size-exclusion chromatography.

9.1.2.1 Limitations of Liquid Chromatography for Quality Control of ^{99m}Tc Pharmaceuticals

In planar radiochromatography (thin-layer chromatography), the detection of the components of any ^{99m}Tc pharmaceutical is performed directly on the chromatographic plate; thus, the total applied ^{99m}Tc activity – theoretically all ^{99m}Tc constituents – can be determined, depending on sample and the system performance.

In the case of column chromatographic techniques, the separated ^{99m}Tc impurities are detected indirectly, after separation, accepting incomplete sample recovery. Some components of a ^{99m}Tc pharmaceutical remain on the column. In fact, reduced, hydrolized ^{99m}Tc activity is commonly retained on the column.

Since determination of radiochemical purity relies on the complete recovery of all constituents of a ^{99m}Tc pharmaceutical applied to the column, calculations based on the peak area of eluted components will overestimate the value of radiochemical purity.

The retention of ^{99m}Tc radiocolloid has been determined indirectly by comparing the precolumn and postcolumn ^{99m}Tc activity, using two detection loops. However, this method depends on the accurate calibration and validation of the system.

Trapping of reducing tin(II) salts (in colloidal form) in the pores of the column material has also been observed to affect radiochemical purity, because the pertechnetate content in the subsequent samples will be reduced, providing false values.

These problems exclude conventional column chromatography from routine use for quality control; however, application of minicolumns has offered an alternative to thin-layer chromatography in the cases of ^{99m}Tc-hexamethylpropylene amine oxime (HMPAO) and others, discussed in the respective monographs (Part 2).

Because of its high resolving capacity, conventional liquid chromatography is an efficient tool in research, when analysis of chemically similar byproducts (secondary, tertiary, etc., complexes), metabolites or products resulting from radiolysis, is required.

9.1.2.2 Gel Permeation or Size-Exclusion Chromatography

This method is based on the accessibility of the pores in the stationary phase for ^{99m}Tc-labeled molecules of different molecular sizes. The sample is eluted from a vertical column packed with porous beads of the gel by gravity or low pressure. Smaller ^{99m}Tc species penetrate the pores and are retained on the column, while larger molecules are excluded and are therefore rapidly eluted from the column. This separation technique has particular application for macromolecules, proteins (serum albumin, immunoglobulins [e.g., monoclonal antibodies and their fragments]), but has also been used for separation of small-molecular-weight ^{99m}Tc-diphosphonate complexes.

Quantification was performed by external scanning of the column and analysis of the activity profile of the retained ^{99m}Tc compounds (mercapto complexes, diphosphonates, etc.), using a modified thin-layer scanner.

Several types of gel column filling material are available (e.g., Sephadex, Sepharose, etc.).

9.1.2.3 High-Performance Liquid Chromatography

J. Imre

High-performance liquid chromatography (HPLC) provides faster separation and higher resolution of the eluted components of a ^{99m}Tc pharmaceutical than does the conventional column chromatography. While in column chromatography the mobile phase passes through the stationary phase by gravity or low pressure, in HPLC the mobile phase is pumped through the column by high pressure up to 6000 psi (~400 bar).

HPLC Instrumentation. The schematic diagram of a radio-HPLC system can be seen in Fig. 9.1.2.1.

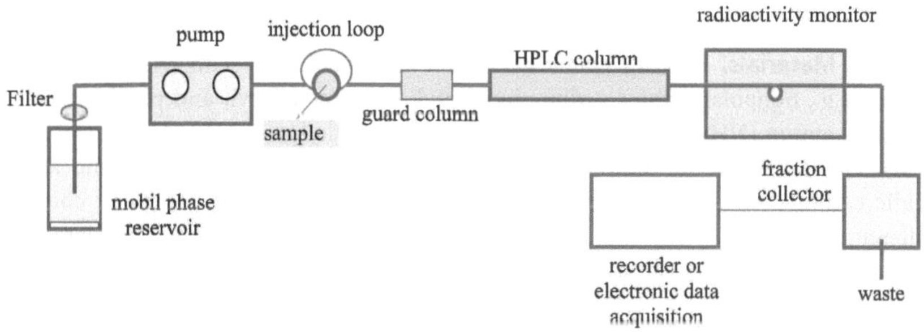

Fig. 9.1.2.1. Components of a radio-high-performance liquid chromatography (HPLC) system

Table 9.1.2.1. Modified silica packing materials with chemically bonded phases

Type of modification	Application
C_2 Dimethyl – RP-2	Reversed-phase and ion-pairing chromatography Shorter retention time than with other RP
C4 Butyl – RP-4	Reversed-phase and ion-pairing chromatography Separation of peptides and proteins
C_8 Octyl – RP-8	Reversed-phase and ion-pairing chromatography Moderately to highly polar (water-soluble compounds)
C_{18} Octadecyl – RP-18	Reversed-phase and ion-pairing chromatography Nonpolar or moderately polar compounds
C_6H_5 Phenyl	Reversed-phase and ion-pairing chromatography Moderately polar compounds, polar aromatics etc.
CN, cyano	Straight- or reversed-phase chromatography
NO_2, nitro	Separation of compounds with double bonds
OH, alkanol – DIOL	Straight- or reversed-phase chromatography
NH_2 amino	Straight-phase, weak anion-exchange, and reversed-phase chromatography
$N(CH_3)_2$ dimethylamino	Weakly basic anion exchanger
Quaternary ammonium	Strongly basic anion exchanger
Sulfonic acid	Strongly acidic cation exchanger

The Eluent Delivery Pump. HPLC pumps deliver the mobile phase from the reservoir to the column. Mainly double piston-actuated pumps accurately deliver solvent over a broad range of flow rates (0.1 μl/min–20 ml/min). Gradient elution (flow rate or composition) can be used in HPLC separations where the use of only one mobile phase (isocratic separation) would not have achieved adequate resolution.

The Sample Application Unit. In the case of radio-HPLC, the more useful system is the injector valve with a sample loop of desired volume, allowing safe filling of the ^{99m}Tc sample.

HPLC Column. In HPLC, a number of chromatographic techniques are readily available. Highly efficient HPLC columns contain small particles of less than 10 μm in diameter (greater than 20 μm for preparative HPLC), tightly packed into 100 to 300 mm length and of small internal diameter (2–5 mm) glass or steel tubes, and the result is good sample resolution and narrow peaks that elute from the column. Columns with high resolving capacity are now commercially available (Table 9.1.2.1) and can be used for several times.

Packing Materials. Packing materials include modified silica with chemically bonded phases, e.g., nonpolar groups – dimethyl, butyl, octyl, octadecyl, and phenyl; and polar groups – amino (NH_2), cyano (CN), nitro (NO_2) dimethylamino [$N(CH_3)_2$], and alkanol (OH). Quaternary ammonium facilitates basic anion exchange, and sulfonic acid the acidic cation exchange, offering separation conditions for a vast variety of chemically different compounds. The use of prefilters and guard columns may considerably improve the life span of a column.

Radioactivity Monitor. The homemade detector systems – where the detector loop is placed into a well-type scintillation detector – are frequently used (De Groot et al.

1986a; Nunn and Fritzberg 1986). Most commercially available radioactivity monitors use flow cells positioned between two photomultipliers with high counting efficiencies. The signal pulses from the photomultipliers are measured in coincidence to suppress noise. The signals are amplified and are then usually subjected to various data processing, integration, etc.

HPLC Applications for ^{99m}Tc Pharmaceuticals. Examples of typical applications of HPLC for separation of ^{99m}Tc-labelled species are shown in Table 9.1.2.2. The most frequently used HPLC techniques (Hnatowich 1986; Millar 1989) are reversed-phase partition chromatography for nonpolar or weakly polar components of ^{99m}Tc pharmaceuticals (HMPAO [Hung et al. 1988; Neirinckx et al. 1987; Weisner et al. 1993], monodentate methoxyisobutyl isocyanide [MIBI; Carvalho et al. 1992; Hung et al. 1991], mercaptoacetyltriglycine [MAG_3; Brandau et al. 1990; Coveney and Robbins 1987; Millar et al. 1990; Shattuck et al. 1994], tetrofosmin [Cagnolini et al. 1998; Graham and Millar 1999; Kelly et al. 1993], iminodiacetic acid [IDAs; Fritzberg and Lewis 1980; Nunn 1983; Nunn et al. 1983], peptides [Vallabhajosula 1986; Zinn et al. 2000], etc.), ion-pair (both ionized and less polar constituents: pertechnetate, diphosphonates; De Groot et al. 1986b; Hoch and Pinkerton 1986; Huigen et al. 1988; Nieuwland et al. 1989; Tanabe et al. 1983, etc.), ion-exchange (ionized forms: pertechnetate, diphosphonates, etc.), and size exclusion chromatography (separation based on molecular size: diphosphonates, human serum albumin [HSA; Vallabhajosula et al. 1982], and antibodies and their fragments [Hnatowich 1986], etc.).

Since HPLC is nondestructive, it can also be used as a preparative technique. The fraction containing the ^{99m}Tc pharmaceutical is formulated and applied in the clinic.

Reversed-Phase Chromatography. At acidic pH, octadecylsilane (ODS or C-18)-coated silica particles are frequently used as a stationary phase, although polymer-based columns (PRP-18) are also popular because of their wider operating pH range (1.0–13.0). The used mobile phases are polar solvents such as water, which is mixed with varying concentrations of miscible organics (e.g., methanol, acetonitrile). Solvent strength can be varied by changing the composition ratios of the mobile phase.

Ion-Pair Chromatography. A number of ^{99m}Tc complexes are ionized by deprotonation at higher pH. It is possible to separate both ionized and nonionized ^{99m}Tc species using reversed-phase chromatography. Ionic charges can be suppressed by manipulation of pH or by the use of an ion-pair reagent. The preferred ion-pairing agents are quaternary ammonium compounds (tetrabutyl ammonium, dodecyltrimethyl ammonium, etc.) for the analysis of anionic ^{99m}Tc complexes, and *N*-alkyl sulfonates for HPLC of cationic ^{99m}Tc compounds dissolved in a solvent similar to that used in reversed phase method.

Ion-Exchange Chromatography. In the case of ^{99m}Tc pharmaceuticals typically weak anion (e.g., amino-modified silica) or cation exchangers are used. The resolution can be affected by the ionic strength and pH of the buffered mobile phase, and gradient elution might be required to hasten elution.

Gel Permeation Chromatography. ^{99m}Tc-labeled proteins on polymer-based column packing can be purified from precursor reagents and unbound radiolabeling. The first eluted macromolecules are separated according to their size from the smaller molecules.

Table 9.1.2.2. High-performance liquid chromatography (HPLC) separation methods applied to ^{99m}Tc pharmaceuticals

Radiopharmaceutical	Column	Isocratic/ Gradient	Solvent(s)	Reference
^{99m}Tc-HMPAO (exametazime)	C-18	Isocratic	For "A" impurity determination: A: acetonitrile B: 0.1 *M* phosphate buffer, pH 3.0 A:B 33:67	Council of Europe 2005
	PRP-1	Gradient	A: 20 m*M* phosphate buffer, pH 7.4 B: tetrahydrofuran 0–25% B over 6 min	Neirinckx 1987
	PRP-1	Gradient	A: 10 m*M* potassium phosphate, pH 7.0 or water containing 1% methanol B: acetonitrile 0–50% B over 5 min	Hung 1988
	PRP-1	Gradient	A: 50 m*M* sodium acetate, pH 5.6 B: tetrahydrofuran 0–100% B over 17 min	Weisner 1993
^{99m}Tc-MIBI (sestamibi)	C-18	Isocratic	For "C" impurity determination: A: acetonitrile B: 50 m*M* ammonium sulfate C: methanol A:B:C, 20:35:45	Ccouncil of Europe 2005
	C-8	Gradient	A: 50 m*M* ammonium sulfate B: methanol 0–95% B over 5 min	Carvalho 1992
	C-18	Isocratic	A: methanol B: 50 m*M* ammonium sulfate C: acetonitrile A:B:C, 45:35:20	Hung 1991
^{99m}Tc-MAG_3 (mertiatide)	C-18	Isocratic with wash	A: ethanol B: 10 m*M* phosphate buffer, pH 6.0 A:B 7:93 After 20 min, wash with methanol:water, 90:10	Council of Europe 2005
	C-18	Isocratic with wash	A: ethanol B: 10 m*M* phosphate buffer, pH 6.5 A:B 5:95 After peak, wash with methanol–water, 90:10	Millar 1990
	C-18	Gradient	A: 10 m*M* potassium phosphate with 1% triethylamine, pH 5.0 B: tetrahydrofuran 0–8% B over 30 min	Shattuck 1994
^{99m}Tc-tetrofosmin	PRP-1	Gradient	A: 10 m*M* phosphate buffer, pH 7.5 B: tetrahydrofuran 0–100% B over 17 min	Kelly 1993
	PRP-1	Isocratic	A: acetonitrile B: 10 m*M* ammonium carbonate A:B, 70:30	Graham 1999
	PRP-1	Isocratic	A: 5 m*M* monopotassium phosphate B: acetonitrile A:B, 50:50	Cagnolini 1998

Table 9.1.2.2 (continued)

Radiopharma-ceutical	Column	Isocratic/Gradient	Solvent(s)	Reference
^{99m}Tc-MDP	Aminex	Isocratic	850 m*M* sodium acetate	Tanabe 1983
	C-18	Isocratic	10 m*M* sodium acetate, 2 m*M* tetrabutylammonium hydroxide, 3% ethyl acetate, pH 6.0	Hoch 1986
^{99m}Tc-EHDP	Aminex	Isocratic	850 m*M* sodium acetate	Huigen 1988
	C-18	Isocratic	50 m*M* EHDP, 10 m*M* sodium acetate, 3 m*M* tetrabutylammonium hydroxide	Nieuwland 1989
^{99m}Tc-IDA	C-18	Isocratic	25 m*M* phosphate buffer pH 6.0	Nunn 1983
	Ultrasphere ODS	Isocratic	A: 10 m*M* phosphate buffer, pH 6.8 B: methanol A:B 50:50	Fritzberg 1980
^{99m}Tc-HSA	Silica gel	Isocratic	100 m*M* phosphate buffer, 100 m*M* sodium chloride, 8 m*M* sodium azide	Council of Europe 2005
	Spherogel-TSK SW	Isocratic	100 *M* phosphate buffer	Vallabhajosula 1982
^{99m}Tc-depreotide	C-18	Gradient	A: 0.1% TFA in water B: 0.1% TFA/90% acetonitrile/H_2O 20% B over 30 min	Vallabhajosula 1996; Zinn 2000

HMPAO hexamethylpropylene amine oxime, *MIBI* monodentate methoxyisobutyl isocyanide, *MAG$_3$* mercaptoacetyltriglycine, *MDP* methylenediphosphonate, *EHDP* etidronate, *IDA* iminodiacetic acid, *HSA* human serum albumin, *TFA* trifluroroacetic acid

The use of a dual detector system, which is a radioactive monitor with a UV detector working at 280 nm, is practical. The mobile phase is usually buffered water.

9.1.2.4 Minicolumns for Routine Quality Control

There is a small-scale version of column chromatography that can compete with the planar chromatographic techniques for routine analysis of certain ^{99m}Tc pharmaceuticals. A small-sized tube or syringe is packed with appropriate stationary phase, forming a short-bed column. Several types of filling material can be used: reversed-phase packing (e.g., C18 modified silica for separation of ^{99m}Tc complexes of hepatoiminodiacetic acids ([HIDAs], MIBI, HMPAO, MAG3, etc.), preswollen gel (e.g., Sephadex for gel filtration of ^{99m}Tc-labeled HSA, immunoglobulin [HIG, monoclonal antibodies]) or aluminium oxide (for adsorption of anionic pertechnetate from inert ^{99m}Tc complexes). The ^{99m}Tc-sample is applied onto the top of the short bed and eluted with a small amount of solvent. The elution may be performed even in a plastic syringe. The separation is fast and simple, and the activity retained on the column or of the collected fractions is measured in an ionization chamber; thus, higher amounts of radioactivity of ^{99m}Tc label may be used than in the case of the planar method. The results depend on the column performance, affected by the tightness of packing and the flow rate; therefore application of standardized, prepacked columns is suggested (e.g., Chromabond, SEP-PAK) at optimal elution speed.

References

Brandau W, Bubeck B, Schober O, Weber E, Taylor DM (1990) $^{99m}Tc-MAG_3$: chemistry and biokinetics of by-products. In: Blaufox MD, Hollenberg NK, Raynaud C (eds) Radionuclides in nephrourology. Karger, Basel pp 11–16

Cagnolini A, Whitener D, Jurisson S (1998) Comparison of the kit performance of three ^{99m}Tc myocardial perfusion agents. Nucl Med Biol 25:435–439

Carvalho PA, Chiu ML, Kronauge JF, Kawamura M, Jones AG, Holman BL, Piwnica-Worms D (1992) Subcellular distribution and analysis of technetium-99m-MIBI in isolated perfused rat hearts. J Nucl Med 33:1516–1522

Council of Europe (2005) European pharmacopeia. Council of Europe, Maisonneuve, Sainte-Ruffine

Coveney JR, Robbins MS (1987) Comparison of technetium-99m MAG3 kit with HPLC-purified technetium-99m MAG3 and OIH in rats. J Nucl Med 28:1881–1887

De Groot GJ, Das HA, deLigny CL (1986a) A system for high performance liquid chromatography of Tc-99m complexes with on-line radiometric detection and data processing. Int J Appl Radiat Isot 37:349–355

De Groot GJ, Das HA, deLigny CL (1986b) Separation of Tc-99m(Sn)HEDP complexes by HPLC and GPC. Int J Appl Radiat Isot 37:23–27

Fritzberg AR, Lewis D (1980) HPLC analysis of Tc-99m iminodiacetate hepatobiliary agents and a question of multiple peaks: concise communication. J Nucl Med 21:1180–1184

Graham D, Millar AM (1999) Artefacts in the thin-layer chromatographic analysis of $^{99}Tc^{m}$-tetrofosmin injections. Nucl Med Commun 20:439–444

Hnatowich DJ (1986) HPLC of radiolabelled antibodies. In: Wieland DM, Tobes MC, Manger TJ (eds) Analytical and chromatographic techniques in radiopharmaceutical chemistry. Springer, Berlin Heidelberg New York

Hoch DJ, Pinkerton TC (1986) Reversed-phase HPLC of ^{99m}Tc methylene diphosphonate bone imaging kits with quantification of pertechnetate. Int J Rad Appl Instrum 37:593–598

Huigen YM, Gelsema WJ, de Ligny C (1988) Effect of the composition of the eluent in the chromatography of ^{99m}Tc-diphosphonate complexes. Int J Rad Appl Instrum 39:381–384

Hung JC, Corlija M, Volkert WA, Holmes RA (1988) Kinetic analysis of technetium-99m D,L-HM-PAO decomposition in aqueous media. J Nucl Med 29:1568–1576

Hung JC, Wilson ME, Brown ML, Gibbons RJ (1991) Rapid preparation and quality control method for technetium-99m-2-methoxy isobutyl isonitrile (technetium-99m-sestamibi). J Nucl Med 32:2162–2168

Kelly JD, Forster AM, Higley B, Archer CM, Booker FS, Canning LR, Chiu KW, Edwards B, Gill HK, McPartlin M (1993) Technetium-99m-tetrofosmin as a new radiopharmaceutical for myocardial perfusion imaging. J Nucl Med 34:222–227

Millar AM (1989) Quality assurance of radiopharmaceuticals. In: Theobald AE (1989) Radiopharmaceuticals and radiotracers in pharmacy. Ellis Horwood, Chichester, UK

Millar AM, Wilkinson AG, McAteer E, Best JJK (1990) $^{99m}Tc-MAG_3$: in vitro stability and in vivo behaviour at different times after preparation. Nucl Med Commun 11:405–412

Neirinckx RD, Canning LR, Piper IM, Nowotnik DP, Picket RD, Holmes RA, Volkert WA, Foster AM, Weisner PS, Marriot JA, Chaplin SB (1987) Tchnetium-99m D,L-HM-PAO: a new radiopharmaceutical for SPECT imaging of regional cerebral blood flow. J Nucl Med 28:191–202

Nieuwland RJ, Das HA, de Ligny CL (1989) Improvement of the reproducibility of ion-pair HPLC of ^{99m}Tc(Sn)EHDP complexes and the influence of the Sn(II) concentration on the composition of the reaction mixture. Int J Rad Appl Instrum 40:153–157

Nunn AD (1983) Structure-distribution relationships of radiopharmaceuticals. Correlation between the reversed-phase capacity factors for Tc-99m phenylcarbamoyl-methyliminodiacetic acids and their renal elimination. J Chromatogr 255:91–100

Nunn AD, Fritzberg AR (1986) Components for the design of a radio-HPLC system. In: Wieland DM, Tobes MC, Manger TJ (eds) Analytical and chromatographic techniques in radiopharmaceutical chemistry. Springer, Berlin Heidelberg New York

Nunn AD, Loberg MD, Conley RA (1983) A structure–distribution–relationship approach leading to the development of Tc-99m-mebrofenin: an improved cholescintigraphic agent. J Nucl Med 24:423–430

Shattuck LA, Eshima D, Taylor AT, Anderson TL, Graham DL, Latino FA, Payne SE (1994) Evaluation of the hepatobiliary excretion of technetium-99m-MAG_3 and reconstitution factors affecting radiochemical purity. J Nucl Med 35:349–355

Tanabe S, Zodda JP, Deutsch E, Heineman WR (1983) Effect of pH on the formation of Tc($NaBH_4$)-MDP radiopharmaceutical analogues. Int J Appl Radiat Isot 34:1577–1584
Vallabhajosula S, Goldsmith SJ, Pollina R, Lipszyc H (1982) Radiochemical analysis of Tc-99m human serum albumin with high pressure liquid chromatography. J Nucl Med 23:326–329
Vallabhajosula S, Moyer BR, Lister-James J, McBride BJ, Lipszyc H, Lee H, Bastidas D, Dean RT (1996) Preclinical evaluation of technetium-99m-labelled somatostatin receptor-binding peptides. JNM 37:1016–1022
Weisner PS, Bower GR, Dollimore LA, Forster AM, Higley B, Storey AE (1993) A method for stabilising technetium-99m exametazime prepared from a commercial kit. Eur J Nucl Med 20:661–666
Zinn KR, Buchsbaum DJ, Chaudhuri TR, Mountz JM, Grizzle WE, Rogers BE (2000) Noninvasive monitoring of gene transfer using a reporter receptor imaged with a high-affinity peptide radiolabeled with ^{99m}Tc or ^{188}Re. J Nucl Med 41:887–895

9.1.3 Electrophoresis

Gy. Jánoki

Electrophoresis is widely used in the separation of inorganic ions and different classes of molecules. The physical principle underlying electrophoresis can be stated as: Particles carrying an electric charge are accelerated when placed in an electric field; this driving force is very rapidly balanced by the friction forces arising in the medium; from that moment, the particles move at a constant speed proportional to their charge. Electric charge carried by molecules originates either from dissociation or from selective adsorption. For polyelectrolytes such as proteins, dissociation of the acidic groups COOH or NH_3^+ constitutes the principal source.

The electromigration rate is directly related to the charge and inversely related to the ionic radius of the complex being separated. It means that the direction of migration of cationic species is toward the cathode while the migration of the anionic species moves toward the anode.

Electrophoresis has been used for quality control of certain radiopharmaceuticals (Belkas and Archimandritis 1979; Pauwels and Feitsma 1979). Paper electrophoresis has been applied successfully for the determination of the relative charge on some ^{99m}Tc complexes.

HPLC, TLC, and ITLC analysis methods available today are more convenient and reproducible than electrophoresis. Despite this, electrophoresis continues to be used as a research tool in radiopharmaceutical studies. The reason for this is that electrophoresis gives information on the charge of ^{99m}Tc complexes that cannot be easily obtained by other techniques. From the standpoint of determining whether a specific radiolabeled complex is anionic, cationic or neutral, electrophoresis is the best technique currently available.

In addition, polyacrylamide gels (PAGE) are used as supports for protein and macromolecular electrophoresis including radiolabeled antibodies and other proteins (Wieme 1965).

References

Belkas EP, Archimandritis S (1979) Quality control of colloid and particulate ^{99m}Tc-labelled radiopharmaceuticals. Eur J Nucl Med 4:375
Pauwels EKJ, Feitsma RIJ (1977) Radiochemical quality control of ^{99m}Tc-labelled radiopharmaceuticals. Eur J Nucl Med 2:97
Wieme RJ (1965) Agar gel electrophoresis. Elsevier, Amsterdam

9.2 Determination of Tin(II)

F. Rakiás

Radiopharmaceutical kits for labeling with ^{99m}Tc eluate contain tin(II)-ion for reduction of sodium pertechnetate to lower valency states, which are chemically reactive. Tin(II) salts are easily oxidized, even by the oxygen in air. Certain chemicals are also assumed to enhance the oxidation of tin(II) to tin(IV). Therefore, determination of tin(II) in radiopharmaceutical kits is an important aspect of quality control.

Classic methods for the determination of tin(II) include the titration with an iodine standard solution, or absorption measurements by spectrophotometric analysis. Both methods have considerable disadvantages for the determination of tin(II) in radiopharmaceutical kits (Rakiás and Zolle 1997).

A new spectrophotometric analysis of tin(II) and tin(IV) in the same matrix was developed by Rakiás et al. (unpublished results); however, the disadvantages of sample preparation and the slow formation of a colored complex for absorption measurements led to the investigation of other methods.

Pulse polarography has been investigated because of its speed and high selectivity. Thus, square wave voltammetry has been extended to the measurement of microgram amounts of tin(II) in radiopharmaceutical kits.

Pulse polarography has offered considerable advantages for the determination of the tin(II) content in a number of radiopharmaceutical kits (Rakias et al. 1988). Square wave voltammetry shows a considerable increase in sensitivity, suitable for the measurement of very small (microgram) amounts of tin(II), as used for the reduction of ^{99m}Tc-Na-pertechnetate when labeling radiopharmaceuticals.

Two kits manufactured by NCPH–"Fredric Joliot Curie" National Research Institute for Radiobiology and Radiohygiene were examined. TromboScint and LeucoScint differ simply by the fact that LeucoScint contains twice the amount of active ingredients, i.e., 0.18 mg HMPAO and 2.28 µg tin(II). LeucoScint is used for labeling leucocytes, and TromboScint is suitable for labeling platelets.

For the determination of tin(II) in kits an EG & G polarographic analyzer (model 384) with a static mercury drop electrode (model 303) and a cabinet reference electrode was used. Model 384 is a microprocessor-based polarographic analyzer with built-in floppy disk memory to store and recall analytical curves. By controlling each step of the analysis, the microprocessor automates polarographic and voltammetric measurements. All experimental parameters may be chosen by the operator. Concentrations are computed automatically and recorded in the range from 0.001 ppb to 9999 ppm.

Square Wave Voltammetry of Tin(II) and Tin(IV). Deoxygenated 1 *N* hydrochloric acid plus 4 *N* ammonium chloride buffer (1:1, v/v) was used as an electrolyte. Sample solutions were prepared by dissolving one kit in the same buffer and adding to the electrolyte solution; recording was started from 0.0–0.6 V.

The tin(IV) maximum appeared at –0.25 V, and the tin(II) maximum at –0.45 V, as the second peak. If only 1 *N* sulfuric acid is used as an electrolyte, tin(II) may be detected selectively at –0.45 V.

The method was validated, and the following parameters were examined.

Linearity of the Instrument. To identify the highest amount of tin(II) that can be measured with a linear response, increasing concentrations of tin(II) up to 5 mg/ml were analyzed.

Linear regression analysis of the data (Booster 1982) showed a correlation coefficient >0.999, indicating linearity of measurements over a wide range of concentration.

Precision of the Method. Measurements were performed with ten samples of each kit. In the case of TromboScint, the average value of Sn(II) was determined as 1.138 µg/kit (theoretically, 1.14 µg Sn(II)/vial) with a standard deviation of 0.050 µg and a coefficient of variation of 4.39%.

In the case of LeucoScint, the average value of Sn(II) was calculated as 2.263 µg/kit (theoretically, 2.28 µg Sn(II)/vial), with a standard deviation of 0.089 µg and a coefficient of variation of 3.93%.

Effect of Sample Volume. Since the vials were filled with 1.0 ml of the dissolved kit content in the production department, there might have been a variation in the amount of tin(II) actually added in this volume. Therefore, the precision of the method by analyzing ten matrix samples to which a homogenous solution of tin(II) chloride containing 1.14 and 2.28 µg of tin(II), respectively, were also investigated. When these known concentrations of tin(II) were analyzed together with the matrix samples, the results were identical excluding a volume effect. A recovery between 92.10 and 106.14% was observed. The average value of Sn(II) in the case of TromboScint was 1.139 µg/kit, with a standard deviation of 0.048 and a coefficient of variation of 4.21%. In the case of LeucoScint, the average value of Sn(II) was 2.266 µg/kit, with a standard deviation of 0.062 and a coefficient of variation of 2.73%.

Selectivity. Although it is stated in the literature that Sn(IV) does not interfere with the measurement of Sn(II), experimental proof adding 2 µg of tin(IV) to the test solution was obtained. When no tin(IV) was added to TromboScint, the average value of Sn(II) was determined as 1.141 µg/kit, with a standard deviation of 0.068 µg and a coefficient of variation of 5.96%. With tin(IV) added to TromboScint, the average value of Sn(II) was determined as 1.126 µg/kit, with a standard deviation of 0.055 µg and a coefficient of variation of 4.88%. Values demonstrating no interference by Sn(IV) were also shown for LeucoScint.

Based on the above-cited results, square wave voltammetry is suitable for the measurement of microgram amounts of tin(II) in radiopharmaceutical kits, with high accu racy (>99%). Impurities causing oxidation to tin(IV) have been shown to have no effect on the recovery. Based on the obtained data, square wave voltammetry has been shown as a reliable and highly sensitive method for the determination of tin(II) in radiopharmaceutical kits.

References

Booster P (1982) Statistische Methoden im Laboratorium [in German]. Elsevier, Amsterdam

Eckelman W, Meinken G, Richards P (1971) Chemical state of Tc-99m in biomedical products. J Nucl Med 12:596–600

Rakiás F, Zolle I (1997) Stannous ion determination: importance and relevance for radiopharmaceutical kits. In: Bergmann H, Kroiss A, Sinzinger H (eds) Radioactive isotope in clinical medicine and research, vol. 22. Birkhauser, Basel, pp 401–409

Rakias F, Kelemen-Küttel I, Tömpe P, Nemeth P, Paal TL (1988) Correlation between the Sn(II)-content of phosphonate, phytate, and gluconate kits and the quality of the obtained images. In: Höfer R, Bergmann H (eds) Radioactive isotope in clinic und research, vol. 18. Schattauer, Stuttgart, pp 417–419

9.3 Sterility Testing of Radiopharmaceuticals

S. R. Hesslewood

It is an integral feature of good manufacturing practice that radiopharmaceuticals for injection that are required to be sterile are prepared under conditions that exclude microbial contamination of the product. There is a higher degree of assurance of sterility of products that are terminally sterilized in their final container than for those that are prepared by aseptic technique. The majority of radiopharmaceuticals in current clinical practice fall into the latter category, and a sterility test is the only analytical method available to demonstrate the absence of microbial contamination. However, the use of commercially available sterile products and starting materials (e.g., kits) with marketing authorization is an important feature in avoiding problems with lack of sterility.

A test for sterility is laid down in the *European Pharmacopeia* for all parenteral products, and two techniques for testing are described, either membrane filtration of the product with subsequent incubation of the filter in suitable culture media (which is the preferred technique), or direct inoculation of the product into the culture medium, followed by incubation at the appropriate temperature for the specified time. Suitable media are described in the *European Pharmacopeia*; although, it is also recognized that other media may be used. In every case, however, it is necessary to demonstrate that the medium is capable of supporting the growth of microorganisms both in the presence and absence of the material to be tested.

9.3.1 Problems in Applying the *European Pharmacopeia* Test to Radiopharmaceuticals

There are specific difficulties in applying the test as written, particularly for radiopharmaceuticals based on technetium-99m, and these difficulties are acknowledged in the general pharmacopeial monograph on radiopharmaceuticals.

First, the batch size of products is often small, and in the case of a technetium product prepared in a hospital or clinic, may consist of only a single vial with a total volume of less than 10 ml.

This makes it impossible to follow the *European Pharmacopeia* requirements for the number of containers to be tested (10% or 4, whichever is the greater), without preparing extra vials of the product specifically for the test, which is economically unrealistic and also imposes an additional radiation burden on to the operators. There are also problems with the volume of the product available for testing. The *European Pharmacopeia* states that if the quantity in the container is between 4 and 20 ml, which is common for technetium products, 2 ml should be used for each culture medium being tested. If followed strictly, this would mean a large reduction in the volume of material available for patient use.

Second, the radioactive nature of the product imposes handling difficulties in performing the test, since quality control departments may not have the necessary facilities for handling radioactive materials, and radiopharmacy departments may not be suitably equipped to perform sterility testing.

One method to avoid these problems would be to let the product decay for a sufficient period of time to allow the level of radiation emitted to fall to a suitably low level to facilitate handling. However, there is published evidence to suggest this technique will decrease the sensitivity of the test, since the number of any viable organisms in the preparation may decrease on storage (Brown and Baker 1986; Stathis et al. 1983; Wind 1985). It is therefore always necessary to perform a sterility test as soon as possible after preparation of the radiopharmaceutical.

Finally, it is recognized that for short-lived radiopharmaceuticals, the long incubation time of the culture media (7 days for the membrane filtration technique, 14 days for direct inoculation) means the result of the sterility test cannot be available before the product is used. In these situations, the test constitutes a control of production techniques and will give valuable information about their suitability.

9.3.2 Recommendations for Sterility Testing of Radiopharmaceuticals

In view of the difficulties in applying the *European Pharmacopeia* test, many variations have been adopted. Whatever the technique used, it is essential to perform a validation to ensure that it would be able to detect the presence of any viable microorganisms in the sample.

If following the *European Pharmacopeia* test, it is often easier to perform the direct inoculation technique. It is suggested that the remnants of a technetium vial are divided equally between two suitable culture media as soon as possible after expiration of the radiopharmaceutical. The inoculated media should then be incubated for not less than 14 days at 20–25 °C for media being used to detect fungi, or 30–35 °C for those used in the detection of bacteria. Depending on the level of radioactivity present, shielding of the culture media may be necessary during some or all of the incubation period. When macroaggregates or microspheres are being tested, it may be necessary to perform a subculture at the end of the incubation mixture, since, dependent on the volume used in the test, the radiopharmaceutical itself may produce turbidity in the culture medium that may be indistinguishable from bacterial growth.

The membrane filtration technique is technically more elaborate and requires that the radiopharmaceutical under test, after aseptic dilution, is passed through a membrane filter with a pore size of 0.45 m, which has been moistened with a sterile nutrient diluent. After filtration, the membrane is either transferred to a suitable culture medium or aseptically cut into two equal parts and one half placed in each of two suitable media. Incubation at the appropriate temperature is required for at least 7 days.

An alternative – although unofficial – technique that has been used is the addition of an equal volume of double-strength culture medium directly to the remnants of the vial of radiopharmaceutical immediately upon its expiration. The advantages and disadvantages of this technique are summarized in Table 9.3.1.

In order to make the test more meaningful, it is necessary to rotate the culture media used and ensure that each product prepared is at some stage tested with each culture me-

Table 9.3.1. Sterility testing with double-strength broth

Advantages	Disadvantages
No delay in testing	Sample size variable
Can be performed in radiopharmacy	Only one culture medium can be used
Container can be easily shielded	Not officially recognized
No dilution of culture medium below normal strength	Not applicable if preservatives need to be diluted out

dia used. After incubation for 14 days, the result is obtained by observing the culture medium for turbidity. A positive control can then be performed by inoculating one of the test organism described in the *European Pharmacopeia* directly into the vial. Again, it is necessary to ensure the test organisms are used in rotation for each product.

9.3.3 Frequency of Testing

It has already been stated that the results of a sterility test are necessarily retrospective and as such, constitute a control of production processes within the radiopharmacy department rather than forming a part of a release procedure for individual products. Sterility testing of every batch prepared, although ideal, is unrealistic in practice. The testing program should ensure that, at some stage, all different types of product prepared are tested on a regular basis. At least one batch should be tested every week. In addition, it is recommended that the remnants of the first eluate of each technetium generator and the final unused eluate should be tested.

9.4 Pyrogen Testing of Radiopharmaceuticals

The *European Pharmacopeia* requires certain radiopharmaceuticals, mainly of biological origin, to comply with a test for pyrogens, which are substances generally arising from bacteria that are capable of inducing fever. The *British Pharmacopeia* general monograph on parenteral products states that compliance is necessary for preparations where the volume in a single injection exceeds 15 ml or is less than 15 ml, but where the label says the product is apyrogenic. The test consists of measuring the rise in body temperature in rabbits, following intravenous administration of the substance and in order to pass the summed response from three rabbits must not exceed 1.15 °C. The *Pharmacopeia* also considers that for some products it may be necessary to allow decay to take place before testing in order that hyperthermia that may be due to the radioactivity of the product is avoided. In view of the sophisticated facilities required for the test and the fact that volumes of radiopharmaceuticals administered are hardly ever above 15 ml, this test is not carried out in hospitals or clinics. If necessary, samples can be submitted to specialist laboratories for testing, although it is most unlikely that this will take place routinely. For short-lived materials, the results are inevitably retrospective to the use of the product, which limits its usefulness.

For some products, a direct measurement of the level of bacterial endotoxins in the preparation is required. The test uses a lysate of amoebocytes from the horseshoe crab,

Table 9.4.1. Products in the *European Pharmacopeia* with endotoxin limits

Product	Endotoxin limit (units ml^{-1})
^{99m} Technetium human albumin	Not greater than 175/V[a]
Iodine 123 MIBG	Not greater than 175/V
Iodine 131 MIBG	Not greater than 175/V
Iodine 131 norcholesterol	Not greater than 175/V
Indium 111 pentetate	Not greater than 14/V

MIBG metaidobenzoguanidine
[a] *V* maximum recommended dose in milliliters

Limulus polyphemus, and is often referred to as *Limulus* amoebocyte lysate (LAL) test. It depends on the fact that endotoxins will produce turbidity, precipitation or, more commonly, gelation of a solution of the lysate within a period of approximately 1 h. The test is described in detail in the *European Pharmacopeia* and requires the performance of control experiments to demonstrate the sensitivity of the lysate using at least three dilutions of an endotoxin standard. It is also necessary to show that the product under test does not contain substances that interfere with the sensitivity of the test. Reagents for performing the test are commercially available and in view of the time scale required, can be used prospectively for short-lived radiopharmaceuticals, although technical expertise in performing the test is essential to avoid obtaining misleading results. Products in the *European Pharmacopeia* for which a bacterial endotoxin test is specified are listed in Table 9.4.1, together with the limits for bacterial endotoxin laid down in the monographs. This limit is the maximum allowable endotoxin concentration in units per ml. The *United States Pharmacopeia* prescribes for a wider range of radiopharmaceuticals, including those based on positron emitters. The maximum limits prescribed are the same as in the *European Pharmacopeia*.

It is known that endotoxins are approximately 1000 times more toxic following intrathecal as opposed intravenous administration and this is recognized in the pharmacopeial monographs for products used intrathecally (e.g. Indium In^{111}-pentetate) since the maximum amount of endotoxin allowed is lower than for intravenous products.

9.4.1 Recommendations for Endotoxin Determinations of Radiopharmaceuticals

The use of commercially available products with marketing authorization that have been tested by the manufacturer will remove the need for routine determination of endotoxins. However, testing should be considered if it is thought that a particular product may be giving rise to problems.

When products are prepared totally within a hospital from raw materials, the hospital assumes responsibility for their safety, quality, and efficacy. In these situations, each batch of a product, especially those intended for intrathecal administration, should be tested.

References

Brown S, Baker MH (1986) The sterility testing of dispensed radiopharmaceuticals. Nucl Med Commun 7:327–336

Stathis VJ, Miller CM, Doerr GF, Coffey JL, Hladik WB (1983) Effect of technetium Tc-99m pertechnetate on bacterial survival in solution. Am J Hosp Pharm 40:634–637

Wind K (1985) Sterility testing. In: Cox PH (ed) Radiopharmacy and radiopharmacology, yearbook 1. Gordon and Breach, London, pp 35–46

Chapter 10

Other Technetium Isotopes: ^{94m}Tc as a Potential Substitute in Positron Emission Tomography Investigations

10

Z. Kovács

10.1 Introduction

Among the neutron-deficient isotopes of Tc, ^{94m}Tc seems to have suitable nuclear properties for application in positron emission tomography (PET) (Browne and Firestone 1986). Considering the relatively new lipophilic Tc-labeled flow tracers used in cardiac perfusion imaging (2-methoxyisobutyl isocyanide [MIBI], teboroxime, tetrofosmin) it would be reasonable to replace ^{99m}Tc with ^{94m}Tc in order to carry out quantitative investigations of biodistribution and clearance with PET. With this replacement, the pharmacokinetics of useful Tc compounds could be studied in a number of cases, which would enhance the introduction of new ^{99m}Tc-labeled pharmaceuticals into clinical practice. The labeling and quality control methods are well known and partly subject of this book. Our task is to review the different methods for the production of ^{94m}Tc and also the separation technique from irradiated targets.

10.2 Methods of Production

The 52 min half-life positron emitter ^{94m}Tc ($\beta^+=70\%$, $E_{\beta^+}=2.5$ MeV; EC=30%, $E_\gamma=871.1$ keV, $I_\gamma=94.2\%$) was first suggested and produced from natural Mo in 1991 (Nickles et al. 1991; Rösch and Beyer 1991). Low-energy (up to 13 MeV) proton and deuteron irradiations were carried out on metal and MoO_3 targets and led to low yields (1.7 and 3.9 Ci/µA at 1-h irradiations) with high contamination from other Tc isotopes. Detailed systematic investigations were carried out in recent years at Jülich (Denzler et al. 1995; Faßbender et al.1995; Qaim et al. 1994; Rösch and Qaim 1993), using highly enriched target materials. Excitation functions were measured by the stacked foil technique (Qaim et al. 1977; Weinreich et al. 1974) for the ^{94}Mo(p,n)^{94m}Tc reaction up to 18.4 MeV (Rösch and Qaim 1993), and the ^{93}Nb(^{3}He,2n) ^{94m}Tc reaction up to 35 MeV (Faßbender et al.1995).

The ^{92}Mo(α,2n)^{94}Ru → ^{94m}Tc and ^{92}Mo(α,pn)^{94m}Tc reaction routes were also studied (Denzler et al. 1995). The excitation functions were measured on 93.9% enriched Mo metal samples (Qaim et al. 1994). The highest yield is given by the ^{94}Mo(p,n) reaction, with an acceptable level of impurity. This is the method of choice at small cyclotrons. However, if higher-energy α-particles are available ($E_\alpha>30$ MeV), the ^{92}Mo(α,2n) reaction can provide extra clean ^{94m}Tc with a considerably higher yield than that given in Table 10.1.

Table 10.1. Comparison of ^{94m}Tc production routes

Nuclear reaction	Optimal energy range (MeV)	Thick target yields at EOB (MBq/μAh)	Major impurities at EOB (%)
$^{94}Mo(p,n)^{94m}Tc$	13 → 7	2000	^{94g}Tc (6)
$^{93}Nb(^{3}He,2n)^{94m}Tc$	18 → 10	34	^{94g}Tc (24) $^{93m,g}Tc$ (14)
$^{92}Mo(\alpha,2n)^{94}Ru \rightarrow {}^{94m}Tc$	26* → 18	35	None
$^{92}Mo(\alpha,pn)^{94m}Tc$	26* → 18	98	^{94g}Tc (30)

*maximum available energy at the compact cyclotron in Jülich

10.3 Methods of Separation

Several methods have been published for the separation of ^{94m}Tc from irradiated $^{94}MoO_3$ target (Nickles et al. 1993 a, b; Rösch et al. 1994). In an online method, a well-focused, vertical 11-MeV proton beam was used for irradiation (Nickles et al. 1993 a). The $^{94}MoO_3$ target was melted by the beam and kept at 800 °C, under continuous surface temperature control. Ninety-five percent of the radiotechnetium, heated out simultaneously, was collected on a cooled quartz tube. One milligram of $^{94}MoO_3$ also was evaporated and adsorbed on the surface, from which the ^{94m}Tc was separated and purified by a wet chemistry method.

Offline thermochromatographic separation was used in another method (Rösch et al. 1994). The irradiated $^{94}MoO_3$ target was placed at the bottom of a vertically arranged thermochromatographic system and heated to 1,090 °C. Both the $^{94}MoO_3$ and the ^{94m}Tc rapidly evaporate in a stream of wet air carrier gas. The $^{94}MoO_3$ condenses in the 800–600 °C zone, and the $H^{94m}TcO_4$ in the 350–250 °C temperature range on the quartz column. The $H^{94m}TcO_4$ layer was dissolved from the surface in NaOH and purified on an alumina column. After purification, no Mo traces or other chemical forms of radiotechnetium could be detected. The yield is 40–45% after a 25-min separation time. The batch yields of ^{94m}Tc amount to about 50 mCi (18 GBq). The final product was of the highest radionuclidic purity reported so far. A method has also been described for the recovery of highly enriched $^{94}MoO_3$, with 95% yield (Rösch et al. 1994). The final chemical form of the radiotechnetium with both separation methods is $^{94m}TcO_4^-$, which is suitable for labeling procedures.

An alternate method is described in (Nickles et al. 1993 b). A Mo foil (0.1 mm thick) was used for irradiation and the radiotechnetium was extracted by an electrochemical etching procedure. With the obtained ^{94m}Tc-pertechnetate, ^{94m}Tc-teboroxime was prepared, which was used for the first preliminary human investigation with this ^{94m}Tc-labeled pharmaceutical (Nickles et al. 1993 b). Another PET study with ^{94m}Tc-2-methoxy isobutyl isonitrile was also reported (Stone et al. 1994).

References

Browne E, Firestone RB (1986) Table of radioactive isotopes. Wiley, New York

Denzler F, Rösch F, Qaim SM (1995) Excitation functions of α-particle induced nuclear reactions on highly enriched ^{92}Mo: comparative evaluation of production routes for ^{94m}Tc. Radiochim Acta 68:13–20

Faßbender M, Novgorodov AF, Rösch F, Qaim SM (1995) Excitation functions of ^{93}Nb(^{3}He,xn)$^{93m,g;94m,g;95m,g;}$ Tc processes from threshold up to 35 MeV: possibility of production of ^{94m}Tc in high radiochemical purity using a thermochromatographic separation technique. Radiochim Acta 65:215–224

Nickles J, Nunn AD, Stone CK, Perlmann SB, Levine RL (1991) Tc-94m flow agents: bridging PET and SPECT. J Nucl Med 32:925

Nickles RJ, Christian BT, Nunn AD, Stone CK (1993 a) Cyclotron production of high-purity ^{94m}Tc by in situ sublimation. J Labelled Compd Radiopharm 23:447

Nickles RJ, Nunn AD, Stone CK, Christian BT (1993 b) Technetium-94m-teboroxime: Synthesis, dosimetry and initial PET imaging studies. J Nucl Med 34:1058–1066

Qaim SM, Stöcklin G, Weinreich R (1977) Excitation functions for the formation of neutron deficient isotopes of bromine and krypton via high-energy deuteron induced reactions on bromine: production of ^{77}Br, ^{76}Br and ^{79}Kr. Int J Appl Radiat Isot 28:947–953

Qaim SM, Rösch F, Scholten B, Stöcklin G, Kovács Z, Tárkányi F (1994) Nuclear data relevant to the production of medically important β^{+}-emitting radioisotopes ^{75}Rb, ^{86}Y, ^{94m}Tc and ^{124}I at a small cyclotron. Dickens JK (ed) Proceedings of International Conference for Science and Technology. Gatlinburg, Tenn., pp 1035–1038

Rösch R, Beyer G-J (1991) Längerlebige Positronenemitter: Einzelnuklide und Generatorsysteme, Produktion am Rossendorfer Zyklotron U-120, radiochemische Studien und Möglichkeiten für die PET, Report ZfK-745 [in German]

Rösch F, Qaim SM (1993) Nuclear data relevant to the production of the positron emitting technetium isotope ^{94m}Tc via the ^{94}Mo(p,n)-reaction. Radiochim Acta 62:115–121

Rösch F, Novgorodov AF, Qaim SM (1994) Thermochromatographic separation of ^{94m}Tc from enriched molybdenum targets and its large scale production for nuclear medical application. Radiochim Acta 64:113–120

Stone CK, Christian BT, Nuckles RJ, Perlman SB (1994) ^{94m}Tc-methoxy isobutyl isonitrile dosimetry and resting cardiac imaging with PET. J Nucl Cardiol 1:425

Weinreich R, Schult O, Stöcklin G (1974) Production of ^{123}I via the ^{127}I(d,6n) ^{123}Xe(β^{+},EC)^{123}I process. Int J Appl Radiat Isot 25:535–543

Chapter 11

The Rules Governing Medicinal Products for Human Use in the European Union

11

A. Verbruggen, I. Zolle

11.1 European Economic Council Directives and Regulations

The peaceful application of nuclear reactors after World War II initiated the production of large quantities of radionuclides, which were introduced to medicine for the investigation of human physiology and disease. The application of radioactive tracers for clinical diagnosis and therapy increased by the year, resulting in the widespread use of radioisotopes in clinical procedures.

Radioactivity has been the concern of physicists, who set up rules and regulations for the protection of the general public and health personnel handling radioactive materials. At most institutions, the commercially available radiopharmaceuticals were managed by a responsible staff of health physicists.

In some European countries, the distribution of radioactive compounds to the users was considered the responsibility of the pharmacist. The first centralized isotope pharmacy was installed in Denmark, taking responsibility for the quality and distribution of radioactive drugs since 1976. Different combinations of primary legislation for non-radioactive drugs (i.e., the Medicines Act) and national administrative arrangements provided legislation for consumer protection regulating the importation, manufacture, and sale or supply of radioactive medicinal products (Appendix A.2).

Requirements for good radiopharmacy practice and specific guidelines for simple procedures such as handling of ready-for-use radiopharmaceuticals and the preparation of ^{99m}Tc-labeled radiopharmaceuticals from generators and kits were presented in a comprehensive report by Kristensen (1979) based on existing good manufacturing practice (GMP) recommendations and standards applied to the manufacture of pharmaceuticals (Appendix A.3). The demand on qualified personnel, and premises for small-scale production in a radiopharmacy posed a considerable challenge.

The European regulatory bodies have three main instruments to reach an approximation of national regulations: directives, guidelines, and European regulations.

- *Directives* are rules addressed to the member states to be translated into the respective national legislation and effectively implemented. Directives are mandatory.
- *Guidelines* are recommendations for the effective implementation of directives by the member states. Guidelines are not mandatory.
- The *European regulations* are mandatory in all European Union countries; they are applied into the national legislation without translation.

In 1989, the European Economic Council issued Directive 89/343/EEC as an amendment to Directive 75/319/EEC, which, for the first time, defines different classes of radiopharmaceutical products and states the requirements for marketing authorization of these products as proprietary medicinal products for human use (Appendix A.1). Implementation of directives into national law provides legislation that is effective in all member states, resulting in a harmonization within the European Union.

The comprehensive legislation regulating all medicinal products in the European Union is organized in seven volumes in *The Rules Governing Medicinal Products in the European Union*:

- Volume 1 contains the directives governing medicinal products for human use in the European Union.
- Volume 2 contains the notice to applicants for marketing authorization of medicinal products.
- Volume 3 contains the guidelines concerning quality, safety, and efficacy.
- Volume 4 contains pharmaceutical legislation of GMP for the manufacture of medicinal products.
- Volumes 2, 3, and 4 present specific guidelines for radiopharmaceutical products.
- Volumes 5, 6, and 7 are dedicated to medicinal products for veterinary use.

Council directives relating to marketing authorization of medicinal products for human use and amending texts are listed in Volume 2B (annex).

Some directives relevant for radiopharmaceuticals as medicinal products are discussed here in chronological order.

Directive 65/65/EEC is the basis of all pharmaceutical regulations. It includes relevant definitions (medicinal product, proprietary medicinal product, officinal and magisterial formula, etc.) and states the registration by law for all proprietary medicinal products. Since 1965, this directive has been amended several times.

Directives 75/318/EEC and 75/319/EEC, amended at different times from the original version, state the documents and requirements for manufacture and marketing authorization. Definition of the professional profile of the qualified person responsible for the quality of medicinal products is presented, and a technical body on medicines, the Committee of Proprietary Medicinal Products (CPMP), is created. Due to technical reasons, some products such as vaccines and sera, homeopathics, hemoderivatives, and radiopharmaceuticals were excluded from the scope of this directive in its original version.

In 1987, Directive 87/22/EEC created the centralized marketing authorization of medicines, particularly those derived from biotechnology in accordance with the CPMP, as a valid procedure for drug registration in all member states. This centralized procedure should be applied to relevant new medicinal products, such as those obtained by biotechnology, monoclonal antibodies, or radionuclides. Therefore, radioactive medicinal products were recognized in this context as drugs, requiring the modification of Directives 65/65/EEC, 75/318/EEC and 75/319/EEC to extend their provisions to the products previously excluded; this amendment was introduced by Directive 89/341/EEC.

After Directive 89/341/EEC had been issued, a block of directives, so-called extension directives, extended the scope of Directives 65/65/EEC, and 75/319/EEC to vaccines and sera, homeopathics, hemoderivatives, and radiopharmaceuticals, giving specific details for them. This means that all pharmaceutical directives must be applied to these products.

One of these extension directives is 89/343/EEC, extending the scope of the pharmaceutical directives to radiopharmaceuticals, defining all radiopharmaceutical products (radiopharmaceuticals ready for use, cold kits, generators, and radionuclide precursors) as drugs for which marketing authorization as proprietary medicinal products for human use is required. However, this authorization shall not be required for radiopharmaceuticals prepared at the time of use, if prepared by using authorized products (cold kits, generators, and radionuclide precursors) and in accordance with the instructions given by the manufacturer. This directive does not derogate in any way the rules on radiation protection.

Directive 91/356/EEC gives the principles of GMP detailed in Volume 4 of *The Rules Governing Medicinal Products in the European Union.* These principles must be applied not only for production at industrial level, but also for small-scale productions, such as in nuclear medicine.

A new block of directives gives details on practical aspects for the rational use of medicinal products. Directive 92/25/EEC concerns the distribution from the producer to the dispenser. According to this directive, distribution of medicinal products must be in accordance with good distribution practice (GDP): only approved products may be distributed, and may be delivered exclusively to persons authorized for supplying the public in the concerned country.

Directive 92/26/EEC states the criteria to be applied for the supply of medicinal products to the public. According to these criteria, radiopharmaceuticals are medicinal products for human use, dispensed only on medical prescription. In most European countries, the only persons authorized to dispense medicinal products are pharmacists, but radioactive medicinal products may constitute an exception to this rule in some countries.

Council Regulation 2309/93 creates the European Agency for the Evaluation of Medicinal Products (EMEA), in London, as administrative authority responsible for medicines in all European countries. The CPMP and the centralized marketing authorization of medicinal products are both assumed by this agency.

In addition to pharmaceutical regulations, the *radioactive* nature of radiopharmaceuticals is also regulated by European Directives. Directives from European Atomic Energy Community (EURATOM) (Directive 84/466 and 84/467) regulate the health protection of patients submitted to ionizing radiation and of the general public and workers against the dangers of ionizing radiation.

Since radiopharmaceutical products are included in the scope of the discussed pharmaceutical directives, regulations applied to nonradioactive medicines equally apply to radiopharmaceuticals.

11.1.1 Application for Marketing Authorization

Marketing authorization for medicinal products, i.e., radiopharmaceuticals, is required for importation, manufacture, and sale in the European Union. Presentation and content of the dossier is contained in the rules governing medicinal products in the European Union: *The Rules Governing Medicinal Products in the European Union, Notice to Applicants, Volume 2B: Medicinal Products for Human Use.*

The requirements for the content of the application dossier are stated in Directive 75/318/EEC as amended, i.e., the documents and data in support of the application for marketing authorization pursuant to Article 4 of Council Directive 65/65/EEC must be organized in a file in four parts, containing:

- Part I, summary of the dossier
- Part II, chemical, pharmaceutical and biological documentation describing the composition and method of preparation of the medicinal product (raw materials, manufacture, characteristics of the labeled compound, quality control, *European Pharmacopeia* standards), and the documentation of preclinical studies performed in animals to determine stability and the elimination kinetics (half-times of removal)
- Part III, toxicopharmacological documentation

- Part IV, clinical documentation of the clinical investigations carried out in humans to demonstrate efficacy and safety

These are the classic requirements for any medicinal product demonstrating quality, safety, and efficacy.

The Summary of Product Characteristics (SPC) is a document specific for any medicinal product. The SPC is included in Part I of the Dossier for Marketing Authorization. After obtaining marketing authorization, the SPC is officially approved and describes the characteristics and uses of the authorized product.

For each approved radiopharmaceutical, the SPC states relevant information concerning pharmaceutical form, approved clinical indications, posology, internal dosimetry and the effective radiation dose resulting from approved diagnostic procedures, detailed instructions for kit labeling, a method to verify the radiochemical purity after labeling before administration, the shelf-life, etc.

These indications and instructions cannot be modified by the user; any relevant modification, such as a new clinical indication, a modification in the posology or a change in the composition or pharmaceutical form, or any modification in the instructions for labeling, requires a new authorization granted by the same authority that authorized the first application.

As a result of regulations by directives and guidelines, all practical aspects related to a radiopharmaceutical are strictly governed by pharmaceutical rules: production, marketing authorization, distribution, clinical investigation, clinical use, etc.

11.1.2 Industrial Production

The industrial producer of a radiopharmaceutical product (radiopharmaceuticals ready for use, generators, radionuclide precursors, cold kits) must be authorized as pharmaceutical laboratory.

The person responsible for the production is a *qualified person*, pharmacist, or any equivalent academic (e.g., chemist, biochemist, biologist) with a pharmaceutical professional profile. This person is responsible for the starting materials, production, quality control, etc., and ensures the characteristics of the final products as stated in the approved requirements and characteristics or stated in the *European Pharmacopeia*. The production must be performed in agreement with GMP.

The quality and conformity of radiopharmaceuticals produced in any country of the European Union are guaranteed by their producer. For radiopharmaceuticals produced outside the European Union, it is necessary that a qualified person within the European Union (re)controls the product and ensures its correct production and quality.

Only officially authorized radiopharmaceuticals may be distributed within the European Union.

11.1.3 Marketing Authorization

The registration process is necessary for production, sale, and use of a medicinal product, including radiopharmaceuticals.

There are two registration procedures, depending on the type of marketing authorization that is intended. National authorization is valid in the particular member state where an application has been submitted. However, if marketing authorization is intended in the entire European Community, a centralized procedure must be applied by submitting an application to EMEA in London. Presentation and content of the dossier is contained in the rules governing medicinal products in the European Union: *The Rules Governing Medicinal Products in the European, Notice to Applicants, Volume 2A: Procedures for Marketing Authorization.* The requirements for the content of the application dossier are stated in Directive 65/65/EEC, as amended, and in Regulation (EEC) No. 2309/93.

The marketing authorization is granted after positive evaluation of the file. The authorization states all characteristics of the medicinal product, specified in the registration file and in the SPC.

Any change in the approved characteristics needs a new authorization.

11.1.4 Sales and Distribution

The distribution of medicinal products is regulated by the rules of GDP. According to that, only approved medicinal products may be distributed, and delivered exclusively to persons authorized for dispensing. In most countries, these persons are the pharmacists. However, in some countries radioactive compounds do constitute an exception to this rule, and have to be delivered directly to the physicians who use them for diagnostic or therapeutic purposes in patients (nuclear medicine physicians and radiotherapists).

The distribution mode is very important for the withdrawal from the market of any product if necessary (lack of the required quality, unexpected adverse effects, etc.).

11.1.5 Pharmacovigilance

Pharmacovigilance is a system for the urgent notification to the health authorities of any adverse effect observed in relationship to any medicinal product. Pharmacovigilance encompasses surveillance of side effects after short-term and long-term use of medicines. During the first 5 years of a new drug being put on the market, pharmacovigilance is particularly important, as comparatively little is known about its safety profile until it has been exposed to a much wider range of patients than is possible through clinical trials.

Pharmacovigilance is obligatory for producers, but in some countries it is obligatory for all involved health professionals as well. Radiopharmaceuticals are included in the scope of pharmacovigilance.

In each country, there is a national system for reporting adverse reactions. However, there is also a European Rapid Alert System, within the activities of the EMEA, to facilitate the transmission of incidents of pharmacovigilance.

11.1.6 Other Aspects

Other aspects concerning the preparation, distribution and use of radiopharmaceuticals are also subject to pharmaceutical regulations, such as pricing, labeling, and advertising, etc.

Despite strict legislation, frequently observed deviations do exist, such as delivery of radiopharmaceuticals to nonauthorized persons, application of nonapproved radiopharmaceuticals, application of radiopharmaceuticals for nonapproved clinical indications or using higher doses than authorized, and labeling of kits under different conditions than recommended by the manufacturer, etc.

11.1.7 SPC for Radiopharmaceutical Products

The authorized sequence of topics for the presentation of product information (package leaflet) according to Guideline III/9163/89, reissued 10/25/1993 are listed in Table 11.1.1.

Table 11.1.1. Authorized sequence of topics for the presentation of product information (package leaflet) according to European Economic Committee Guideline III/9163/89

Heading no.	Name
1	Name of the Product (Trade Name)
2	Qualitative and Quantitative Composition
	2.1 Active Substance(s)
	2.2 Physical Characteristics
3	Pharmaceutical Form
4	Clinical Particulars
	4.1 Clinical Indications
	4.2 Posology and Method of Administration
	4.3 Contraindications
	4.4 Special Warnings and Special Precautions for Use
	4.5 Interactions with Other Medicaments
	4.6 Pregnancy and Lactation
	4.7 Effects on the Ability to Drive and Use Machines
	4.8 Undesirable Effects
	4.9 Overdose
5	Pharmacological Properties
	5.1 Pharmacodynamic Properties
	5.2 Pharmacokinetics Properties
	5.3 Preclinical Safety Data
	5.4 Radiation Dosimetry
6	Pharmaceutical Particulars
	6.1 List of Excipient(s)
	6.2 Incompatibilities
	6.3 Shelf-life
	6.4 Special Precautions for Storage
	6.5 Nature of Contents of Container
	6.6 Instructions for Use/Handling (Labeling and QC Methods)
	6.7 Holder of the Marketing Authorization
7	Marketing Authorization Holder
8	Marketing Authorization Number
9	Date of First Authorization/Renewal

11.2 The *European Pharmacopeia*

A. Verbruggen

11.2.1 General

A pharmacopeia (from the Greek words *pharmakon* [drug] and *poeioo* [to work, to manufacture] is a collection of standardized specifications that define the quality of pharmaceutical preparations, their constituents, or even their containers. It is an official book stating not only the necessary characteristics of medicinal products, described in monographs, but also including monographs on methods of analysis, reagents, pharmaceutical forms, and other related items. The specifications described in pharmacopeias are not recommendations but are obligatory.

Within the European Union, the official pharmacopeia is the *European Pharmacopeia* (*Ph. Eur.*) from the Council of Europe. In most countries it is used as such, in some other countries it constitutes the basis for the national version of the pharmacopeia. The official texts of the *Ph. Eur.* are published in English and French. Translations in other languages may be prepared by the signatory states of the European Pharmacopeia Convention. In case of doubt or dispute, the English and French versions are alone authoritative.

The monographs of the *Ph. Eur.* are legally enforced in the countries being signatories to the Convention on the Elaboration of a European Pharmacopeia. The terminology used in the marketing authorization dossiers must be identical to that of the *Ph. Eur.* Likewise, raw materials, preparations, dosage forms of medicines and, if necessary, containers, must comply with the requirements of the *Ph. Eur.*

The *Ph. Eur.* was inaugurated in 1964 through a convention elaborated under the aegis of the Council of Europe. The eight founder countries were Belgium, France, Federal Republic of Germany, Italy, Luxembourg, the Netherlands, Switzerland, and the United Kingdom. It was considered that the free movement of medicines, for reasons of public health as well as international trade, requires that manufacturing and quality control standards be unified for pharmaceutical substances for human and veterinary use and that these standards keep pace with scientific progress. For this reason, the initiative was taken for unifying national pharmacopeias. Adherence to the *Ph. Eur.* should guarantee that the citizens of Europe receive the same quality of medicines.

As of start of 2006, the European Pharmacopeia Convention has been signed by 35 parties including the European Union and the following countries: Austria, Belgium, Bosnia and Herzegovina, Bulgaria, Croatia, Cyprus, the Czech Republic, Denmark, Estonia, Finland, the Former Yugoslav Republic of Macedonia, France, Germany, Greece, Hungary, Ireland, Iceland, Italy, Latvia, Lithuania, the Grand-Duchy of Luxembourg, Malta, Norway, the Netherlands, Portugal, Serbia and Montenegro (formerly Yugoslavia), Romania, the Slovak Republic, Slovenia, Spain, Sweden, Switzerland, Turkey, and the United Kingdom. In addition, 16 observers participate at a lower level, namely the WHO, four European states (Albania, Georgia, Poland, Ukraine), and 11 non-European states (Algeria, Australia, Canada, China, United States [Food and Drug Administration], Madagascar, Malaysia, Marocco, Senegal, Syria, Tunisia).

The quality standards of the *Ph. Eur.* have, therefore, an impact on the quality of medicines, which goes far beyond the European region. In absence of a description of a particular medicinal product in the *Ph. Eur.*, other relevant pharmacopeias may be considered, such as the *United States Pharmacopeia* (*USP*).

The new European Directives 2001/82/EC and 2001/83/EC on medicines for human use and veterinary use maintain the mandatory character of the *Ph. Eur.* monographs in the preparation of dossiers for marketing authorization of medicines, which was instituted in the first directive, Directive 75/318/EEC in 1975. It means that the monographs of the *Ph. Eur.* must therefore be updated to keep pace with products on the market, with scientific progress, and with regulatory developments.

11.2.2 Monographs on Radiopharmaceuticals in the *Ph. Eur.*

The *Ph. Eur.* defines requirements for purity and activity for a wide range of radiopharmaceutical products, whether they are used for therapeutic or diagnostic purposes. As of the beginning of 2006, there are:

- Two general monographs:
  - Radiopharmaceutical preparations (01/2005:0125). This monograph describes the principles and methods to be applied in the respective paragraphs of the individual monographs, e.g., how to measure the half-life of a radionuclide.
  - 5.7. Table of physical characteristics of radionuclides mentioned in the *Ph. Eur.* (01/2005:50700)
- 56 monographs on a specific radiopharmaceutical, of which:
  - 46 monographs on non–positron emission tomography (PET) radiopharmaceuticals
    - 18 ^{99m}Tc preparations
    - 11 radiopharmaceuticals labeled with iodine-123 or iodine-131
    - Preparations labeled with ^{32}P, ^{89}Sr, ^{3}H, ^{51}Cr, ^{57}Co, ^{58}Co, ^{67}Ga, ^{111}In, ^{81m}Kr, ^{201}Tl, or ^{133}Xe
  - 10 monographs on radiopharmaceuticals labeled with a positron emitting radionuclide

The requirement to guarantee the quality and to analyze radiopharmaceuticals according to the specifications set in these monographs applies to every producer within the countries adhering to the *Ph. Eur.* More specifically, this means that:

- For radiopharmaceuticals delivered to the (radio)pharmacist, nuclear medicine physician, or radiotherapist in a ready-to-use form (such as ^{201}TlCl, ^{18}FDG, ^{131}I-capsules, ^{123}I-iodide for radiolabeling, etc.), the manufacturer has to perform the analyses as described in the monographs.
- Radiopharmaceuticals prepared in a nuclear medicine department using licensed labeling kits (mostly ^{99m}Tc-labeled preparations) need not to be analyzed after their preparation according to the *Ph. Eur.* monograph, but rather, follow the instructions in the package leaflet. In most cases, this means a rather simple check of the radiochemical purity and in some cases, of the pH. However, if for any reason a full analysis is required (e.g., in case of serious side effects), these preparations must meet all requirements of the monograph.
- There are no monographs on labeling kits. Part of the quality control of licensed or locally produced labeling kits is that the manufacturer (industrial or in radiopharmacy) verifies whether their labeling according to the instructions results in a prep-

aration that meets the requirements of the corresponding *Ph. Eur.* monograph on the radiolabeled preparation.
- Up to now unlicensed radiopharmaceuticals (e.g., ^{11}C-raclopride, ^{18}F-fluoride for bone scintigraphy, etc.) the manufacturer, i.e., the local radiopharmacy, has to perform a full analysis according to the monograph of the *Ph. Eur.*

In its introductory general statements, the *Ph. Eur.* mentions that a preparation is not of *Ph. Eur.* quality unless it complies with *all* the requirements stated in the monograph. This does not imply that performance of all the tests in a monograph is necessarily a prerequisite for a manufacturer in assessing compliance with the *Ph. Eur.* before release of a product. The manufacturer may obtain assurance that a product is of *Ph. Eur.* quality from data derived, for example, from validation studies of the manufacturing process and from in-process controls. Parametric release *in circumstances deemed appropriate by the competent authority* is thus not precluded by the need to comply with the *Ph. Eur.* "Competent authority" means the national, supranational, or international body or organization vested with the authority for making decisions concerning the issue in question. It may, for example, be a national pharmacopeia authority, a licensing authority, or an official control laboratory. In practice, this means that the local pharmaceutical inspector has the right to decide whether or not particular tests of a monograph have to be performed before release of the product, unless the monograph specifies the reverse (as it is the case with the tests for sterility and bacterial endotoxins of radiopharmaceutical preparations for injection labeled with a short-lived radionuclide). As a result, there is no uniform rule with respect to the possibility of applying a parametric release of radiopharmaceuticals throughout Europe and a variety of attitudes in this respect exists. Harmonization of this situation in the European Union would be more than welcome.

The introductory general statements of the *Ph. Eur.* also clarify that the tests and assays described are the official methods upon which the standards of the *Ph. Eur.* are based. *With the agreement of the competent authority* (again), alternative methods of analysis may be used for control purposes, provided that the methods used enable an unequivocal decision to be made as to whether compliance with the standards of the monographs would be achieved if the official methods were used. In the event of doubt or dispute, the methods of analysis of the *Ph. Eur.* are alone authoritative.

11.2.3 Elaboration of New *Ph. Eur.* Monographs on Radiopharmaceutical Preparations

Monographs on radiopharmaceutical preparations are elaborated by Expert Group 14 of the European Pharmacopeia Commission, i.e., the group on radioactive compounds. Groups of experts are appointed by the Commission (the organ that takes all final decisions) for a renewable period of 3 years. The experts appointed to such groups are chosen for their personal competence on the proposal of their national delegations. The competent authority of each member state may propose one appointed member for each group of experts, taking account of the competence of the person proposed for the work involved. In addition, subject to approval by the Commission, specialists in a particular field (e.g., PET radiopharmaceuticals) may assist the appointed experts when supplementary technical advice is needed. At this moment, Group 14 has one expert from 11 European countries, three additional specialists in the field of PET-radiopharmaceuticals, a scientific officer from the European Pharmacopoeia Commission and a president.

The elaboration of a new monograph (or the revision of an existing monograph) on a radiopharmaceutical preparation requires several steps according to a strict procedure:

- The European Pharmacopeia Commission agrees on the elaboration of a new monograph or the revision of an existing monograph by Group 14; proposals concerning the introduction of monographs, general chapters, and other texts into the *Ph. Eur.* may be made by the chair of the Commission, a national delegation, or a group of experts through the intermediary of its chair.
- The Group designates a rapporteur and, if necessary, a corapporteur for the monograph to be elaborated.
- The European directorate for the quality of medicines (EDQM), in charge of the secretariat of the European Pharmacopeia Commission, is responsible initially for finding and compiling information, and it asks the manufacturers and suppliers for samples for analysis and in some cases, for use as reference substances. In the case of radioactive compounds, the samples are sent directly to the experts, as the laboratory of the EDQM is not allowed to handle radioactivity.
- The expert rapporteur initiates the required work, if necessary, with the collaboration of manufacturer(s), user(s), the corapporteur, and possibly other members of Group 14.
- Based on the laboratory results, the rapporteur elaborates a first draft according to the *Technical Guide for the Elaboration of Monographs* and the *Style Guide* of the *Ph. Eur.*

  The *Technical Guide* explains the principles that govern the elaboration of European monographs, and it can also be used in preparing registration files. It is continually being revised and updated. As an example, the *Technical Guide* specifies that, preferentially, liquid chromatography (LC) must be used for control of organic impurities, and thin-layer chromatography should be reserved for control of specific impurities that cannot conveniently be controlled by LC or gel chromatography.

  The aim of the *Style Guide* is to provide the means of drafting clear unambiguous texts, with similar requirements presented in the same way in every monograph. It is considered that a uniform style is of great help in conveying information in an easily understandable and unambiguous manner. An analyst who has already carried out a test prescribed in the *Ph. Eur.* will find it easier to set up and carry out a similar test presented in the same way.
- The draft monograph is discussed within Group 14 in one or mostly several sessions. The Group examines the text and if necessary, carries out additional experimentation which may involve the participation of the EDQM laboratory (only for work not involving radioactivity).
- Once a consensus is reached within the group of experts, the monograph (in English and French) is subjected to an international public survey procedure which takes 6 months and involves the national pharmacopeia authorities of the member states and all the readers of *Pharmeuropa.*

  Pharmeuropa is the official journal of the EDQM (four issues per year) and aims to be a forum for the publication of draft texts or monographs elaborated by the *Ph. Eur.* groups of experts, and the regular presentation and communication of general policy. The issues also include information on forthcoming conferences and

training courses, public inquiries, news and updates on texts undergoing international harmonization, and scientific notes. It allows the users in the field to share their experience and give their views and input on monographs and other texts being elaborated by *Ph. Eur.* groups of experts.
- The comments received during the public survey are discussed and examined within Group 14, and the monograph is amended if necessary.
- Once the monograph has gone through these various steps, it is submitted to the European Pharmacopeia Commission for adoption.
- The Commission also proposes the date for common entry into force, which is ratified by a resolution of the Public Health Committee of the Council of Europe.

Apart from the elaboration of new monographs, existing texts are regularly updated, taking into account the changes in marketed products, scientific progress, comments from users, and authorities. Such revisions are made at the request either of a public health authority or of an industrialist, one of the delegations or the chair of the group of experts.

11.3 European Union Legislation Concerning New Drug Development

A. Verbruggen

11.3.1 EU Commission Directives and Guidelines

On 1 May 2004, EU Directive 2001/20/EC (European Parliament and of the Council of the European Union) came into force in the European countries, laying down the principles of good clinical practice (GCP) in the conduct of clinical trials on medicinal products for human use. At this moment, the provisions of this directive have been transposed into national laws in most European member states, while complementary guidance documents on the submission, content, and format of clinical trial applications (CTAs) have been issued by the EU Commission (European Commission 2003a, b; European Medicines Agency 2004). In addition, EU Directive 2005/28/EC (European Commission 2005) was published on 8 April 2005, which not only defined the principles and detailed guidelines for GCP regarding investigational medicinal products for human use, but also set out the requirements for authorization of the manufacturing or importation of such products.

As a consequence of the new legislation for clinical trials, documentation on the quality and preclinical (i.e., toxicological) data of investigational medicinal products, including radiopharmaceuticals, needs to be submitted to obtain approval from the national health authorities in the member states prior to initiating a clinical study in humans. Furthermore, all clinical trials which started in the European Union after 1 May 2004 need to be recorded in the European EudraCT database (European Commission 2003b).

11.3.2 Clinical Trials

Clinical trials are studies to evaluate the effectiveness, pharmacokinetics, and safety of medications or medical devices by monitoring their effects on large groups of people. Clinical research trials may be conducted by government health agencies, researchers affiliated with a hospital or university, independent researchers, or private industry. The use of an approved medicinal product in nonapproved conditions (different administration route, dose, clinical indications, etc.) requires also a new clinical trial.

All clinical trials must be carried out according to a protocol approved by an ethical committee and according to the GCP rules. Participants must be volunteers. Clinical trials with radiopharmaceuticals must be performed in suitable patients rather than in healthy people in order to avoid any unnecessary exposure to radiation. In addition, administrative authorization is needed in some countries, as well as other specific requirements. Only results obtained by authorized clinical trials may be used in support of a clinical indication for an application for marketing authorization.

According to the pharmaceutical legislation and European guidelines related to CTAs, a written approval should also be obtained for the clinical data (i.e., study protocol) related to a specific CTA. However, although evaluation and approval of the study protocol are generally being performed by the independent (leading) ethics' committee, some exceptions do exist in certain countries. In any case, for multicenter clinical trials, approval for the quality, safety, and efficacy of the investigational drug product should be sought in every country involved in the trial.

Since Directive 2001/20/EC was translated into country-specific national laws, the quality requirements for medicinal products such as radiopharmaceuticals used in clinical studies often differ to a certain extent from one country to another. In fact, the same holds true for the current legislation regarding preparation, dispensing, and use of radiopharmaceuticals, radioactive precursors, or radionuclides that are routinely used in hospitals for diagnostic or therapeutic purposes. In addition, the scope of the national legislation, which is based on the definition of an investigational medicinal product, may also differ from the scope of the corresponding EU legislation and therefore may have substantial impact on the clinical research in each member state.

11.3.3 CTA Requirements

Evidently, the existing differences in the EU countries relating to the regulatory documentation required for CTAs using radiopharmaceuticals often provide additional administrative burden, in particular in the case of multicentric studies, and are often experienced by (non)academic investigators as "overregulation" from the national health authorities. In addition, the currently available chemical-pharmaceutical quality requirements for investigational radiopharmaceutical products are covered by the general EMEA draft guideline regarding investigational medicinal products in clinical trials (European Medicines Agency 2004), which provides only a limited level of detailed guidance for ready-to-use compounds (e.g., ^{67}Ga-citrate), kit-based radiopharmaceuticals (e.g. ^{99m}Tc-mertiatide), or PET radiopharmaceuticals (e.g., ^{18}F-fludeoxyglucose). Subsequently, this lack of detailed information often leads to considerable uncertainty and misinterpretation by professionals in this particular research field who are regularly faced with substantial difficulties when seeking approval from the national authorities to allow the start of a clinical study (e.g., due to submission of insufficient or inappropriate quality data).

Another point of concern relates to the impact of the new legislation on clinical trials performed by noncommercial research centers. Many academic research centers developing new radiopharmaceuticals might not have the necessary regulatory experience or time to compile a CTA according to the current European requirements, and often do not have the facilities or budget to produce small batches of radioligands for clinical use under full GMP conditions. Moreover, a lot of confusion still seems to exist at the academic level on the liabilities of the investigator and sponsor. Evidently, the currently existing difficulties and uncertainties experienced by academics impede the broader development and clinical use of new radiopharmaceuticals in Europe, and of course, the marketing of such investigational drug products.

An even bigger lack of clarity is experienced by professionals in the field concerning the regulations to be followed for the manufacturing, dispensing, and use of radiopharmaceuticals or radionuclides that are being prepared in hospital radiopharmacies or centralized radiopharmacies for daily clinical practice in hospitals. In most cases, it concerns formulations for intravenous use that are prepared in house due to the short half-life of the used radioisotope, under the responsibility of a qualified person (QP). As a consequence, a variety of procedures are being used across European countries, since a lot of uncertainty still exists on the requirements with respect to preparation conditions, quality of starting materials, analyses of the finished product to be performed, as well as on the technical profile of the qualified person.

Acknowledgment

The authors thank Dr. Christophe Lahorte for the useful discussions and his valuable information, which contributed significantly to the content of this chapter.

Appendix

A.1 List of European Economic Council Directives and Regulations Applied to Radiopharmaceuticals

These documents are available in the current version at http://dg3.eudra.org/eudralex

Council Directive 65/65/EEC of 26 January 1965, on the approximation of provisions laid down by law, regulation, or administrative action relating to medicinal products

Council Directive 75/318/EEC of 20 May 1975, on the approximation of the laws of member states relating to analytical, toxicopharmacological and clinical standards, and protocols in respect of the testing of medicinal products

Council Directive 75/319/EEC of 20 May 1975, on the approximation of provisions laid down by law, regulation, or administrative action relating to medicinal products

Directive 87/22/EEC of 22 December 1986, on the approximation of national measures relating to the placing on the market of high-technology medicinal products, particularly those derived from biotechnology (centralized marketing authorization of medicines)

Council Directive 89/341/EEC of 3 May 1989, amending Directives 65/65/EEC, 75/318/EEC and 75/319/EEC, on the approximation of provisions laid down by law, regulation, or administrative action relating to proprietary medicinal products

Council Directive 89/343/EEC of 3 May 1989, extending the scope of Directives 65/65/EEC and 75/319/EEC, and laying down additional provisions for radiopharmaceuticals

Commission Directive 91/356/EEC of 13 June 1991, laying down the principles and guidelines of good manufacturing practice for medicinal products for human use
Council Directive 92/25/EEC of 31 March 1992, on the wholesale distribution of medicinal products for human use
Council Directive 92/26/EEC of 31 March 1992, concerning the classification for the supply of medicinal products for human use
Council Regulation (EEC) No 2309/93 of 22 July 1993, laying down Community procedures for the authorization and supervision of medicinal products for human and veterinary use and establishing a European Agency for the Evaluation of Medicinal Products
Committee of Proprietary Medicinal Products (CPMP) Note for guidance III/9163/89, reissued 25 October 1993, summary of the product characteristics
Council Directive 84/466/EURATOM of 3 September 1984, laying down basic measures for the radiation protection of persons undergoing medical examination or treatment
Council Directive 84/467/Euratom of 3 September 1984, amending Directive 80/836/EURATOM, regarding the basic safety standards for the health protection of the general public and workers against the dangers of ionizing radiation

A.2 Good Practice in the Manufacture of Pharmaceuticals (from Kristensen 1979)

Basic standards of good manufacturing practice for pharmaceutical products (1972) Report PHI/72, EFTA Secretariat, Geneva
Good practice in the manufacture and quality control of drugs (1977) World Health Organization, Resolution WHO 28.65, WHO Chronicle 31 (Suppl.)
Guidelines for the manufacture of sterile products, annex to the basic standards of good manufacturing practice for pharmaceutical products (1973) Report PHI/73, EFTA Secretariat, Geneva
Guide to good pharmaceutical manufacturing practice (1977) Department of Health and Social Security (HMSO), London
Manual on radiation sterilization of medical and biological materials (1973) Technical Reports Series No. 149. International Atomic Energy Agency, Vienna
Hygiene Recommendations (1972) Drug manufacturers association, Stockholm (Dec. 1972)
United States Food and Drug Administration, Code of Federal regulations, Title 21, Part 210–211

A.3 Good Radiopharmacy Practice (from Kristensen 1979)

Kristensen K (1975) The quality of radiopharmaceuticals. A review of current problems in quality factors in nuclear medicine. 13th international meeting of the Society of Nuclear Medicine. Copenhagen, pp 135.1–135.17
Kristensen K (1979) Preparation and control of radiopharmaceuticals in hospitals. Technical Reports Series No. 194. International Atomic Energy Agency, Vienna
Kristensen K, Müller T, Rhodes BA (1977) Good radiopharmacy practice, quality control in nuclear medicine. Mosby, St. Louis, Mo., pp 268–275
Good pharmaceutical manufacturing practice applied to the hospital preparation of radiopharmaceuticals (1977) Dept. of Health and Social Security, notes for guidance issued by the Medicines Inspectorate, London
Good practice in the manufacture of radiopharmaceuticals (1971) A proposal put forward by a group of pharmacists from the Nordic countries. Arch Pharm Chem 78:1002–1009
Guidelines for the preparation of radiopharmaceuticals in hospitals (1975) British Institute of Radiology, London, Special Report No. 11
Hospital Physicists' Association (1976) The hospital preparation of radiopharmaceuticals. Scientific Report Series 16
Isotope Pharmacy (1976) On the preparation of Tc-99m labelled radiopharmaceuticals. Rep. IA-5, Copenhagen
Trott NG (1978) The preparation of radiopharmaceuticals in hospitals. In: Höfer R, Bergmann H (eds) Radioactive isotopes in clinical medicine and research, vol. 13. Egermann, Vienna, pp 517–528

References

European Commission (2003a) Detailed guidance for the request for authorization of a clinical trial on a medicinal product for human use to the competent authorities, notification of substantial amendments and declaration of the end of the trial. (http://pharmacos.eudra.org/F2/pharmacos/docs/Doc2005/10_05/CA_14-2005.pdf)

European Commission (2003b) Detailed guidance on the European clinical trials database (EudraCT Database). http://pharmacos.eudra.org/F2/pharmacos/docs/Doc2003/april/cp-guidance-eudract_ 230403.pdf)

European Commission (2005) Commission Directive 2005/28/EC of 8 April 2005 laying down principles and detailed guidelines for good clinical practice as regards investigational medicinal products for human use, as well as the requirements for authorisation of the manufacturing or importation of such products.

European Medicines Agency (2004) CPMP/CHMP/QWP185401 Draft guideline on the requirements to the chemical and pharmaceutical quality documentation concerning investigational medicinal products in clinical trials. (http://www.emea.eu.int/Inspections/docs/18540104en.pdf)

European Parliament and the Council of the European Union (2001) Directive 2001/20/EC of the European Parliament and of the Council of 4 April 2001 on the approximation of the laws, regulations and administrative provisions of the Member States relating to the implementation of good clinical practice in the conduct of clinical trials on medicinal products for human use. (http://pharmacos.eudra.org/F2/eudralex/vol-1/DIR_2001_20/DIR_2001_20_EN.pdf)

Kristensen K (1979) Preparation and control of radiopharmaceuticals in hospitals. Technical Reports Series No. 194. International Atomic Energy Agency, Vienna

Part II

Chapter 12

Monographs of ^{99m}Tc Pharmaceuticals

12

12.1 ^{99m}Tc-Pertechnetate

I. Zolle and P.O. Bremer

Chemical name	**Chemical structure**
Sodium pertechnetate Sodium pertechnetate ^{99m}Tc injection (fission) (*Ph. Eur.*) Technetium Tc 99m pertechnetate injection (*USP*) ^{99m}Tc(VII)-Na-pertechnetate	Pertechnetate anion ($^{99m}TcO_4^-$)
Physical characteristics	**Commercial products**
$E\gamma = 140.5$ keV (IT) $T_{1/2} = 6.02$ h	^{99}Mo/^{99m}Tc generator: GE Healthcare Bristol-Myers Squibb Mallinckrodt/Tyco

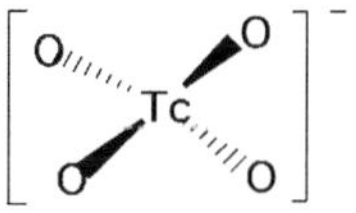

Preparation

Sodium pertechnetate ^{99m}Tc is eluted from an approved ^{99}Mo/^{99m}Tc generator with sterile, isotonic saline. Generator systems differ; therefore, elution should be performed according to the manual provided by the manufacturer. Aseptic conditions have to be maintained throughout the operation, keeping the elution needle sterile. The total eluted activity and volume are recorded at the time of elution. The resulting ^{99m}Tc activity concentration depends on the elution volume.

Sodium pertechnetate ^{99m}Tc is a clear, colorless solution for intravenous injection. The pH value is 4.0–8.0 (*Ph. Eur.*).

Description of Eluate

^{99m}Tc eluate is described in the *European Pharmacopeia* in two specific monographs depending on the method of preparation of the parent radionuclide ^{99}Mo, which is generally isolated from fission products (Monograph 124) (Council of Europe 2005 a), or produced by neutron activation of metallic ^{98}Mo-oxide (Monograph 283) (Council of Europe 2005 b). Sodium pertechnetate ^{99m}Tc injection solution satisfies the general requirements of parenteral preparations stated in the *European Pharmacopeia* (Council of Europe 2004).

The specific activity of ^{99m}Tc-pertechnetate is not stated in the *Pharmacopeia*; however, it is recommended that the eluate is obtained from a generator that is eluted regularly, every 24 h. Details for obtaining carrier-free technetium from the ^{99}Mo/^{99m}Tc generator and the resulting radionuclidic purity of eluates are discussed in Chap. 5, Sect. 5.1.1.

Pertechnetate anion ($^{99m}TcO_4^-$) is stable in aqueous solutions. It is chemically not reactive; its ability to form ligand complexes depends on the reduction to lower valence states (Steigman and Eckelman 1992; Steigman and Richards 1974; Steigman et al. 1975). The major use of this short-lived radionuclide in nuclear medicine to date is for the ad hoc preparation of ^{99m}Tc pharmaceuticals, which is performed with kits containing stannous ion as a reducing agent (Johannsen and Narasimhan 1992; Lin and Winchell 1972; Lin et al. 1971).

Clinical Applications

^{99m}Tc(VII)-pertechnetate is used after intravenous injection:

- Thyroid scintigraphy
  - Determination of technetium uptake and morphology
  - Diagnosis and localization of hot/cold nodules
- Salivary gland scintigraphy
  - To asses salivary gland function and duct status
- Imaging of gastric mucosa
  - To diagnose ectopic gastric mucosa (Meckel's diverticulum)
- Brain scintigraphy
  - Visualization of brain lesions when the blood-brain barrier (BBB) is defective
- Lachrimal duct scintigraphy
  - To evaluate nasolachrimal drainage
- In vivo labeling of RBC
  - Regional blood pool imaging
  - First-pass cardiac radionuclide angiography (ejection fraction, wall motion)
  - Detection of occult gastrointestinal bleeding

Sodium pertechnetate Tc-99m was introduced for scanning the thyroid (Harper 1964) and for brain scanning (McAfee et al. 1964; Quinn 1965), primarily because of its physical properties; other applications followed (Harper et al. 1966). The striking similarity of the heptavalent anion with iodide, however, has made pertechnetate an excellent radionuclide for thyroid scanning and for the study of thyroid physiology (Andros et al. 1965; Kusic et al. 1990).

In vivo labeling of red blood cells (RBC) with sodium ^{99m}Tc-pertechnetate (Callahan et al. 1982) is performed subsequent to pretreatment of RBC in vivo with a stannous reducing agent, using stannous pyrophosphate cold kits (TechneScan PYP, AngioCis) as well as Amerscan Stannous Agent.

Time of Examinations

- Thyroid scintigraphy is performed 20 min after intravenous injection.
- Salivary gland scintigraphy should begin immediately after intravenous injection and at regular intervals up to 60 min.
- Meckel's diverticulum scintigraphy should commence immediately after intravenous injection and at regular intervals up to 30 min.
- For brain scintigraphy, sequential images are taken immediately within the first minute after intravenous injection; static imaging is performed 1-h later.
- Lachrimal duct scintigraphy's dynamic imaging should begin immediately after tracer application for 10 min.
- For in vivo RBC labeling, the blood pool scintigraphy should start 10 min after intravenous injection of a bolus of ^{99m}Tc-pertechnetate: cardiac dynamic imaging should begin immediately, and abdominal imaging should also begin immediately and at various times up to 24 h.

Recommended Activities for Indications. The activity range for intravenous administration in patients (70 kg) is:

- Thyroid scintigraphy: 75 MBq
- Salivary gland scintigraphy: 40 MBq
- Meckel's diverticulum: 185 MBq
- Brain scintigraphy: 550 MBq, after blocking thyroid and choroid plexus to avoid nonspecific uptake of ^{99m}Tc-pertechnetate
- Lachrimal duct scintigraphy: 2–4 MBq instilled into each eye
- In vivo RBC labeling: 740 MBq, after pretreatment with a stannous agent

Pediatric Dose. The amount of radioactivity for infants and children administered for myocardial scintigraphy is based on body weight, using the scaling factors given in Appendix 1 (Table A1.2). The minimum value of the scaling factor is 0.1, since activities below 10% of the value of the adult radioactivity have resulted in poor image quality.

Additional Information

Iodinated contrast agents or iodine-containing medication interfere with ^{99m}Tc-pertechnetate thyroid imaging.

Persistent vascular activity of ^{99m}Tc-pertechnetate has been observed when scintigraphy has been performed following a bone scan, due to radiolabeled RBC that had retained stannous ions (Ancri et al. 1977; Montelibano et al. 1979).

Several drugs interfere with the normal biodistribution of ^{99m}Tc-pertechnetate (Hladik et al. 1987). Thus, cancer chemotherapeutic agents (methotrexate) can affect brain scintigraphy; atropine, isoprenaline, and analgesics interfere in abdominal imaging; iodine and other blockers (perchlorate, perrhenate) can modify thyroid uptake.

Quality Control

Radiochemical Purity. Sodium pertechnetate [^{99m}Tc] must satisfy the requirements stated in the *European Pharmacopeia* (Council of Europe 2004a). More than 95% of ^{99m}Tc activity must be present as pertechnetate anion.

Paper Chromatography. The *European Pharmacopeia* describes descending paper chromatography using methanol/water (80:20 v/v) as solvent; developing time is 2 h. The ^{99m}Tc-pertechnetate anion migrates with an R_f value of 0.6. More than 95% of the measured radioactivity corresponds to an R_f of 0.6; less than 5% are detected at the start.

Recommended Methods by the Manufacturer

Thin-layer chromatography		
Stationary phase:	Gelman ITLC-SG (fiberglass), 2×9.5 cm	
Solvent:	Saline (0.9% NaCl)	
Developing time:	10 min	
R_f values:	^{99m}Tc reduced, hydrolized:	0.0–0.1
	^{99m}Tc-Na-pertechnetate:	0.9–1.0 (>99%)

Thin-layer chromatography on silica gel plates offers reliable results faster; ^{99m}Tc-pertechnetate migrates with the solvent (saline) front; reduced, hydrolized activity remains at the start. The results of radiochemical purity generally exceed 99%.

Radionuclidic Purity. The *European Pharmacopeia* and national regulatory agencies recommend determination of ^{99}Mo in the primary eluate to assure high quality of generator eluates (DIN 6854).

Generators are eluted after shipment, before administration of eluates to patients. The primary eluate contains the highest concentration of chemical impurities and of carrier ^{99}Tc (decay product). Also, parent ^{99}Mo is highest in the first eluate (Hammermayer et al. 1986).

Less than 0.1% of the total ^{99m}Tc activity is due to parent ^{99}Mo, and not more than 0.01% is due to other radionuclidic impurities. These limits are however, never observed with the available generator systems.

Determination of ^{99}Mo Impurity. Determination of ^{99}Mo should be performed with a sample of the fresh eluate (37 MBq) by γ-spectrometry. A lead absorber 6-mm thick is placed between the sample and the NaI-detector. The fraction of γ-radiation measured at 740 keV (^{99}Mo) should not exceed the reading obtained with a reference (^{99}Mo) of 37 kBq.

The purity of generator eluates is directly related to the performance of a generator system (see Chap. 5, Sect. 5.1.1).

Pharmacokinetic Data

The monovalent anion pertechnetate is actively transported into the thyroid in a manner similar to iodide; however, it is not metabolized and is released from the thyroid unchanged. It has been demonstrated that the active transport of iodide and pertechnetate share the same carrier mechanism (Wolff and Maurey 1962). Uptake in the thyroid gland is between 1.5 and 2% of the injected activity within 20 min (Andros et al. 1965).

Assuming an activity of 74 MBq for thyroid scintigraphy, the concentration of pertechnetate in blood has been calculated as 2.5×10^{-12} *M*, based on a volume of distribution of 20 l. This concentration is far below the value required to saturate the carrier-mediated uptake mechanism. However, iodinated contrast agents or iodine-containing medication do affect serum levels of iodide (up to 3×10^{-4} *M*) and thus would interfere with ^{99m}Tc-pertechnetate thyroid imaging (Loberg 1979).

In blood, 70–80% of ^{99m}Tc-pertechnetate is bound to proteins. Perchlorate has been shown to displace pertechnetate from plasma protein-binding sites (Oldendorf et al. 1970). The unbound fraction is preferentially concentrating in the thyroid gland and other related structures, such as salivary glands, gastric mucosa, choroid plexus, and mammary tissue. ^{99m}Tc-pertechnetate is selectively excluded from the cerebrospinal fluid (Andros et al. 1965).

Elimination of ^{99m}Tc-pertechnetate from plasma after intravenous or oral administration (same subject) showed disappearance curves with an initial fast elimination of radioactivity; 50–60% are cleared with a half-time of 15 min; the remainder is eliminated more slowly, with half-times of approximately 3 h. After oral administration, the highest value of ^{99m}Tc activity in blood was reached within 30 min; the activity level was approximately one half of the radioactivity measured at the same and subsequent times after intravenous injection (Andros et al. 1965). The rates of disappearance for both whole blood and plasma were shown to be the same (Prince et al. 1980).

Pertechnetate is excreted by the kidneys, but other pathways may be relevant in specific circumstances, such as saliva, gastric juice, milk, sweat, etc. (Hays 1973). Lactating women secrete 10% of pertechnetate in milk (Ahlgren et al. 1985). Pertechnetate crosses the placental barrier.

A major difference between the kinetics of pertechnetate and iodide in humans is the excretion pattern; only pertechnetate is excreted in the feces (Andros et al. 1965; Hays and Berman 1977). Renal excretion by glomerular filtration is observed in the first 24 h after administration; 25–30% of the injected activity is recovered in the urine. During the following 48–72 h, fecal excretion predominates and may amount to 35% of the injected radioactivity. A total of approximately 60% of the administered radioactivity is recovered in urine and feces in 72 h; approximately 40% is retained in the body, mainly in the digestive tract. The whole-body biological half-time is estimated to be 53 h (Andros et al. 1965). Tubular reabsorption of pertechnetate was determined as 86.5% of the filtered amount. ^{99m}Tc-pertechnetate is excreted unchanged (Dayton et al. 1969).

In certain clinical situations, when thyroidal uptake of ^{99m}Tc-pertechnetate should be avoided, pretreatment with an oral dose of potassium perchlorate is used to inhibit uptake. Perchlorate anion shows greater affinity for the transporter than does iodide and is, therefore, a competitive inhibitor of the thyroid iodide trap (Wolff and Maurey 1962). Like pertechnetate, perchlorate is concentrated in the thyroid and is not metabolized. Assuming 100% resorption of an oral dose of 300 mg of $KClO_4$ and an initial distribution volume of 20 l, the resulting concentration of perchlorate in blood is approximately 10^{-4} *M*, sufficient for saturation of the thyroid trapping mechanism (Loberg 1979).

The effect of pretreatment with perchlorate and iodide on pertechnetate pharmacokinetics has been studied (Oldendorf et al. 1970; Prince et al. 1980; Welch et al. 1969).

^{99m}Tc-pertechnetate cannot pass through the intact BBB, but in areas of the brain where structural defects permit diffusion, uptake has been used as an indicator of vascular and neoplastic brain lesions (Jhingram and Johnson 1973).

^{99m}Tc-pertechnetate has an affinity for RBC that have been treated (in vivo) with a reducing agent, causing "stannous loading". Approximately 95% of the administered ^{99m}Tc activity is taken up by RBC. Unbound pertechnetate is excreted by the kidneys. Approximately 15% of the activity is excreted in the urine during the first day (Porter et al. 1983). Radioactivity is removed from blood with a half-time of 60 h by renal excretion (International Commission on Radiological Protection 1987).

Radiation Dose

The radiation exposure after the intravenous administration of ^{99m}Tc-pertechnetate depends on the thyroid status, and whether a blocking agent has been administered. The thyroid gland, stomach wall, small intestine, upper and lower intestinal wall, and urinary bladder wall are the most exposed organs.

The effective (whole body) dose equivalent for pertechnetate ^{99m}Tc is 0.013 mSv/MBq (International Commission on Radiological Protection 1987). The effective dose in adults (70 kg) resulting from 75 MBq of intravenously injected ^{99m}Tc-pertechnetate for thyroid scintigraphy is approximately 1 mSv. The absorbed radiation dose to the thyroid (without a blocking agent) resulting from an intravenous injection of 75 MBq of ^{99m}Tc-pertechnetate corresponds to 1.7 mGy.

The effective dose in adults (70 kg) resulting from 40 MBq of intravenously injected ^{99m}Tc-pertechnetate for salivary gland scintigraphy is approximately 0.5 mSv.

Lachrimal duct scintigraphy using 4 MBq of ^{99m}Tc-pertechnetate corresponds to an effective dose of 0.05 mSv. The absorbed radiation dose to the optical lens is given as 0.038 mGy/MBq. Using 4 MBq of ^{99m}Tc-pertechnetate, the absorbed radiation dose to the lens corresponds to 0.15 mGy.

Diagnosis of ectopic gastric mucosa using 185 MBq of intravenously injected ^{99m}Tc-pertechnetate corresponds to an effective dose of approximately 2.4 mSv.

The effective dose in adults (70 kg) resulting from 550 MBq of intravenously injected ^{99m}Tc-pertechnetate for brain scintigraphy (after blocking thyroid and choroid plexus) is approximately 2.9 mSv. The effective (whole body) dose equivalent (with blocking agent) is 0.0053 mSv/MBq.

^{99m}Tc-Labeled Erythrocytes. The effective (whole body) dose equivalent for ^{99m}Tc-labeled erythrocytes is 0.0085 mSv/MBq (International Commission on Radiological Protection 1987). The effective dose in adults (70 kg) resulting from 740 MBq of intravenously injected ^{99m}Tc-pertechnetate for angioscintigraphy is approximately 6.3 mSv. The absorbed radiation dose to the heart resulting from an intravenous injection of 740 MBq of ^{99m}Tc-pertechnetate corresponds to 17.0 mGy. The absorbed radiation dose to the kidneys is corresponding to 7.4 mGy.

Storage and Stability

Storage. Sodium pertechnetate ^{99m}Tc injection is stored at room temperature with shielding.

Stability. Sodium pertechnetate ^{99m}Tc injection may be used up to 6 h after elution if the generator had been eluted within 24 h.

References

Ahlgren L, Ivarsson S, Johansson L, Mattsson S, Nosslin B (1985) Excretion of radionuclides in human breast milk after administration of radiopharmaceuticals. J Nucl Med 26:1085–1090

Ancri D, Lonchampt M, Basset J (1977) The effect of tin on the tissue distribution of Tc-99m sodium pertechnetate. Radiology 124:445–450

Andros G, Harper PV, Lathrop KA, McCardle RJ (1965) Pertechnetate-99m localization in man with application to thyroid scanning and the study of thyroid physiology. J Clin Endocrinol Metab 25:1067–1076

Callahan RJ, Froelich JW, McKusick KA, Leppo J, Strauss WH (1982) A modified method for the in vivo labeling of red blood cells with Tc-99m: concise communication. J Nucl Med 23:315–318

Council of Europe (2005 a) Sodium pertechnetate Tc-99m injection (fission), monograph 124. European pharmacopeia. Council of Europe, Maisonneuve, Sainte-Ruffine

Council of Europe (2005 b) Sodium pertechnetate Tc-99m injection, monograph 283. European pharmacopeia. Council of Europe, Maisonneuve, Sainte-Ruffine

Council of Europe (2004) Preparations for parenteral use, monograph 520. European pharmacopeia. Council of Europe, Maisonneuve, Sainte-Ruffine

Dayton DA, Maher FT, Elveback LR (1969) Renal clearance of technetium (^{99m}Tc) as pertechnetate. Mayo Clin Proc 44:549–551

Hammermaier A, Reich E, Bögl W (1986) Chemical, radiochemical, and radionuclidic purity of eluates from different commercial fission ^{99}Mo/^{99m}Tc-generators. Eur J Nucl Med 12:41–46

Harper PV, Beck R, Charleston D, Lathrop KA (1964) Optimization of scanning method using ^{99m}Tc. Nucleonics 22:50–54

Harper PV, Lathrop KA, Gottschalk A (1966) Pharmacodynamics of some technetium-99m preparations. In: Andrews GA, Knisely RM, Wagner HN Jr (eds) Radioactive pharmaceuticals. AEC Symposium Series Conf 651111 1966, pp 335–357

Hays MT (1973) ^{99m}Tc-pertechnetate transport in man: absorption after subcutaneous and oral administration; secretion into saliva and gastric juice. J Nucl Med 14:331–335

Hays MT, Berman M (1977) Pertechnetate distribution in man after intravenous infusion: a compartmental model. J Nucl Med 18:898–904

Hladik III WB, Ponto JA, Lentle BC, Laven DL (1987) Iatrogenic alterations in the biodistribution of radiotracers as a result of drug therapy: reported instances. In: Hladik WB III, Saha GB, Study KT (eds) Essentials of nuclear medicine science. Williams & Wilkins, Baltimore, pp 189–219

International Commission on Radiological Protection (1987a) Pertechnetate. In: Annals of the ICRP, radiation dose to patients from radiopharmaceuticals, biokinetic models and data. ICRP publication 53, vol. 18, no. 1–4. Pergamon, Oxford, pp 197–200

International Commission on Radiological Protection (1987b) Technetium-labelled erythrocytes. In: Annals of the ICRP, radiation dose to patients from radiopharmaceuticals, biokinetic models and data. ICRP publication 53, vol. 18, no. 1–4. Pergamon, Oxford, pp 209–210

Jhingram SG, Johnson PC (1973) Radionuclide angiography in the diagnosis of cerebrovascular disease. J Nucl Med 14:265–268

Johannsen B, Narasimhan DVS eds (1992) Preparation of kits for ^{99m}Tc-radiopharmaceuticals. International Atomic Energy Agency: IAEA-TECDOC-649

Kusic Z, Becker DV, Saenger EL, Paras P, Gartside P, Wessler T, Spaventi S (1990) Comparison of technetium-99m and iodine-123 imaging of thyroid nodules: correlation with pathological findings. J Nucl Med 31:393–399

Lin SM, Winchell HS (1972) A "kit" method for the preparation of a technetium-tin(II) colloid and a study of its properties. J Nucl Med 13:58–65

Lin SM, Winchell HS, Shipley BA (1971) Use of Fe(II) or Sn(II) alone for technetium labelling of albumin. J Nucl Med 12:204–211

Loberg MD (1979) Radiotracer distribution by active transport: the implications of nonlinear kinetics. In: Colombetti LG (ed) Principles of radiopharmacology, vol. III. CRC Press, Boca Raton, pp 43–59

McAfee JG, Fueger GF, Stern HS, Wagner Jr, HN, Migita T (1964) ^{99m}Tc-pertechnetate for brain scanning. J Nucl Med 5:811–827

Montelibano EB, Ford DR, Sayle BA (1979) Altered Tc-99m pertechnetate distribution in a thyroid scan after Tc-99m pyrophosphate administration. Clin Nucl Med 4:277–278

Oldendorf WH, Sisson WB, Iisaka Y (1970) Compartmental redistribution of ^{99m}Tc-pertechnetate in the presence of perchlorate ion and its relation to plasma protein binding. J Nucl Med 11:85–88

Porter WC, Dees SM, Freitas JE, Dworkin HJ (1983) Acid-citrate-dextrose compared with heparin in the preparation of in vivo/in vitro technetium-99m red blood cells. J Nucl Med 24:383–387

Prince JR, Bancroft S, Dukstein WG (1980) Pharmacokinetics of pertechnetate administered after pretreatment with 400 mg of potassium perchlorate: concise communication. J Nucl Med 21:763–766

Quinn JL (1965) ^{99m}Tc-pertechnetate for brain scanning. Radiology 84:354–355

Steigman J, Eckelman WC (1992) The chemistry of technetium in medicine. Nuclear Science Series, NAS-NS-3204 National Academy Press, Washington, D.C.

Steigman J, Richards P (1974) Chemistry of technetium 99m. Sem Nucl Med 4:269–279

Steigman J, Meinken G, Richards P (1975) The reduction of pertechnetate-99 by stannous chloride–I. The stoichiometry of the reaction in HCl, in a citrate buffer and in a DTPA buffer. Int J Appl Radiat Isot 26:601–609

United States Pharmacopeial Convention (2005) Official Monographs: USP 28, technetium Tc-99m pertechnetate injection (sodium). United States Pharmacopeia (USP) 28–national formulary (NF) 23, p 1861

Welch MJ, Adatepe M, Potchen EJ (1969) An analysis of technetium ($^{99m}TcO_4^-$) kinetics: the effect of perchlorate and iodide pretreatment. Int J Appl Radiat Isot 20:437–445

Wolff J, Maurey JR (1962) Thyroid iodide transport. III. Comparison of iodide with anions of periodic group VIIA. Biochim Biophys Acta 57:422–426

12.2 ^{99m}Tc-Labeled Human Serum Albumin

12.2.1 ^{99m}Tc-Albumin (HSA)

I. Zolle and Gy. Jánoki

Chemical name

Human serum albumin (HSA)

Technetium ^{99m}Tc albumin injection (*Ph. Eur.*)

Technetium Tc 99m albumin injection (*USP*)

^{99m}Tc-HSA

Kit components		Commercial products	
Human serum albumin	10 mg	Albumoscint	Nordion
Stannous chloride dihydrate	0.02 mg	TechneScan HSA	Mallinckrodt/Tyco
Sodium chloride q.s.		VasculoCis (TCK-2)	CIS Bio

Preparation

Commercial kits contain the freeze-dried, sterile formulation in a multidose vial, sealed under a nitrogen atmosphere. The lyophilized preparation is readily soluble in ^{99m}Tc-pertechnetate injection and in saline. For labeling, the vial is placed into a lead-shielded container. Aseptically sterile ^{99m}Tc-pertechnetate should be injected into the vial in a volume of 1–8 ml, with an activity up to 2.22 GBq (60 mCi). Before removing the syringe, 2–5 ml of gas should be withdrawn from the space above the solution to normalize the pressure inside the vial. The shielded vial should be agitated gently to dissolve the lyophilized material. The reaction should proceed at room temperature for about 20 min, with occasional agitation.

^{99m}Tc-human serum albumin (HSA) is a clear, pale yellow solution for intravenous injection. The pH value is 2.5–3.5 (2.0–6.5, *European Pharmacopeia*).

Description of the Kit

^{99m}Tc-labeled albumin is a product derived from HSA, which is a natural constituent of blood. HSA is isolated from donor blood and complies with the purity standards stated in the *European Pharmacopeia* (monographs 255 and 853) (Council of Europe 2001a, b) in accordance with EU and WHO requirements for biological substances (Council of Europe 1982; World Health Organization 1994).

Several methods for labeling HSA with technetium-99m in the reduced state (Steigman et al. 1975) have been described. Originally, ferric ion and ascorbic acid were used as a reducing system; labeling of HSA was performed at acidic pH (Stern et al. 1965; Persson and Liden 1969). Adjustment of the pH to 7.8 was essential in order to produce ferrous ion, the active principle for the reduction of Tc(VII)-pertechnetate (Zolle et al. 1973). An alternate approach used reduction of pertechnetate with concentrated HCl to produce Tc(V) in the dry state and subsequent addition of albumin for labeling at acidic pH (Williams and Deegan 1971). Additional insight into the labeling mechanism was obtained by the electrolytic reduction of pertechnetate using a zirconium crucible as the anode and a platinum wire as the cathode (Benjamin 1969; Benjamin et al. 1970). Zirconium as the anode was eventually replaced by tin (Narasimhan and Mani 1975). Anodic dissolution of Zr resp. Sn ions was made responsible for the reduction of pertechnetate and labeling of HSA at acidic pH. A sterile electrolytic kit procedure was introduced for the preparation of ^{99m}Tc-HSA by Dworkin and Gutkowski (1971). Further studies have demonstrated that high labeling yields were obtained by reduction with either Fe(II) or Sn(II) alone (Lin et al. 1971). The advantages of using stannous ion in the production of ^{99m}Tc-pharmaceuticals had been demonstrated by the one-step kit preparation of ^{99m}Tc-HSA (Eckelman et al. 1971). Freeze-drying of the nonradioactive components has considerably enhanced the stability of kits containing stannous ion (Deutsch and Redmond 1972).

Clinical Applications

Intravenous injection: Cardiac blood pool imaging (static)
First-pass ventriculography
Gated equilibrium ventriculography
Regional circulatory imaging

Angiocardiography by first-pass imaging of right ventricular (RV) and left ventricular (LV) function is performed in patients with coronary artery disease. A bolus of ^{99m}Tc-HSA or ^{99m}Tc-red blood cells (RBC) is injected intravenously in a small volume (Berger et al. 1979; Philippe et al. 1988).

Gated radionuclide ventriculography or equilibrium (gated) radionuclide angiocardiography is performed to evaluate LV function after the radiotracer has become distributed throughout the vascular space (Strauss et al. 1971).

Time of Examination. Immediately or shortly after intravenous injection, depending on the type of examination.

Recommended Activities for Indications

Blood pool imaging:	111–185 MBq (3–5 mCi)
Angiocardiography:	370–740 MBq (10–20 mCi) as a bolus of 1–2 ml
Gated ventriculography:	185–925 MBq (5–25 mCi) (or ^{99m}Tc-RBC)
Circulation and blood flow:	18.5–185 MBq (0.5–5 mCi)

Additional Information

^{99m}Tc-HSA is administered by intravenous injection, intrathecal application is contraindicated. ^{99m}Tc-HSA has originally been used to image the placenta; however, this indication is no longer accepted. The preparation of ^{99m}Tc-HSA must not be injected into individuals hypersensitive to protein.

Quality Control

Radiochemical Purity. The *European Pharmacopeia* requires thin-layer chromatography (TLC) (distance 10–15 cm) for the identification of impurities using methylethylketone (MEK) as solvent. The radioactivity corresponding to ^{99m}Tc-HSA must not be less than 95% (Council of Europe 2005).

Here, a similar procedure was used for analysis. A sample of approximately 1,000 counts/s is applied to silica gel fiberglass sheets and developed in acetone. Parallel to the labeled product, sodium ^{99m}Tc-pertechnetate is also analyzed. After development, the dried strips are cut into 1-cm pieces and measured in an NK-350 type scintillation well counter coupled with an automatic sample sorter. ^{99m}Tc-HSA remained at the start (R_f=0.0), ^{99m}Tc-pertechnetate moved with the solvent front (R_f=0.9–1.0).

Thin-layer chromatography		
Stationary phase:	Gelman ITLC-SG (fiberglass), 2×9.5 cm	
Solvent:	Acetone	
Developing time:	5 min	
R_f values:	^{99m}Tc-HSA:	0.0–0.1
	^{99m}Tc reduced, colloidal:	0.0–0.1
	^{99m}Tc-Na-pertechnetate:	0.9–1.0 (<5%)

A quick test for determination of unbound ^{99m}Tc-pertechnetate in labeled albumin preparations has been described using precipitation with trichloroacetic acid (TCA) and separation by membrane filtration. Free ^{99m}Tc-pertechnetate is efficiently separated from ^{99m}Tc-HSA (Lamson et al. 1974). ^{99m}Tc-HSA shows high in vitro stability (Benjamin 1969; Stern et al. 1965).

Pharmacokinetic Data

Following intravenous injection, ^{99m}Tc-HSA is distributed homogenously in the vascular compartment. It does not concentrate in the thyroid, salivary, and gastric glands. In pregnant women receiving a placental scan, 50–75% of the radioactivity was measured in the blood 30 min after injection. The elimination from blood has been described by two half-times, namely 6 h and 3 days (McAfee et al. 1964; Stern et al. 1966).

A comparison with ^{131}I-albumin (^{131}I-IHSA) in experimental animals showed that ^{99m}Tc-HSA parallels the disappearance of ^{131}I-albumin, being eliminated slightly faster. The tissue distribution in pregnant rabbits was also similar, showing slightly lower blood levels and a much lower concentration within the fetus. ^{99m}Tc-HSA accumulates in the kidneys, which showed the highest tissue concentration. Though at low levels, an increase of radioactivity was seen also in the stomach and gut (McAfee et al. 1964; Stern et al. 1966).

In another extensive study (184 samples of maternal plasma and 98 samples of fetal plasma), the disappearance curve indicated an initial rapid decrease with a half-time of 2 h and a slow elimination corresponding to a half-time of 35 h. Initial plasma clearance was 16% in the first hour and 51% in 6 h, the incremental loss being 38% per day. The major loss from the albumin pool resulted from urinary excretion (32% in the first 24 h). Fecal excretion was low, with a maximum of 5.8% in 48 h. Free pertechnetate in maternal plasma (17 samples) was 2–21.2%, with a mean value of 13.6% (Herbert et al. 1969).

Transplacental passage of activity was evident a few minutes after maternal injection; after 2.5 h, fetal plasma activity was in equilibrium with maternal plasma, showing 4.3% of the maternal plasma concentration, totally as free ^{99m}Tc-pertechnetate. However, the total plasma content in the fetus never exceeded 0.15% of the maternal administered activity. Urine immediately obtained after birth showed high concentrations of free ^{99m}Tc-pertechnetate. Passage of ^{99m}Tc-HSA across the placenta is minimal (Herbert et al. 1969).

The fetal plasma equilibrium value of 4.3% of maternal plasma concentration was approximately twice that of IHSA reported by Hibbard and Herbert (1960); also, accumulation in the liquor was high (no liquor activity was detected with IHSA).

^{99m}Tc-HSA was labeled by the method of Stern et al. (1965). Preparations showed high in vitro stability and contained virtually no free radionuclide when injected. Yet, no explanation has been given for the rapid liberation of technetium-99m from albumin after injection reflected in the rapid initial plasma clearance, the excretion pattern, and the evidence of fetal uptake of pertechnetate.

The results of clinical studies performed by McAfee et al. (1964) and by Herbert et al. (1969) are well documented; they are, however, at variance with respect to the excretion of activity in urine and uptake in the fetus. McAfee et al. (1964) recovered less than 0.5% of the injected radioactivity in urine or feces, and 0.4% of the maternal radioactivity in two infants delivered approximately 1 and 4 h after intravenous injection to the mother.

Evidence that ^{99m}Tc-HSA prepared by electrolytic kit labeling shows high in vivo stability has been provided by measuring the whole-blood disappearance in mice over a period of 90 min (Dworkin and Gutkowski 1971) and in patients for 1 h (Callahan et al. 1976).

Radiation Dose

In analogy with iodine-labeled albumin, absorbed dose calculations are based on the elimination of radioactivity from blood by three half-times, namely 6.8 h (0.40), 1.29 days (0.22), and 19.4 days (0.38, Takeda and Reeve 1963). Uniform distribution of ^{99m}Tc-HSA outside the blood pool and rapid renal excretion of the released radionuclide is assumed (International Commission on Radiological Protection 1987).

The effective (whole body) dose equivalent for ^{99m}Tc-labeled albumin is 0.0079 mSv/MBq (International Commission on Radiological Protection 1987). The effective dose in adults (70 kg) resulting from 740 MBq (20 mCi) of intravenously injected ^{99m}Tc-HSA for angioscintigraphy is approximately 5.8 mSv. Blood pool imaging with 185 MBq (5 mCi) of ^{99m}Tc-HSA is delivering 1.5 mSv.

The absorbed radiation dose to the heart resulting from an intravenous injection of 740 MBq of ^{99m}Tc-HSA corresponds to 14.8 mGy, and to the kidneys, 6.0 mGy.

The absorbed radiation dose per unit activity (GBq) of administered ^{99m}Tc-HSA is shown for selected organs in adults (70 kg) in Table 12.2.1.1. Calculations are based on a biological half-time of 6.8 h for the elimination from blood and whole body (expressed as mGy/GBq and rad/mCi).

Table 12.2.1.1. ^{99m}Tc-human serum albumin (HSA)-Absorbed organ dose

Organs	mGy/GBq	rad/mCi
Adrenals	8.3	0.022
Heart	20.0	0.054
Kidney	8.1	0.022
Liver	7.3	0.019
Spleen	14.0	0.038

Storage and Stability

Storage. Kits should be stored at 2–8 °C. ^{99m}Tc-HSA injection solution should be kept in the refrigerator.

Stability. ^{99m}Tc-HSA injection solution should be used within 6 h after labeling.

References

Benjamin PP (1969) A rapid and efficient method of preparing ^{99m}Tc-human serum albumin: its clinical applications. Int J Appl Radiat Isot 20:187–194

Benjamin PP, Rejali A, Friedell H (1970) Electrolytic complexation of ^{99m}Tc at constant current: its application in nuclear medicine. J Nucl Med 11:147–154

Berger HJ, Matthay RA, Pytlik L, Gottschalk A, Zaret BL (1979) First-pass radionuclide assessment of right and left ventricular performance in patients with cardiac and pulmonary disease. Semin Nucl Med 9:275–295

Callahan RJ, McKusick KA, Lamson III M, Castronovo FP, Potsaid MS (1976) Technetium-99m-human serum albumin: Evaluation of a commercially produced kit. J Nucl Med 17:47–49

Council of Europe (1992) Guide for the preparation, use and quality assurance of blood components. Council of Europe, Strasbourg

Council of Europe (2004 a) Human albumin injection, monograph 255. European pharmacopeia. Council of Europe, Maisonneuve, Sainte-Ruffine

Council of Europe (2004 b) Human plasma for separation, monograph 853. European pharmacopeia. Council of Europe, Maisonneuve, Sainte-Ruffine

Council of Europe (2005) Technetium ^{99m}Tc albumin injection. European pharmacopeia 5.0, monograph 640. Council of Europe, Maisonneuve, Sainte-Ruffine

Deutsch ME, Redmond ML (1972) Unitary freeze-dried kits for preparation of technetium-labeled human serum albumin. J Nucl Med 13(Abstr):426–427

Dworkin HJ, Gutkowski RF (1971) Rapid closed system production of ^{99m}Tc-albumin using electrolysis. J Nucl Med 12:562–565

Eckelman WC, Meinken G, Richards P (1971) ^{99m}Tc-Human serum albumin. J Nucl Med 12:707–710

Herbert RJT, Hibbard BM, Sheppard MA (1969) Metabolic behaviour and radiation dosimetry of ^{99m}Tc-albumin in pregnancy. J Nucl Med 10:224–232

Hibbard BM, Herbert RJT (1960) Fetal radiation dose following administration of radio-iodinated albumin. Clin Sci 19:337–344
International Commission on Radiological Protection (1987) Technetium-labelled albumin (HSA). In: Annals of the ICRP. Radiation dose to patients from radiopharmaceuticals, biokinetic models and data. ICRP publication 53, vol 18, no 1–4. Pergamon, Oxford, p 173
Lamson III M, Callahan RJ, Castronovo FP, McKusick KA, Potsaid MS (1974) A rapid index of free activity in preparations of ^{99m}Tc-albumin. J Nucl Med 15:1061–1062
Lin SM, Winchell HS, Shipley BA (1971) Use of Fe(II) or Sn(II) alone for technetium labelling of albumin. J Nucl Med 12:204–211
McAfee JG, Fueger GF, Baggish MS, Holzman GB, Zolle I (1964) ^{99m}Tc labeled serum albumin for scintillation scanning of the placenta. J Nucl Med 5:936–946
Narasimhan DVS, Mani RS (1975) Electrolytic preparation of ^{99m}Tc-human serum albumin using tin electrodes. Radiochim Radioanal Lett 20:307–316
Persson RBR, Liden K (1969) ^{99m}Tc-labelled human serum albumin: a study of the labelling procedure. Int J Appl Radiat Isot 20:241–248
Philippe L, Mena I, Sarcourt J, French WJ (1988) Evaluation of valvular regurgitation by factor analysis of first-pass angiography. J Nucl Med 29:159–167
Steigman J, Meinken G, Richards P (1975) The reduction of pertechnetate-99 by stannous chloride I. The stoichiometry of the reaction in HCl, in a citrate buffer and in a DTPA buffer. Int J Appl Radiat Isot 26:601–609
Stern HS, Zolle I, McAfee JG (1965) Preparation of ^{99m}Tc-labelled serum albumin. Int J Appl Radiat Isot 16:283–288
Stern HS, McAfee JG, Zolle I (1966) Technetium-99m–albumin. In: Andrews GA, Knisely RM, Wagner HN Jr (eds) Radioactive pharmaceuticals. AEC symposium series no. 6 (CONF-651111), Oak Ridge, Tenn., pp 359–382
Strauss HW, Zaret BL, Hurley PJ, Natarajan TK, Pitt P (1971) A scintiphotographic method for measuring left ventricular ejection fraction in man without cardiac catheterization. Am J Cardiol 28:575–580
Takeda Y, Reeve EB (1963) Studies of the metabolism and distribution of albumin with autologous ^{131}I-albumin in healthy men. J Lab Clin Med 61:183202
United States Pharmacopeial Convention (2005) Official Monographs: Technetium Tc 99m albumin injection. United States Pharmacopeia (USP) 28–national formulary (NF) 23, p 1849
Williams MJ, Deegan T (1971) The process involved in the binding of technetium-99m to human serum albumin. Int J Appl Radiat Isot 22:767–774
World Health Organization (1994) Requirements for the collection, processing and quality control of blood, blood components and plasma derivatives, requirements for biological substances No. 27, WHO, Technical Report Series No. 840
Zolle I, Oniciu L, Höfer R (1973) Contribution to the study of the mechanism of labelling human serum albumin (HSA) with Technetium-99m. Int J Appl Radiat Isot 24:621–626

12.2.2 ^{99m}Tc-Albumin Macroaggregates (MAA) (Size Range: 10–50 μm)

I. Zolle and Gy. Jánoki

Chemical name

Macroaggregated human serum albumin (MAA)

Technetium 99m Macrosalb injection (*Ph. Eur.*)

^{99m}Tc-MAA

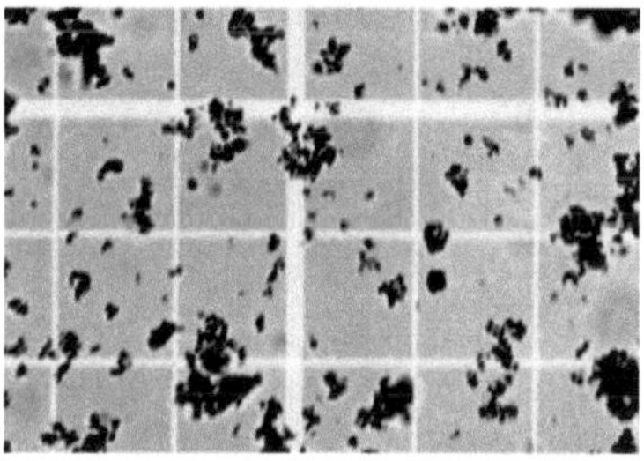

Albumin Macroaggregates
(Small square 50×50 μm, 150-fold)

Kit components

Macroaggregated albumin	0.5–2 mg
Human albumin (HSA)	0.5 mg
Stannous chloride (Dihydrate)	1.2–2.0 mg

Commercial products

TechneScan MAA	Mallinckrodt/Tyco
MAAScint	Nordion
PulmoCis (TCK-8)	CIS Bio
Macrotec Sorin	GE Healthcare
MAASol Sorin	GE Healthcare

Preparation

Commercial kits contain the lyophilized, sterile components in a multidose vial, sealed under a nitrogen atmosphere. Labeling is carried out by adding aseptically a volume of 2–10 ml of sterile ^{99m}Tc-pertechnetate to the vial with an activity up to 3.7 GBq. The manufacturer's instructions should be followed. The lyophilized material will dissolve by agitating the reaction vial. The reaction should be allowed to proceed for 5–20 min, with occasional agitation. ^{99m}Tc-macroaggregated albumin (MAA) is a pale-white suspension ready for intravenous injection. The pH of the suspension is 3.5–7.5.

Description of the Kit

Macroaggregates are obtained by aggregation of human serum albumin (HSA). HSA is isolated from donor blood and complies with the purity standards stated in the *European Pharmacopeia,* monographs 255 and 853) (Council of Europe 2004 a, b) in accordance with EU and WHO requirements for biological substances (Council of Europe 1992; World Health Organization 1994).

Aggregation of albumin is carried out at acidic pH under controlled conditions, yielding macroaggregates with a particle size distribution between 10 and 90 μm. Larger aggregates are separated by filtration of the crude product. Typically, kits contain a suspension of MAA with a size distribution of 10–50 μm. Generally, 2 mg of macroaggregates correspond to 1.5–2 million albumin particles.

Lactose, glucose (anhydrous), HSA, succinic acid, sodium acetate, sodium phosphate, dibasic sodium phytate (anhydrous), as well as Tween-80 or PVP-40 may serve as stabilizers. The kit contains no preservatives.

The first suspensions of radioalbumin were introduced for photoscanning the liver, spleen, lung, and other organs, because they contained particles in the colloidal range (≤1 μm) and between 10 and 20 μm (Taplin et al. 1964 a, b). At the time, solutions of iodinated albumin (0.1% radioalbumin, pH 5.5) with a specific activity of 1 mCi of ^{131}I/mg of HSA) had been provided by E. R. Squibb & Sons for investigative purposes, and were heat-aggregated in-house for 20 min, at different temperatures in a water bath (60–80 °C). At Johns Hopkins, these preparations (10–100 μm) were evaluated in dogs in which pulmonary embolism had been produced experimentally, extending lung perfusion scanning to the first human studies with ^{131}I-MAA (Wagner et al. 1964 a, b). In the early seventies, one-step kit preparations with technetium-99m became available and have contributed to the worldwide application of ^{99m}Tc-MAA (Chandra et al. 1973; Monroe et al. 1974; Robbins et al. 1976; Subramanian et al. 1972; Taplin and MacDonald 1971).

Clinical Applications

Intravenous injection: Lung perfusion scintigraphy
Radionuclide venography

Lung perfusion imaging with labeled, biodegradable particles has become an important diagnostic tool for the diagnosis of regional perfusion defects observed in patients with pulmonary disease (i.e., emphysema, chronic obstructive disease, pulmonary hypertension, fibrosis, acute arterial obstruction) (Bell and Simon 1976; Saenger et al. 1985; Tow et al. 1966). The mechanism of lung retention of particles like MAA is known as capillary blockade.

Even though a perfusion scan will localize an obstructed pulmonary artery, accurate diagnosis of pulmonary embolism requires an additional ventilation study with a radioactive gas or aerosol (Agnew 1991; Gottschalk et al. 1993 a, b; Taplin and Chopra 1978; Tow and Wagner 1967; Wagner 1995; Wagner et al. 1968).

^{99m}Tc-MAA has also been used for the evaluation of deep vein thrombosis by venous blood flow studies (Dibos 1995; Vlahos et al. 1976). In order to visualize the deep venous system, a special technique is used for injection into veins on the dorsum of each foot.

Time of Examination

- Lung perfusion scintigraphy: immediately after intravenous injection
- Scintigraphy of the lower extremities: shortly after bilateral intravenous injection

Recommended Activities for Indications

- Lung scintigraphy (adults): 37–185 MBq (1–5 mCi)
- Scintigraphy of the lower extremities: 130–150 MBq (3.5–4.0 mCi).

The lung scanning dose. The number of aggregates to be administered for a lung scan lies between 250,000 and 700,000. The *United States Pharmacopeia* (*USP*) limits the protein concentration to 1 mg of MAA per 37 MBq (1 mCi) of Tc-99m at the time of administration (United States Pharmacopeia Convention 2000). The number of macroaggregates in 1 mg has been calculated as approximately 700,000 particles, which is the upper limit for intravenous injection in adults. A sufficient number of particles need to be administered to avoid nonuniform spatial distribution of radioactivity in lung regions (Heck and Duley 1974).

Calculation of safety factors. Clinical safety factors for particulate injections are derived from lethal dose (LD_{50}) values or the minimum lethal dose (MLD) expressed in units of weight (mg/kg body weight [BW]) or as the number of particles per gram of BW. Accordingly, the amount of particles injected is given in milligrams or as the number of particles. BW is based on 70 kg (average adult).

$$\text{Safety factor (SF)} = \frac{\text{LD}_{50} \text{ or MLD (mg/kg)} \times 70 \text{ (kg) BW}}{\text{Injected dose (mg or number of particles)}}$$

Based on SF derived from toxicity studies, the minimum amount of MAA for a lung scan in adults (70 kg) has been derived as 10 μg/kg BW, resulting in a lung scanning dose (0.7 mg) at least a 1,000 times below the minimum toxic dose in dogs (20 mg/kg) (Taplin and MacDonald 1971). If 1 mg of MAA is injected, an SF of 1,400 would apply. This safety margin also applies to an intravenous dose of 700,000 aggregates.

Pediatric Dose. The amount of radioactivity for infants and children administered for lung scintigraphy is based on body weight, using the scaling factors given in Appendix 1 (Table A1.2). The number of macroaggregates in 1 ml of the injection solution should be considered. Not more than 165,000 particles should be injected in children up to 1 year of age, and not more than 50,000 aggregates in newborns (Davis and Taube 1979; Heyman 1979).

Additional Information

Macroaggregates of albumin must not be injected in patients with a history of hypersensitivity to human albumin.

Careful consideration should be given to limiting the number of aggregates injected when studying patients with known severe pulmonary hypertension (Vincent et al. 1968). Also, in patients with right-to-left cardiac shunts, the number of aggregated albumin particles administered for a lung scan should be reduced to the minimum, because shunted macroaggregates are distributed throughout the entire systemic circulation and may cause microembolism in the brain and kidneys (Taplin and MacDonald 1971).

^{99m}Tc-MAA is injected intravenously with the patient in supine position. Prior to injection, the syringe should be inverted repeatedly to resuspend sedimented macroaggregates; they should be injected slowly over a period of at least 30 s to normalize different phases of the respiratory cycle (Wagner 1995).

Aspiration of blood must be avoided; if blood is drawn into the syringe, formation of larger aggregates with coagulated blood occurs. The syringe should not be back flushed.

^{99m}Tc-MAA should not be injected through a lying catheter (butterfly) because of the occasional observation of "hot spots" in the lung.

Quality Control

Radiochemical Purity. The *European Pharmacopeia* (Council of Europe 2005) requires membrane filtration (3 μm) for the determination of radiochemical purity. Unbound radioactivity is in the filtrate; not less than 90% of the total radioactivity is measured on the filter.

The *USP* permits centrifugation of a sample obtained from a well mixed injection solution. Less than 10% of the total radioactivity is measured in the supernatant.

Paper chromatography (*USP*) is recommended for the identification of impurities, using 70% methanol as solvent. Free ^{99m}Tc-sodium pertechnetate is measured at R_f=0.6, and ^{99m}Tc-MAA is identified at the origin. The radiochemical purity of ^{99m}Tc-MAA should not be less than 90%.

Paper chromatography (*USP*)		
Stationary phase:	Whatman No. 1 paper	
Solvent:	Methanol:water, 70:30 (v/v)	
R_f values:	^{99m}Tc reduced, hydrolized:	0.0–0.1
	^{99m}Tc-MAA:	0.0–0.1 (>90%)
	^{99m}Tc-Na-pertechnetate:	0.6–0.7

Thin-layer chromatography (recommended by the manufacturer)		
Stationary phase:	Gelman ITLC-SG (fiberglass), 2×9.5 cm	
Solvent:	Acetone (resp. methylethylketone [MEK])	
R_f values:	^{99m}Tc reduced, hydrolized:	0.0–0.1
	^{99m}Tc-MAA:	0.0–0.1
	^{99m}Tc-Na-pertechnetate:	0.9–1.0

Thin-layer chromatography offers the advantage of rapid development (5–10 min).

Quantification of labeled components. Each chromatogram is measured (TLC linear analyzer or gamma counter), and the regional radioactivities are expressed as a percentage of the total recovered counts:

$$^{99m}\text{Tc-MAA}\ (\%) = 100 - F$$

where F (%) = ^{99m}Tc-Na-pertechnetate (free).

Results of ^{99m}Tc-MAA analysis *(12 samples)*

Labeling and stability	20 min (%)	4 h (%)
Paper chromatography		
^{99m}Tc-macroaggregates	98.8±0.06	99.2±0.04
^{99m}Tc-Na-pertechnetate	1.2±0.05	0.8±0.06
Thin-layer chromatography		
^{99m}Tc-macroaggregates	99.4±0.01	99.6±0.01
^{99m}Tc-pertechnetate	0.6±0.02	0.4±0.01

Particle Size Distribution. The method described in the *European Pharmacopeia* is based on the microscopic examination of 5,000 albumin aggregates, using a suitable counting chamber such as a hemocytometer.

Macroaggregates show a size distribution of diameters between 10 and 90 μm. Typically, 90% of a suspension is within 10–50 microns. Less than ten aggregates are bigger than 75 μm; none is larger than 100 μm. The number of macroaggregates in 1 ml of the injection solution should also be determined.

Pharmacokinetic Data

Following intravenous injection, more than 90% of the technetium-99m MAA is extracted during the first pass and retained in lung capillaries (8.2 ± 1.5 μm) and arterioles (25 ± 10 μm) (Taplin and MacDonald 1971). Organ selectivity is directly related to particle size. Albumin aggregates smaller than 8 μm pass the pulmonary capillary bed and are taken up in the reticuloendothelial system. With a particle diameter above 15 μm, aggregates are retained in the lung capillaries by a purely mechanical process. Distribution of aggregated albumin in the lung is a function of regional pulmonary blood flow (Wagner et al. 1964 a).

Macroaggregates are sufficiently fragile for the capillary microocclusion to be temporary. Erosion and fragmentation reduce the particle size, facilitating removal of aggregates from the lung (Taplin and MacDonald 1971). Subsequently, the fragments are accumulated in the liver by phagocytosis (Chandra et al. 1973; Robbins et al. 1976).

The elimination of radioactivity from the lung is described by half-times between 4 and 6 h (Taplin and MacDonald 1971). Experimental data indicate an initial fast component, which has been interpreted as rapid release of unbound ^{99m}Tc activity (Malone et al. 1983).

Accumulation in the liver is assumed to amount to 25%, with an uptake half-time of 6 h and an elimination half-time of 5 days (International Commission on Radiological Protection 1987); considerable uptake in the liver has been reported (Chandra et al. 1973; Robbins et al. 1976).

Excretion of released pertechnetate in the urine is reported as 40 ± 14% in 24 h, and an additional 9.0 ± 3.8% up to 48 h. Intestinal activity has been shown to increase slowly up to 24 h (Malone et al. 1983).

Radiation Dose

Lung, liver, and the bladder wall are the most exposed organs. The effective (whole body) dose equivalent is 0.012 mSv/MBq (International Commission on Radiological Protection 1987). Elimination from the lung is assumed with half-times of 6 h (0.85) and 3 days (0.15). The liver takes up a fraction of 0.25, with an uptake half-time of 6 h and an elimination half-time of 5 days. Radionuclide released from the lung is primarily excreted by the kidneys (Malone et al. 1983).

The effective (whole body) dose in adults (70 kg) resulting from an intravenous injection of 185 MBq of ^{99m}Tc-MAA for lung scintigraphy is approximately 2.2 mSv. The absorbed radiation dose to the lung resulting from an intravenous injection of 185 MBq of ^{99m}Tc-MAA for a lung scan corresponds to 12.4 mGy, and to the liver, approximately 3.0 mGy.

The effective (whole body) dose resulting from bilateral venography injecting a total of approximately 185 MBq (5 mCi) of ^{99m}Tc-MAA was estimated as 1.35 mSv (Malone et al. 1983).

Storage and Stability

Storage. Kits should be stored at 2–8 °C, and ^{99m}Tc-MAA injection solution should be kept at 2–8 °C, with adequate shielding.

Stability. ^{99m}Tc-MAA injection solution should be used within 6 h after labeling.

References

Agnew JE (1991) Characterizing lung aerosol penetration. J Aerosol Med 4:237–250
Bell WR, Simon TL (1976) A comparative analysis of pulmonary perfusion scans with pulmonary angiograms: From a national cooperative study. Am Heart J 92:700–706
Chandra R, Shannon J, Braunstein P, Durlov OL (1973) Clinical evaluation of an instant kit for preparation of ^{99m}Tc-MAA for lung scanning. J Nucl Med 14:702–705
Council of Europe (1992) Guide for the preparation, use and quality assurance of blood components. Council of Europe, Strasbourg
Davis MD, Taube RA (1979) Re: toxicity and safety factors associated with lung perfusion studies with radiolabeled particles (letter to the editor). J Nucl Med 20:1099
Dibos PE (1995) Deep venous thrombosis. In: Wagner HN Jr, Szabo S, Buchanan JW (eds) Principles of nuclear medicine, 2nd edn. Saunders, Philadelphia, pp 872–880
Council of Europe (2004a) Human albumin injection, monograph 255. European pharmacopeia. Council of Europe, Maisonneuve, Sainte-Ruffine
Council of Europe (2004b) Human plasma for separation, monograph 853. European pharmacopeia. Council of Europe, Maisonneuve, Sainte-Ruffine
Council of Europe (2005) Technetium ^{99m}Tc macrosalb injection. European pharmacopeia 5.0, monograph no 296. Council of Europe, Maisonneuve, Sainte-Ruffine
Gottschalk A, Juni JE, Sostman HD, Coleman RE, Thrall J, McKusick KA, Froelich JW, Alavi A (1993a) Ventilation-perfusion scintigraphy in the PIOPED study. Part I. Data collection and tabulation. J Nucl Med 34:1109–1118
Gottschalk A, Sostman HD, Coleman RE, Juni JE, Thrall J, McKusick KA, Froelich JW, Alavi A (1993b) Ventilation-perfusion scintigraphy in the PIOPED study. Part II. Evaluation of the scintigraphic criteria and interpretations. J Nucl Med 34:1119–1126

Heck LL, Duley JW Jr (1974) Statistical considerations in lung imaging with Tc-99m-albumin particles. Radiology 113:675–679
Heyman S (1979) Toxicity and safety factors associated with lung perfusion studies with radiolabeled particles (letter to the editor). J Nucl Med 20:1098–1099
International Commission on Radiological Protection (1987) Technetium-labelled macroaggregated albumin. In: Annals of the ICRP. Radiation dose to patients from radiopharmaceuticals, biokinetic models and data. ICRP publication 53, vol. 18, no. 1–4. Pergamon, Oxford, pp 223–224
Malone LA, Malone JF, Ennis JT (1983) Kinetics of technetium 99m-labelled macroaggregated albumin in humans. Br J Radiol 56:109–112
Monroe LA, Thompson WL, Anderton NS, Burdine JA (1974) Evaluation of an improved ^{99m}Tc-stannous aggregated albumin preparation for lung scanning. J Nucl Med 15:192–194
Robbins PJ, Feller PA, Nishiyama H (1976) Evaluation and dosimetry of a ^{99m}Tc-Sn-MAA lung imaging agent in humans. Health Phys 30:173–176
Saenger EL, Buncher CR, Specker BL, McDvitt RA (1985) Determination of clinical efficacy: nuclear medicine as applied to lung scanning. J Nucl Med 26:793–806
Subramanian G, Arnold RW, Thomas FD, McAfee JG (1972) Evaluation of an instant ^{99m}Tc-labeled lung scanning agent. J Nucl Med 13(Abstr):790
Taplin GV, Chopra SK (1978) Lung perfusion-inhalation scintigraphy in obstructive airway disease and pulmonary embolism. Radiol Clin N Am 16:491–513
Taplin GV, MacDonald NS (1971) Radiochemistry of macroaggregated albumin and newer lung scanning agents. Sem Nucl Med 1:132–152
Taplin GV, Johnson DE, Dore EK, Kaplan HS (1964a) Suspensions of radioalbumin aggregates for photoscanning the liver, spleen, lung and other organs. J Nucl Med 5:259–275
Taplin GV, Johnson DE, Dore EK, Kaplan HS (1964b) Lung photoscans with macroaggregates of human serum radioalbumin. Health Phys 10:1219–1227
Tow DE, Wagner HN Jr (1967) Recovery of pulmonary arterial blood flow in patients with pulmonary embolism. N Engl J Med 276:1053–1059
Tow DE, Wagner HN Jr, Lopez-Majano V, Smith E, Migita T (1966) Validity of measuring regional pulmonary arterial blood flow with macroaggregates of human serum albumin. Am J Roentgenol 96:664–676
United States Pharmacopeial Convention (2005) Official Monographs: Technetium Tc 99m albumin aggregated injection. United States Pharmacopeia (USP) 28–national formulary (NF) 23, p 1850
Vincent WR, Goldberg SJ, Desilets D (1968) Fatality immediately following rapid infusion of macroaggregates of ^{99m}Tc-albumin (MAA) for lung scan. Radiology 91:1181–1184
Vlahos L, MacDonald AF, Causer DA (1976) Combination of isotope venography and lung scanning. Br J Radiol 49:840–851
Wagner HN Jr (1995) Regional ventilation and perfusion. In: Wagner HN Jr, Szabo S, Buchanan JW (eds) Principles of nuclear medicine, 2nd edn. Saunders, Philadelphia, pp 887–895
Wagner HN Jr, Sabiston DC Jr, Iio M, McAfee JG, Langan JK (1964a) Regional pulmonary blood flow in man by radioisotope scanning. JAMA 187:601–603
Wagner HN Jr, Sabiston DC Jr, McAfee JG, Tow DE, Stern HS (1964b) Diagnosis of massive pulmonary embolism in man by radioisotope scanning. N Engl J Med 271:377–384
Wagner HN Jr, Lopez-Majano V, Langan JK, Joshi RC (1968) Radioactive xenon in the differential diagnosis of pulmonary embolism. Radiology 91:1168–1174
World Health Organization (1994) Requirements for the collection, processing and quality control of blood, blood components and plasma derivatives, requirements for biological substances No. 27, Technical Report Series No. 840

12.2.3 ^{99m}Tc-Albumin Microspheres (HAM) (Size Range: 10–50 μm)

I. Zolle

Chemical name

Human serum albumin microspheres (HAM)
HSA-microspheres

Technetium ^{99m}Tc microspheres injection (*Ph. Eur.*)

^{99m}Tc-HSA microspheres

^{99m}Tc-HAM

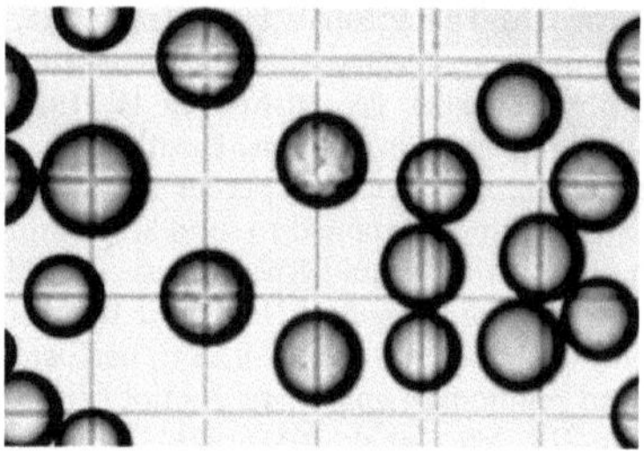

Albumin microspheres (40 μm)
(Small square 50×50 μm, 150-fold)

Kit components(1)		Kit components(2)	
HSA-microspheres	10.0 mg	HSA-microspheres	2.5 mg
Stannous chloride·2 H_2O	1.3 mg	Stannous chloride·2 H_2O	0.1 mg
Pluronic F68	0.25 mg	Tween-80	0.6 mg
Sodium chloride	9.0 mg		

Preparation

Commercial kits contain the lyophilized, sterile components including preformed albumin microspheres in a multidose vial, sealed under a nitrogen atmosphere. Labeling with ^{99m}Tc eluate is carried out under aseptic conditions by injecting a volume of 2–10 ml of eluate corresponding to a ^{99m}Tc activity of maximal of 5.5 GBq (5–150 mCi). The reaction is allowed to proceed at room temperature for 15 min. ^{99m}Tc-human albumin microspheres (HAM) is a sterile, pyrogen-free suspension suitable for intravenous injection. The pH of the injection solution is 4.0–9.0 (*European Pharmacopeia*).

Description of the Kit

Albumin microspheres are obtained by heat denaturation of HSA in vegetable oil (Zolle et al. 1970). HSA (25% solution) used for preparation of microspheres complies with the purity standards stated in the *European Pharmacopeia* (Council of Europe 2004 a, b) in accordance with EU and WHO requirements for biological substances (Council of Europe 1992; World Health Organization 1994).

The size distribution of albumin particles depends mainly on the degree of dispersion of HSA in oil; generally, a size distribution between 12 and 45 μm was produced.

Fractions with distinct size ranges were obtained by sieving (Rhodes et al. 1969; Zolle et al. 1970). This permits a close estimate of the number of microspheres in 1 mg.

Commercial kits may contain 10 mg of microspheres, corresponding to 800,000–1,600,000 microspheres per vial (kit formulation 1) or 2.5 mg of microspheres, corresponding to 300,000–500,000 microspheres per vial (kit formulation 2). Typically, kits contain albumin microspheres with a size distribution between 10 and 50 μm. Tween-80 or Pluronic F68 serve as surfactants to avoid aggregation of microspheres.

Originally, albumin microspheres containing iron (ferric hydroxide) were used for labeling with ^{113m}In or ^{99m}Tc; these labeling procedures included heating in a water bath (Rhodes et al. 1969). Later on, a kit preparation of ^{99m}Tc-HAM, using iron loaded microspheres in the presence of sodium thiosulfate at acidic pH (and heating) was introduced (Bolles et al. 1973; Krejcarek et al. 1974). A comparison of different labeling techniques has been reported (Mayron and Kaplan 1975).

Preformed albumin microspheres are labeled with high radiochemical yield either by using a commercial kit or by electrolytic reduction by using a tin electrode. No heating or pH adjustments are required. The specific activity of ^{99m}Tc-HAM should not be less than 185 MBq (5 mCi) per 1 million microspheres according to the *European Pharmacopeia* (Council of Europe 2005).

^{99m}Tc-eluate used for labeling the preformed microspheres should be obtained from a generator by daily elution in order to minimize ^{99}Tc carrier, and should comply with specifications stated in the *European Pharmacopeia.*

Clinical Applications

Intravenous injection:	Pulmonary perfusion scintigraphy
	Determination of right-to-left shunts
Arterial injection:	Regional perfusion in other organs

^{99m}Tc-albumin microspheres satisfy the requirements of particles for systemic application with the indicator fractionation technique, namely, uniform size and shape, available in desired sizes, extracted completely in a single passage through capillary beds, and metabolized within hours after injection (Rhodes et al. 1969, 1971; Wagner et al. 1968, 1969; Zolle and Kropf 1982).

When injected into a systemic artery, microspheres indicate regional blood flow in that organ. Myocardial blood flow was measured in dogs (Fortuin et al. 1971; Weller et al. 1972) . ^{99m}Tc-HAM has also been used for measuring regional blood perfusion in the heart (Martin et al. 1973) and the brain in monkeys (Alm 1975). This method has been used for the measurement of the shunted blood through patent arteriovenous connections in the leg after femoral artery injection (Rhodes et al. 1969, 1973).

The size and number of microspheres can be controlled, and labeling is performed with high specific activity; as a result, ^{99m}Tc-albumin microspheres of distinct particle size have been used to quantify the shunted blood flow reaching the lungs (Strauss et al. 1969; Wagner et al. 1969) and arteriovenous shunts in brain tumors (Bergmann et al. 1973).

Time of Examination

- Lung perfusion scintigraphy: immediately after intravenous injection
- Regional organ perfusion: immediately after arterial injection

Recommended Activities for Indications

Lung scintigraphy:	75–185 MBq, (2–5 mCi) injected intravenously
	≤0.010 mg microspheres/kg body weight (BW)
	≤250,000 microspheres (Food and Drug Administration [FDA])
Regional organ scintigraphy:	75–111 MBq, (2–3 mCi) injected intraarterially

The lung scanning dose. The number of microspheres to be administered for a lung scan lies between 100,000 and 250,000. The lung scanning dose normalized to BW is 10 μg/kg, keeping the patient dose generally below 1 mg of microspheres. The FDA has limited the number of microspheres used for lung scintigraphy to 250,000 particles. However, a sufficient number of particles need to be administered to avoid nonuniform spatial distribution of radioactivity in lung regions (Heck and Duley 1974).

Calculation of safety factors. Clinical safety factors for particulate injections are derived from LD_{50} values or the minimum lethal dose (MLD) expressed in units of weight (mg/kg BW) or as the number of particles per gram of BW. Accordingly, the amount of particles injected is given in mg or as the number of particles. BW is based on 70 kg (average adult).

$$\text{Safety factor (SF)} = \frac{\text{LD}_{50} \text{ or MLD (mg/kg)} \times 70 \text{ (kg) BW}}{\text{Injected dose (mg or number of particles)}}$$

The sublethal toxicity of microspheres (15–30 μm) in the lung was determined as 20 mg/kg BW in dogs (Bolles et al. 1973). Based on a patient dose of 0.7–1 mg of microspheres (15–30 μm) for lung perfusion scintigraphy, SFs between 1,400 and 2,000 are derived. Theoretically, 1 mg of microspheres with a diameter of 20 μm contains 238,732 spheres. A sample of microspheres ranging between 12 and 29 μm in diameter actually showed 234,800–246,000 spheres in 1 mg by Coulter measurements (Zolle et al. 1970).

Pediatric Dose. The amount of radioactivity for infants and children administered for lung scintigraphy is based on body weight, using the scaling factors given in Appendix 1 (Table A1.2). In addition, the number of microspheres in 1 ml of injection solution should also be considered. Only a fraction of the number of microspheres used in adults should be administered in children (Davis and Taube 1978; Heyman 1979).

Additional Information

The use of ^{99m}Tc-albumin microspheres in patients with a history of hypersensitivity to human albumin is contraindicated.

^{99m}Tc-albumin microspheres should not be injected together with other drugs or components to avoid aggregates.

For intravenous injection ^{99m}Tc-HAM microspheres should be homogenously suspended to avoid in vivo aggregates. For this reason, the aspiration of blood into the syringe must be avoided.

^{99m}Tc-HAM should not be injected through a lying catheter (butterfly) because of the occasional observation of "hot spots" in the lung.

Quality Control

Radiochemical Purity. ^{99m}Tc-albumin microspheres are described in the *European Pharmacopeia* (Council of Europe 2005). Membrane filtration (3-μm pore size) is used to determine unbound radioactivity in the filtrate. Not less than 95% of the total radioactivity is measured on the filter.

An alternate method for laboratory use is centrifugation of a homogenously suspended sample of ^{99m}Tc-albumin microspheres for 2 min at 2,000 rpm and separation of the supernatant. Both test tubes are measured in a gamma counter, and the radioactivity is expressed as a percentage of the sum of recovered counts. Less than 5% of the total radioactivity is measured in the supernatant.

Recommended Methods

Paper chromatography is recommended by the manufacturer, using 80% methanol as solvent. Free ^{99m}Tc-sodium pertechnetate is measured at an R_f of 0.6 and ^{99m}Tc-HAM at the origin (R_f=0.0). The radiochemical purity of ^{99m}Tc-HAM should not be less than 95%.

Recommended methods for the determination of radiochemical purity

Paper chromatography		
Stationary phase:	Whatman No. 1 paper strips, 2×9.5 cm	
Solvent:	Methanol:water, 80:20 (v/v)	
Developing time:	10 min	
R_f values:	^{99m}Tc-HAM:	0.0–0.1 (>95%)
	^{99m}Tc-pertechnetate:	0.6–0.7
Thin-layer chromatography		
Stationary phase:	Gelman ITLC-SG (fiberglass), 2×9.5 cm	
Solvent:	Acetone	
Developing time:	5 min	
R_f values:	^{99m}Tc HAM:	0.0–0.1
	^{99m}Tc reduced, colloidal:	0.0–0.1
	^{99m}Tc-pertechnetate:	0.9–1.0 (<5%)

Quantification of labeled components. Each chromatogram is measured (TLC linear analyzer or gamma counter), and the regional radioactivities are expressed as a percentage of the total recovered counts:

$$^{99m}\text{Tc-HAM}\ (\%) = 100 - F$$

where F (%) = ^{99m}Tc-Na-pertechnetate (free).

Results of analysis *(12 samples)*

Results were obtained using paper chromatography and methanol-water as solvent.

Labeling and stability	15 min (%)	3 h (%)
^{99m}Tc-HSA microspheres	99.1 ± 0.05	98.6 ± 0.20
Free ^{99m}Tc-pertechnetate	0.9 ± 0.05	1.4 ± 0.21

Particle Size Distribution. The method described in the *European Pharmacopeia* is based on the microscopic examination of 5,000 albumin microspheres, using a suitable counting chamber such as a hemocytometer.

Microspheres show a homogenous distribution of diameters between 10 and 50 μm. Less than ten microspheres are bigger than 75 μm, and none is larger than 100 μm.

The number of microspheres in 1 ml of the injection solution should also be determined.

Pharmacokinetic Data

^{99m}Tc-albumin microspheres are trapped in the first capillary bed they encounter. When injected intravenously, this is the lung (Zolle and Kropf 1982; Zolle et al. 1970). More than 95% of the injected radioactivity is extracted by the pulmonary capillaries and arterioles in a single passage and distributed in the lung according to regional pulmonary arterial blood flow. The size of the microspheres and the diameter of blood vessels affect the measurement of blood flow. Capillary sizes may vary from organ to organ, and a fraction of the blood may pass through arteriovenous shunts. Under these conditions, ^{99m}Tc-HAM bypass the pulmonary circulation when injected for lung scanning. Microspheres passing through transpulmonary shunts are distributed throughout the systemic circulation. The brain is selected for measuring the fractional radioactivity of microspheres lodged in the cerebral microcirculation (Strauss et al. 1969).

^{99m}Tc-albumin microspheres are metabolized in the capillary lumen (Zolle and Kropf 1982; Zolle et al. 1970). Lung removal rates depend on the size and density of microspheres. Larger microspheres are removed from the lung more slowly. Microspheres with a particle diameter of 12–44 μm are extracted from blood almost completely (≥98%); when prepared at 146 °C, HSA microspheres are eliminated from lung capillaries with a biological half-time of 7.2 h (Zolle and Kropf 1982; Zolle et al. 1970).

Toxicity and Safety Factors. Hemodynamic effects caused by capillary blockade show a direct relationship between the size and number of microspheres injected (Mishkin and Brashear 1971). To avoid an overdose of microspheres, the parameters causing an increase in pulmonary arterial pressure were systematically investigated (Allen et al. 1978; Davis and Taube 1978; Harding et al. 1973; Heyman 1979).

Calculations derived from microspheres with uniform size, having diameters of 20, 28, 40, and 60 μm, have indicated a similar proportion of pulmonary vessels blocked, namely 0.2–0.3%, when 1 mg of these microspheres was injected. However, when calculations were based on the same number of microspheres (i.e., 100,000), the percentage of blocked vessels increased considerably with the diameter of the microspheres (Harding et al. 1973).

Table 12.2.3.1 Influence of particle size on safety factors expressed for total number (100,000) and total mass (mg) of particles injected

Particle diameter (µm)	No. of particles per mg	Safety factors	
		Per 10^5 particles injected	Per mg total mass injected
13.5	776,247	62,380	8,094
15.0	565,884	33,600	5,938
15.8	484,206	33,600	6,939
25.7	112,513	16,630	14,781
28.0	87,002	16,630	19,114
45.4	20,410	4,010	19,647
90.7	2,560	360	14,062

HSA microspheres with a size distribution between 20 and 40 µm contain approximately 90,000 microspheres per milligram. Smaller HSA microspheres with a size distribution between 15 and 20 µm contain approximately 500,000 microspheres per milligram (Table 12.2.3.1).

^{99m}Tc-albumin microspheres (≤1 mg) are well tolerated without complications (Rhodes 1971; Stang et al. 1975). One case of adverse reaction to ^{99m}Tc-HAM has been reported (Littenberg 1975). Hypersensitivity reactions are rare and caused mainly by agents used for suspension of ^{99m}Tc-albumin microspheres. Intracoronary injection of up to 200,000 microspheres had no significant effect upon cardiac function and does not cause myocardial damage (Weller 1975).

Sublethal pulmonary toxicity. Albumin microspheres (15–30 µm) were repeatedly administered in dogs and monkeys showing transient symptoms such as diarrhea, emesis, or decreased activity with amounts of 20 mg/kg BW (Bolles et al. 1973). Single doses up to 40 mg/kg BW were tolerated in mice, rabbits, and dogs without complications (Rhodes et al. 1969).

Acute pulmonary toxicity. The first signs of pulmonary toxicity are an abrupt increase in the pulmonary artery pressure and a sharp fall in the femoral artery pressure (Bolles et al. 1973). The lethal dose (LD_{50}) of microspheres (15–30 µm) in mice is 72.2 ± 8.3 mg/kg BW and in rats, 43.8 ± 5.0 mg/kg BW.

Radiation Dose

The lung and bladder wall are the most exposed organs. The effective (whole body) dose equivalent is 0.011 mSv/MBq (International Commission on Radiological Protection 1987). Elimination from the lung is assumed with half-times of 1.8 h (0.60) and 1.5 days (0.40). Radionuclide released from the lung is primarily excreted by the kidneys (Blau et al. 1982).

The effective (whole body) dose in adults (70 kg) resulting from an intravenous injection of 185 MBq of ^{99m}Tc-HAM for lung scintigraphy is 2.0 mSv.

The absorbed radiation dose to the lung resulting from an intravenous injection of 185 MBq of ^{99m}Tc-HAM is 10.7 mGy.

Storage and Stability

Storage. The lyophilized kit is stored at 2–8 °C.

Stability. The ^{99m}Tc-HAM injection solution is stable for 6 h after preparation.

References

Allen DR, Ferens JM, Cheney FW, Nelp WB (1978) Critical evaluation of acute cardiopulmonary toxicity of microspheres. J Nucl Med 19:1204–1208

Alm A (1975) Radioactively labelled microspheres in regional cerebral blood flow determinations. A study on monkeys with 15- and 35-μm spheres. Acta Physiol Scand 95:60–65

Bergmann H, Böck F, Brenner H, Höfer R (1973) Quantitative evaluation of arterio-venous shunts in brain tumours. In: Medical radioisotope scintigraphy 1972, vol. II, International Atomic Energy Agency, Vienna, p 487

Blau M, Wicks R, Thomas SR, Lathrop KA (1982) MIRD dose estimate report no. 10. Radiation absorbed dose from albumin microspheres labelled with technetium-99m. J Nucl Med 23:915–917

Bolles TF, Kubiatowicz DO, Evans RL, Grotenhuis IM, Nora JC (1973) Tc-99m-labelled albumin (human) microspheres (15–30 micron). Their preparation, properties and uses. In: Radiopharmaceuticals and labelled compounds, vol. 1. International Atomic Energy Agency, Vienna, p 151–167

Council of Europe (1992) Guide for the preparation, use and quality assurance of blood components. Strasbourg

Council of Europe (2004 a) Human albumin injection, monograph 255. European pharmacopeia. Council of Europe, Maisonneuve, Sainte-Ruffine

Council of Europe (2004 b) Human plasma for separation, monograph 853. European pharmacopeia. Council of Europe, Maisonneuve, Sainte-Ruffine

Council of Europe (2005) Technetium ^{99m}Tc microspheres injection. European pharmacopeia 5.0, monograph no 570. Council of Europe, Maisonneuve, Sainte-Ruffine

Davis MA, Taube RA (1978) Pulmonary perfusion imaging: acute toxicity and safety factors as a function of particle size. J Nucl Med 19:1209–1213

Fortuin NJ, Kaihara S, Becker LC, Pitt B (1971) Regional myocardial blood flow in the dog studied with radioactive microspheres. Cardiovasc Res 5:331–336

Harding LK, Horsfield K, Singhal SS, Cumming G (1973) The proportion of lung vessels blocked by albumin microspheres. J Nucl Med 14:579–581

Heck LL, Duley JW Jr (1974) Statistical considerations in lung imaging with Tc-99m-albumin particles. Radiology 113:675–679

Heyman S (1979) Toxicity and safety factors associated with lung perfusion studies with radiolabeled particles. J Nucl Med 20:1098–1099

International Commission on Radiological Protection (1987) Technetium-labelled albumin microspheres (1987) In: Annals of the ICRP, Radiation dose to patients from radiopharmaceuticals, Biokinetic models and data. ICRP publication 53, vol 18, no 1–4. Pergamon, Oxford, pp 227–228

Krejcarek GE, Bradford KL, Bolles TF (1974) Instant labelling human serum albumin microspheres. J Nucl Med 15(Abstr):509

Littenberg RL (1975) Anaphylactoid reaction to human albumin microspheres. J Nucl Med 16:236 237

Martin LG, Larose JH, Sybers RG, Tyras DH, Symbas PN (1973) Myocardial perfusion imaging with Tc-99m-albumin microspheres. Radiology 107:367–370

Mayron LW, Kaplan E (1975) A comparison of four techniques for labelling albumin microspheres with technetium-99m. Int J Nucl Med Biol 2:74–80

Mishkin FS, Brashear RE (1971) Pulmonary and systemic blood pressure responses to large doses of albumin microspheres. J Nucl Med 12:251–252

Rhodes BA (1971) Low probability of allergic reaction to albumin microspheres. J Nucl Med 12:649–650

Rhodes BA, Zolle I, Buchanan JW, Wagner HN Jr (1969) Radioactive albumin microspheres for studies of the pulmonary circulation. Radiology 92:1453–1460

Rhodes BA, Stern HS, Buchanan JW, Zolle I, Wagner HN Jr (1971) Lung scanning with Tc-99m-microspheres. Radiology 99:613–621
Rhodes BA, Greyson ND, Siegel ME, Giargiana FA Jr, White RI Jr, Williams GM, Wagner HN Jr (1973) The distribution of radioactive microspheres after intra-arterial injection in the legs of patients with peripheral vascular disease. Am J Roentgenol Radium Ther Nucl Med 118:820–826
Stang PC, Roelands JF, Cohen P (1975) Immunologic relationships of human serum albumin, macroaggregated albumin and albumin microspheres. In: Subramanian G, Rhodes BA, Cooper JF, Sodd VJ (eds) Radiopharmaceuticals. Society of Nuclear Medicine, New York, p 292
Strauss HW, Hurley PJ, Rhodes BA, Wagner HN Jr (1969) Quantification of right-to-left transpulmonary shunts in man. J Lab Clin Med 74:597–607
Wagner HN Jr, Stern HS, Rhodes BA, Reba RC, Hosain F, Zolle I (1968) Design and development of new radiopharmaceuticals. In: Medical radioisotope scintigraphy, vol. II. IAEA Vienna, pp 3–24
Wagner HN Jr, Rhodes BA, Sasaki Y, Ryan JP (1969) Studies of the circulation with radioactive microspheres. Invest Radiol 4:374–386
Weller DA (1975) Toxicity of particles on intracoronary injection. In: Subramanian G, Rhodes BA, Cooper JF, Sodd VJ (eds) Radiopharmaceuticals. Society of Nuclear Medicine, New York, p 370
Weller DA, Adolph RJ, Wellman HN, Carrol RG, Kim O (1972) Myocardial perfusion scintigraphy after intracoronary injection of Tc-99m-labeled human albumin microspheres. Toxicity and efficacy for detecting myocardial infarction in dogs, preliminary results in man. Circulation 46:963–975
World Health Organization (1994) Requirements for the collection, processing and quality control of blood, blood components and plasma derivatives, requirements for biological substances No. 27. WHO Technical Report Series No. 840
Zolle I, Kropf G (1982) Factors affecting the trapping and clearance of microspheres. In: Anghilieri L (ed) General processes of radiotracer localization, vol. 2. CRC Press, Boca Raton, pp 15–38
Zolle I, Rhodes BA, Wagner HN Jr (1970) Preparation of metabolizable radioactive human serum albumin microspheres for studies of the circulation. Int J Appl Rad Isot 21:155–167

12.3 ^{99m}Tc-Labeled Colloids

12.3.1 ^{99m}Tc-Labeled Microcolloids

12.3.1.1 ^{99m}Tc-Tin Colloid (Size Range: 0.2–0.8 μm)

I. Zolle

Chemical name		**Listed trade names**	
Colloidal tin hydroxide		Amerscan Hepatate II	GE Healthcare (1)
Technetium ^{99m}Tc tin colloid injection (*Ph. Eur.*)		Livoscint	Bristol-Myers Squibb (2)
^{99m}Tc-tin colloid			
Kit components$^{(1)}$		**Kit components**$^{(2)}$	
Tin(II)-fluoride	0.125 mg	Tin(II)-chloride dihydrate	0.3 mg
Sodium fluoride	1.0 mg	Sodium fluoride	0.9 mg
Poloxamer 188	0.5 mg	Sodium chloride	4.5 mg

Preparation

The kit contains the lyophilized, sterile ingredients in a multidose vial. Labeling with ^{99m}Tc-pertechnetate is carried out under aseptic conditions by adding a suitable volume of sterile ^{99m}Tc eluate, up to 3.7 GBq (100 mCi) to the reaction vial. The reaction is allowed to proceed at room temperature for 20 min. ^{99m}Tc-tin colloid is a sterile, pyrogen-free, opalescent solution suitable for intravenous injection. The pH is 4.0–6.0 on pH paper.

Description of the Kit

^{99m}Tc-tin colloid is a hydrolysis product and easily formed at pH > 3, with the reduced technetium-99m-oxide (Lin and Winchell 1972). Poloxamer 188 is added to stabilize the colloid. No heating or pH adjustments are required. Once formed, the particle size distribution is unaffected by time.

The ^{99m}Tc-tin colloid shows a particle size distribution between mainly 0.2 and 0.8 μm (Lin and Winchell 1972; Whateley and Steele 1985).

Factors causing low colloid formation (low specific activity) are primarily related to pH, inadequate reaction time, or a defect of kit formulation. ^{99m}Tc eluate used for colloid preparation should be obtained from a generator by daily elution in order to minimize ^{99}Tc carrier (Ponto et al. 1987).

Clinical Applications

Intravenous injection:	Liver and spleen scintigraphy

Time of Examination
- 10–15 min after the intravenous injection

Recommended Activities for Indications

Liver and spleen scintigraphy:	75–185 MBq (2–5 mCi), injected intravenously, slowly 200 MBq maximal activity for single-photon emission computed tomography (SPECT) 0.2 mg tin colloid/kg body weight The manufacturer's instructions should be followed.

Pediatric Dose. The amount of radioactivity for infants and children administered for liver/spleen scintigraphy is based on body weight, using the scaling factors given in Appendix 1, Table A1.2. The minimum value of the scaling factor is 0.1, since activities below 10% of the value of the adult radioactivity have resulted in poor image quality.

Additional Information

^{99m}Tc-tin colloid should not be injected together with other drugs or components.

Diffuse pulmonary accumulation of radiocolloids may result from elevated plasma levels of aluminum in patients with antacid therapy, because the ^{99m}Tc-tin colloid coprecipitates with larger aggregates of aluminum phosphate (Ponto et al. 1987). A number of clinical conditions associated with pulmonary uptake of ^{99m}Tc-labeled colloids have been reported (Hladik et al. 1987a).

Hepatotoxic substances interfere with phagocytosis, causing transient changes, such as inhomogeneous or irregular distribution of radiocolloids in the liver/spleen or a shift of uptake from liver to spleen and/or bone marrow. Anesthetics (halothane) affect both phagocytic and catabolic reticuloendothelial system (RES) function, resulting in an increased uptake in the spleen (Hladik et al. 1987b). The plasma expander dextran or protein deficiency may also reduce uptake of ^{99m}Tc-labeled colloids by the Kupffer cells (Hodges 1987).

Androgen therapy may stimulate the phagocytic activity of the RES, so that macrophages are released from storage sites and trapped in the lung. In the lung, these phagocytic cells extract the ^{99m}Tc-labeled colloid (Hladik et al. 1987b).

Quality Control

Radiochemical Purity. The *European Pharmacopeia* requires thin-layer chromatography on silica gel fiberglass sheets and a migration distance 10–15 cm for the identification of impurities, using 0.9% sodium chloride solution (saline) as solvent. Free ^{99m}Tc-sodium pertechnetate is measured at an R_f of 1.0, and ^{99m}Tc-tin colloid is identified at the origin. The radiochemical purity of ^{99m}Tc-tin colloid should not be less than 95% (Council of Europe 2005)

The analysis of radiocolloids is based on the determination of free ^{99m}Tc-Na-pertechnetate, since colloidal activity remains at the start. Hydrolized ^{99m}Tc activity cannot be distinguished from the ^{99m}Tc-tin colloid.

Recommended Methods

Thin-layer chromatography using saline ore acetone as solvent

Thin-layer chromatography (*Ph. Eur.*)		
Stationary phase:	Gelman ITLC-SG (fiberglass), 2×10 cm	
Solvent:	Saline	
Developing time:	10 min	
R_f values:	^{99m}Tc tin colloid:	0.0–0.1
	^{99m}Tc-reduced, colloidal:	0.0–0.1
	^{99m}Tc-pertechnetate:	0.9–1.0 (< 5%)

TLC using saline is described for: ^{99m}Tc-tin colloid (Amerscan Hepatate II).

TLC using acetone has been in clinical use for: ^{99m}Tc-tin colloid, ^{99m}Tc–Nanocoll and ^{99m}Tc-(Re)-sulphide colloids.

Quantification of labeled components. Each chromatogram is measured (TLC linear analyzer or gamma counter), and the regional radioactivities are expressed as a percentage of the total recovered counts:

$$^{99m}\text{Tc-(tin) colloid (\%)} = 100 - F$$

where F (%) = ^{99m}Tc-Na-pertechnetate (free).

Results of analysis *(12 samples)*
Results were obtained using thin-layer chromatography and saline as solvent.

Labeling and stability	15 min (%)	3 h (%)
^{99m}Tc-tin colloid	99.0 ± 0.37	98.7 ± 0.35
^{99m}Tc-Na-pertechnetate	1.0 ± 0.39	1.3 ± 0.39

Bioassay of Colloidal Suspensions. To assure a particle size in the colloidal range (≤1 µm), the *European Pharmacopeia* recommends a physiological test in mice. At least 80% of the intravenously injected radiocolloid should localize in the liver and spleen of three mice. Lung uptake should not exceed 5%.

Pharmacokinetic Data

Intravenously injected colloids distribute within the body according to the phagocytic function of the reticuloendothelial system (RES). Size has considerable influence on the biodistribution of colloidal particles. With a particle size of 0.3–0.6 µm, 80–90% of the radioactivity is seen in the liver, with 5–10% seen in the spleen and 5–9% in the bone marrow (Adams et al. 1980; Colombetti 1974; Lin and Winchell 1972; Nelp 1975; Whateley and Steele 1985). Larger colloidal particles show increased splenic uptake, whereas smaller particles localize in the bone marrow (Schuind et al. 1984; Subramanian and McAfee 1970).

Increased splenic uptake has been seen with decreased liver function, i.e., cirrhosis (Atkins et al. 1975). Extrahepatic accumulation of ^{99m}Tc-labeled colloids is also seen in the bone marrow in the case of hyperplastic bone marrow and with certain hematological disorders (Höfer and Egert 1963; Höfer et al. 1964; Nelp and Bower 1969). Occasionally, uptake has been observed in the lung and kidneys (Klingensmith et al. 1976).

In healthy persons, the colloid is rapidly removed from the blood by phagocytosis, mainly in the liver. The colloid disappearance rate has been used to estimate liver blood flow (Vetter et al. 1954). The clearance half-time of colloidal radiogold in male and female patients without liver or circulatory disorders was measured as 2.57 and 2.64 min, respectively; patients with liver cirrhosis showed a considerable increase of clearance half-times (mean 7.37 and 7.29 min).

Inorganic colloids may remain within the macrophages indefinitely. However, ^{99m}Tc-tin colloid is eliminated from the fixed macrophages of the liver and spleen with half-times of approximately 71 and 37 h, respectively (Lin and Winchell 1972).

The metabolism of soluble bivalent tin was studied in rats after intravenous injection of $[^{113}Sn]SnCl_2 \cdot 2H_20$ (30 µCi, no carrier added). It was apparent that tin(II)-chloride is rapidly cleared from the circulation, accumulating mainly in the liver (56.2% after 24 h) and spleen (6.9% after 24 h). The elimination from the liver followed an effective half-time of 85 days, from the spleen, of 50 days (Marciniak 1981).

Toxicity studies. Fourteen-day studies in mice and dogs (intravenous injection) produced no significant signs of toxicity at levels of 3,108 and 490 times the maximum human dose (≤0.2 mg/kg).

The acute chemical toxicity of $SnCl_2 \cdot 2H_2O$ was determined after the intravenous injection of 1–12 mg Sn^{2+}/kg body weight in rats. The animals were observed for 30 days for signs of toxicity or death. The LD_{50} value of $SnCl_2 \cdot 2H_2O$ was determined in rats as 7.83 mg Sn^{2+}/kg body weight. Lethality was observed with doses exceeding 3 mg Sn^{2+}/kg body weight; death occurred within 24 h after the intravenous injection (Marciniak 1981).

Radiation Dose

The liver, spleen, and red marrow are the most exposed organs. Calculations of the absorbed radiation dose resulting from liver and spleen scintigraphy are based on technetium-labeled colloids (International Commission on Radiological Protection 1987). The effective dose equivalent is 0.014 mSv/MBq. The effective whole-body dose in adults (70 kg) resulting from an intravenous injection of 185 MBq of ^{99m}Tc-tin colloid is 2.6 mSv.

The effective dose equivalent per ICRP 62 has been replaced by the quantity effective dose. Values per unit administered activity were published in Addendum 1 (International Commission on Radiological Protection 1991). Values calculated accordingly are slightly lower than the effective whole body doses presented here.

Storage and Stability

Storage. The lyophilized kit is stored at 2–8 °C.

Stability. ^{99m}Tc-tin colloid injection is stable for 6 h after the preparation.

References

Adams FG, Horton PW, Selim SM (1980) Clinical comparison of three liver scanning agents. Eur J Nucl Med 5:237–239

Atkins HL, Cloutier RJ, Lathrop KA et al (1975) Studies of the reticuloendothelial system (RES). III. Blockade of the RES in man. J Nucl Med 16:108A–108B

Colombetti LG (1974) ^{99m}Tc-Sn-colloid for liver dynamic studies. Radiobiol Radiother 15:47–53

Council of Europe (2005) Technetium ^{99m}Tc tin colloid injection. European pharmacopeia 5.0, monograph no 689. Council of Europe, Maisonneuve, Sainte-Ruffine

Hladik WB, III, Gregorio N, Braun TL, Stathis VJ, Ponto JA (1987 a) Radiopharmaceutical information and consultation services. In: Hladik WB, III, Saha GB, Study KT (eds) Essentials of nuclear medicine science. Williams & Wilkins, Baltimore, p 392

Hladik WB, III, Ponto JA, Lentle BC, Laven DL (1987b) Iatrogenic alterations in the biodistribution of radiotracers as a result of drug therapy: reported instances. In: Hladik WB, III, Saha GB, Study KT (eds) Essentials of nuclear medicine science. Williams & Wilkins, Baltimore, pp 189–219

Hodges R (1987) Iatrogenic alterations in the biodistribution of radiotracers as a result of drug therapy: theoretical considerations. In: Hladik WB, III, Saha GB, Study KT (eds) Essentials of nuclear medicine science. Williams & Wilkins, Baltimore, pp 165–188

Höfer R, Egert H (1963) Radiogoldverteilung im Knochenmark [in German]. Bull Schweiz Akad Med Wiss 19:36–46

Höfer R, Ogris E, Pfeiffer G (1964) Kolloidspeicherung im Knochenmark bei Bluterkrankungen. Wien Z Inn Med 45:330–335

International Commission on Radiological Protection (1987) Technetium-labeled colloids. In: Annals of the ICRP, radiation dose to patients from radiopharmaceuticals, biokinetic models and data. ICRP publication 53, vol 18, no 1-4. Pergamon, Oxford, pp 179–183

International Commission on Radiological Protection (1991) In: Annals of the ICRP, radiological protection in biomedical research, ICRP Publication 62, vol 22, no 3. Pergamon, Oxford, pp 25–28

Klingensmith WC III, Tsan MF, Wagner HN Jr (1976) Factors affecting the uptake of ^{99m}Tc-sulfur colloid by the lung and kidney. J Nucl Med 17:681

Lin MS, Winchell HS (1972) A "kit" method for the preparation of a technetium-tin(II) colloid and a study of its properties. J Nucl Med 13:58–65

Marciniak M (1981) Bivalent tin metabolism and toxicity after intravenous injection in rats. Acta Physiol Pol 32:2

Nelp WB (1975) An evaluation of colloids for RES function studies. In: Subramanian G, Rhodes BA, Cooper JF, Sodd VJ (eds) Radiopharmaceuticals. Society of Nuclear Medicine, New York, pp 349–355

Nelp WB, Bower RE (1969) The quantitative distribution of the erythron and the RE cell in the bone marrow organ of man. Blood 34:276–280

Ponto JA, Swanson DP, Freitas JE (1987) Clinical manifestations of radio-pharmaceutical formulation problems. In: Hladik WB III, Saha GB, Study KT (eds) Essentials of nuclear medicine science. Williams & Wilkins, Baltimore, pp 268–289

Schuind F, Schoutens A, Verhas M, Verschaeren A (1984) Uptake of colloids by bone is dependent on bone blood flow. Eur J Nucl Med 9:461–463

Subramanian G, McAfee JG (1970) Stannous oxide colloid labeled with ^{99m}Tc or ^{113}In for bone marrow imaging. J Nucl Med 11:365–366

Vetter H, Falkner R, Neumayr A (1954) The disappearance rate of colloidal radiogold form the circulation and its application to the estimation of liver blood flow in normal and cirrhotic patients. J Clin Invest 33:1594

Whateley TL, Steele G (1985) Particle size and surface charge studies of a tin colloid radiopharmaceutical for liver scintigraphy. Eur J Nucl Med 19:353–357

12.3.1.2 ^{99m}Tc-Rhenium Sulfide Colloid (Size Range: 0.3–1.0 µm)

I. Zolle

Chemical name	**Listed trade names**
Technetium ^{99m}Tc colloidal rhenium sulfide injection (*Ph. Eur.*) Technetium Tc 99m sulfur colloid injection (*USP*) ^{99m}Tc-(Re)-sulfide colloid ^{99m}Tc-sulfur colloid	HepatoCis (TCK-1) (1) Sulfotec Sorin (2)

Kit components$^{(1)}$		**Kit components**$^{(2)}$	
Potassium perrhenate	4.17 mg	Potassium perrhenate	4.17 mg
Sodium thiosulphate·5 H_2O	10 mg	Sodium thiosulphate·5 H_2O	10 mg
Gelatin	166 mg	Gelatin	100 mg
Water for injection	6 ml		

Preparation

Each kit consists of one vial with two syringes, one syringe containing 1 ml of 4.6 N HCl, the other 2 ml of 0.65 M citrate buffer (pH 12.0).

Each vial contains a sterile, pyrogen-free solution of the ingredients. After adding 5 ml of sterile ^{99m}Tc eluate under aseptic conditions to the vial, the reaction is acidified by adding 1 ml of 4.6 N HCl (syringe 1). The reaction vial is placed into a boiling water bath for 5 min. After cooling, 2 ml of 0.65 M citrate buffer solution (syringe 2) are injected into the vial. The resulting pH should be between 3.5 and 6.0 on pH paper. The manufacturer's instructions should be followed. ^{99m}Tc-(Re)-sulfide colloid is a sterile, pyrogen-free, light brown, solution (14 ml), suitable for intravenous injection.

Description of the Kit

^{99m}Tc-(Re)-sulfide colloid is formed at acidic pH in the boiling water bath. The reaction is based on the formation of colloidal sulfur and technetium heptasulfide (Tc_2S_7) at acidic pH. Rhenium is used as a carrier (Larson and Nelp 1966; Patton et al. 1966). Ten milligrams of sodium thiosulfate pentahydrate generate 2 mg of colloidal sulfur (Stern et al. 1966). The added ^{99m}Tc eluate should not be less than 5 ml, the ^{99m}Tc activity not less than 370 MBq (10 mCi).

In principle, ^{99m}Tc-sulfur colloid and ^{99m}Tc-(Re)-sulfide colloid are prepared using similar conditions (Patton et al. 1966; Larson and Nelp 1966); however, the original formulation did not contain K-perrhenate carrier. Gelatin was used as a stabilizer (Stern et al. 1966; Haney et al. 1971).

^{99m}Tc-(Re)-sulfide colloid and ^{99m}Tc-sulfur colloid, respectively, show a particle size distribution mainly between 0.3 and 0.8 µm.

Factors causing low colloid formation (low specific activity) are primarily related to pH, incorrect order of mixing, low heating temperature, heating a large volume, inadequate boiling time, or a defect of kit formulation. ^{99m}Tc eluate used for colloid preparation should be obtained from a generator by daily elution in order to minimize ^{99}Tc carrier (Ponto et al. 1987). A flocculent precipitate is formed in the presence of Al^{3+} cation (1 µg/ml) (Haney et al. 1971; Ponto et al. 1987).

Clinical Applications

Intravenous injection:	Liver and spleen scintigraphy
	Bone marrow scintigraphy
Oral application:	Digestive transit scintigraphy
	Gastroduodenal motor activity

Time of Examination

Liver and spleen scintigraphy:	15–20 min after the intravenous injection
Bone marrow scintigraphy:	1 h after the intravenous injection
Digestive transit scintigraphy:	Dynamic imaging is started immediately after oral administration of test meal.

Recommended Activities for Indications

Liver and spleen scintigraphy:	75–150 MBq (2–4 mCi), injected intravenously, slowly
	200 MBq maximal activity for single-photon emission computer tomography (SPECT)
	≤0.2 mg sulfur colloid/kg body weight
Bone marrow scintigraphy:	150 MBq (400 MBq maximum activity)
Digestive transit scintigraphy:	37 MBq (1 mCi) in 30 ml water, followed by 300 ml water
	The manufacturer's instructions should be followed.

Pediatric Dose. The amount of radioactivity for infants and children administered for liver and spleen, and bone marrow scintigraphy is based on body weight, using the scaling factors given in Appendix 1, Table A1.2. The minimum value of the scaling factor is 0.1, since activities below 10% of the value of the adult radioactivity have resulted in poor image quality.

Additional Information

^{99m}Tc-(Re)-sulfide colloid should not be injected together with other drugs or components.

Diffuse pulmonary accumulation of radiocolloids may result from elevated plasma levels of aluminum in patients with antacid therapy, because the ^{99m}Tc-sulfur colloid coprecipitates with larger aggregates of aluminum phosphate (Ponto et al. 1987). A number of clinical conditions associated with pulmonary uptake of ^{99m}Tc-labeled colloids have been reported (Hladik et al. 1987 a).

Hepatotoxic substances interfere with phagocytosis, causing transient changes such as inhomogeneous or irregular distribution of radiocolloids in the liver and spleen or a shift of uptake from liver to spleen and/or bone marrow. Anesthetics (halothane) affect both phagocytic and catabolic reticuloendothelial system (RES) function, resulting in an increased uptake in the spleen (Hladik et al. 1987 b). The plasma expander dextran or protein deficiency may also reduce uptake of ^{99m}Tc-labeled colloids by the Kupffer cells (Hodges 1987).

Androgen therapy may stimulate the phagocytic activity of the RES, so that macrophages are released from storage sites and trapped in the lung. In the lung these phagocytic cells extract the ^{99m}Tc-labeled colloid (Hladik et al. 1987 b).

Renal accumulation of ^{99m}Tc-sulfur colloid has been reported in transplant patients during rejection and during episodes of acute tubular necrosis. Kanamycin is known to cause acute tubular necrosis (Hodges 1987).

Gastric emptying has been studied using solid meals to which the labeled colloid (^{99m}Tc-(Re)-sulfide colloid or ^{99m}Tc-sulfur colloid) was added (Stacher and Bergmann 1992). After 6 h of fasting, the patient is allowed to eat the test meal. From the plot of percent activity versus time, the half-time of gastric emptying is determined. Depending on the consistency of the test meal (liquid or solid), a shorter (10–15 min) or longer (50–80 min) half-time is measured. Meal size and composition must be standardized. Delayed emptying is associated with a number of disease states.

Orally administered colloids are not absorbed from the gastrointestinal tract.

Quality Control

Radiochemical Purity. The *European Pharmacopeia* requires paper chromatography (distance 10–15 cm) for the identification of impurities using 0.9% sodium chloride solution (saline) as solvent. Free ^{99m}Tc-pertechnetate is measured at R_f 0.6 and ^{99m}Tc-(Re)-sulfide colloid is identified at the origin.

The radiochemical purity of ^{99m}Tc-(Re)-sulfide colloid and ^{99m}Tc-sulfur colloid, respectively, should not be less than 92% (Council of Europe 2005 a, b).

The analysis of radiocolloids is based on the determination of free ^{99m}Tc-Na-pertechnetate, since colloidal activity remains at the start. Hydrolized ^{99m}Tc activity cannot be distinguished from the ^{99m}Tc-(Re)-sulfide colloid.

Paper chromatography (*Ph. Eur.*)		
Stationary phase:	Whatman No. 1 paper, 2×10 cm	
Solvent:	Saline (0.9% NaCl)	
Developing time:	10 min	
R_f values:	^{99m}Tc-(Re)-sulfide colloid:	0.0–0.1 (>92%)
	^{99m}Tc reduced, hydrolized	0.0–0.1
	^{99m}Tc-pertechnetate:	0.6–0.7

Recommended Methods

Thin-layer chromatography on Gelman silica gel fiberglass sheets using acetone as solvent has been recommended for the identification of free ^{99m}Tc-pertechnetate at the solvent front (R_f=1.0) and ^{99m}Tc-(Re)-sulfide colloid at the origin.

This method has been in clinical use for: ^{99m}Tc-(Re)-sulfide colloid, ^{99m}Tc-tin colloid, and ^{99m}Tc-Nanocoll.

Quantification of labeled components. Each chromatogram is measured (TLC linear analyzer or gamma counter), and the regional radioactivities are expressed as a percentage of the total recovered counts:

$$^{99m}\text{Tc-(Re)-sulfide colloid } (\%) = 100 - F$$

where F (%) = ^{99m}Tc-Na-pertechnetate (free).

Results of analysis *(12 samples)*
Results were obtained using thin-layer chromatography and acetone as solvent.

Labeling and stability	15 min (%)	3 h (%)
^{99m}Tc-(Re)-sulfide colloid	97.3±0.47	96.7±0.57
^{99m}Tc-pertechnetate	2.7±0.38	3.3±0.97

Bioassay of Colloidal Suspensions. To assure a particle size in the colloidal range (≤1 μm), the *European Pharmacopeia* recommends a physiological test in mice. At least 80% of the intravenously injected radiocolloid should localize in the liver and spleen of three mice. Lung uptake should not exceed 5%.

Pharmacokinetic Data

Intravenously injected colloids distribute within the body according to the phagocytic function of the RES. Size has considerable influence on the biodistribution of colloidal particles. With a particle size of 0.3–0.6 μm, 80–90% of the radioactivity is seen in the liver, 4–8% in the spleen, and 3–5% in the bone marrow (Adams et al. 1980; Nelp 1975). Larger colloidal particles show increased splenic uptake, whereas smaller particles localize in the bone marrow (Nelp and Bower 1969; Schuind et al. 1984). Increased splenic uptake has been seen with decreased liver function, i.e., cirrhosis (Atkins et al. 1975). Extrahepatic accumulation of ^{99m}Tc-labeled colloids is also seen in the bone marrow in the case of hyperplastic bone marrow and with certain hematological disorders (Höfer and Egert 1963; Höfer et al. 1964). Occasionally, uptake has been observed in the lung and kidneys (Klingensmith et al. 1976).

^{99m}Tc-(Re)-sulfide colloid is rapidly extracted from the blood circulation by phagocytosis, approximately 94% in one passage. The colloid disappearance rate has been used to estimate liver blood flow (Vetter et al. 1954). The clearance half-time of colloidal radiogold in male and female patients without liver or circulatory disorders was measured as 2.57 and 2.64 min, respectively; patients with liver cirrhosis showed a considerable increase of clearance half-times (means of 7.37 and 7.29 min). For the measurement of plasma clearance, a high radiochemical purity of the colloid is essential (Alavi 1982).

Inorganic colloids may remain within the macrophages indefinitely. It has been reported, however, that ^{99m}Tc-labeled sulfur colloid may be metabolized and excreted (Warbick-Cerone and Phythian 1982). However, the urinary excretion (in 48 h) measured in three patients was less than 4% of the injected radioactivity (Larson and Nelp 1966).

No signs of toxicity were observed with large doses of ^{99m}Tc-sulfur colloid (40 mg sulfur/kg) after intravenous injection in mice over a period of 4–7 days (Haney et al. 1971).

Radiation Dose

The most exposed organs are the liver, spleen, and red marrow. Calculations of the absorbed radiation dose resulting from liver and spleen scintigraphy are based on technetium-labeled colloids (International Commission on Radiological Protection 1987). The effective dose equivalent is 0.014 mSv/MBq. The effective whole-body dose in adults (70 kg) resulting from an intravenous injection of 185 MBq of ^{99m}Tc-(Re)-sulfide colloid is 2.6 mSv.

Calculations of the absorbed dose resulting from the oral application of ^{99m}Tc-(Re)-sulfide colloid are based on ICRP Publication 53 (International Commission on Radiological Protection 1987b) for nonabsorbable markers. The effective dose in adults (70 kg) resulting from 37 MBq (1 mCi) of orally administered ^{99m}Tc-(Re)-sulfide colloid is approximately 1 mSv.

The effective dose equivalent per ICRP Publication 62 has been replaced by the quantity effective dose. Values per unit administered activity were published in Addendum 1 (International Commission on Radiological Protection 1991). Values calculated accordingly are slightly lower than the effective whole-body doses presented here.

Storage and Stability

Storage. The kit is stored at room temperature.

Stability. The ^{99m}Tc-(Re)-sulfide colloid injection is stable for 6 h after preparation.

References

Adams FG, Horton PW, Selim SM (1980) Clinical comparison of three liver scanning agents. Eur J Nucl Med 5:237–239

Alavi A (1982) Detection of gastrointestinal bleeding with ^{99m}Tc-sulfur colloid. Semin Nucl Med 12:126–138

Council of Europe (2005 a) Technetium ^{99m}Tc colloidal rhenium sulphide injection. European pharmacopeia 5.0, monograph, no 126. Council of Europe, Maisonneuve, Sainte-Ruffine

Council of Europe (2005 b) Technetium ^{99m}Tc sulfur colloid injection. European pharmacopeia 5.0, monograph, no 131. Council of Europe, Maisonneuve, Sainte-Ruffine

Haney TA, Ascanio I, Gigliotti JA, Gusmano EA, Bruno GA (1971) Physical and biological properties of a ^{99m}Tc-sulfur colloid preparation containing disodium edetate. J Nucl Med 12:64–68

Hladik WB III, Gregorio N, Braun TL, Stathis VJ, Ponto JA (1987a) Radiopharmaceutical information and consultation services. In: Hladik WB, III, Saha GB, Study KT (eds) Essentials of nuclear medicine science. Williams & Wilkins, Baltimore, p 392

Hladik WB III, Ponto JA, Lentle BC, Laven DL (1987b) Iatrogenic alterations in the biodistribution of radiotracers as a result of drug therapy: reported instances. In: Hladik WB III, Saha GB, Study KT (eds) Essentials of nuclear medicine science. Williams & Wilkins, Baltimore, pp 189–219

Hodges R (1987) Iatrogenic alterations in the biodistribution of radiotracers as a result of drug therapy: theoretical considerations. In: Hladik WB III, Saha GB, Study KT (eds) Essentials of nuclear medicine science. Williams & Wilkins, Baltimore, pp 165–188

Höfer R, Egert H (1963) Radiogoldverteilung im Knochenmark [in German]. Bull Schweiz Akad Med Wiss 19:36–46

Höfer R, Ogris E, Pfeiffer G (1964) Kolloidspeicherung im Knochenmark bei Bluterkrankungen [in German]. Wien Z Inn Med 45:330–335

International Commission on Radiological Protection (1987a) Technetium-labelled colloids (1987) In: Annals of the ICRP, radiation dose to patients from radiopharmaceuticals, biokinetic models and data. ICRP publication 53, vol 18, no 1–4. Pergamon, Oxford, pp 179–183

International Commission on Radiological Protection (1987b) Technetium-labelled non-absorbable markers (1987) In: Annals of the ICRP, radiation dose to patients from radiopharmaceuticals, biokinetic models and data. ICRP publication 53, vol. 18, no. 1–4. Pergamon, Oxford, pp 225–226

International Commission on Radiological Protection (1991) In: Annals of the ICRP, Radiological protection in biomedical research, ICRP Publication 62, vol 22, no 3. Pergamon, Oxford, pp 25–28

Klingensmith WC III, Tsan MF, Wagner HN Jr (1976) Factors affecting the uptake of ^{99m}Tc-sulfur colloid by the lung and kidney. J Nucl Med 17:681

Larson SM, Nelp WB (1966) Radiopharmacology of a simplified technetium-99m-colloid preparation for photoscanning. J Nucl Med 7:817–826

Nelp WB (1975) An evaluation of colloids for RES function studies. In: Subramanian G, Rhodes BA, Cooper JF, Sodd VJ (eds) Radiopharmaceuticals. Society of Nuclear Medicine, New York, pp 349–355

Nelp WB, Bower RE (1969) The quantitative distribution of the erythron and the RE cell in the bone marrow organ of man. Blood 34:276–280

Patton, DP, Garcia, EN, Webber, MM (1966) Simplified preparation of technetium-99m-sulfide colloid for liver scanning. Am J Roentgenol Radium Ther Nucl Med 97:880

Ponto JA, Swanson DP, Freitas JE (1987) Clinical manifestations of radio-pharmaceutical formulation problems. In: Hladik WB III, Saha GB, Study KT (eds) Essentials of nuclear medicine science. Williams & Wilkins, Baltimore, pp 268–289

Schuind F, Schoutens A, Verhas M, Verschaeren A (1984) Uptake of colloids by bone is dependent on bone blood flow. Eur J Nucl Med 9:461–463

Stacher G, Bergmann H (1992) Scintigraphic quantitation of gastrointestinal motor activity and transport: oesophagus and stomach. Eur J Nucl Med 19:815–823

Stern HS, McAfee JG, Subramanian G (1966) Preparation, distribution and utilization of technetium-99m-sulfur colloid. J Nucl Med 7:665–675

United States Pharmacopeial Convention (2005) United States Pharmacopeia (USP) 28 official monographs: Technetium Tc 99m sulfur colloid injection. United States Pharmacopeial Convention, Rockville, Md, p 1865

Vetter H, Falkner R, Neumayr A (1954) The disappearance rate of colloidal radiogold form the circulation and its application to the estimation of liver blood flow in normal and cirrhotic patients. J Clin Invest 33:1594

Wagner HN Jr, Iio M (1964) Studies of the reticuloendothelial system (RES). III. Blockade of the RES in man. J Clin Invest 42:1525–1532

Warbick-Cerone A, Phythian JR (1982) The role of endocytosis in the localization of radiotracers. In: Anghileri LJ (ed) General processes of radiotracer localization, vol 1. CRC Press, Boca Raton, p 173

12.3.1.3 ^{99m}Tc-Albumin Microcolloid (Size Range: 0.2–2.0 μm)

I. Zolle

Chemical name		**Listed trade names**	
Human albumin microaggregates		AlbuRes	Solco (1)
Technetium albumin colloid injection (*USP*)		Microlite	DuPont Merck (2)
^{99m}Tc-albumin colloid			
^{99m}Tc-HSA-microcolloid			
Kit components$^{(1)}$		**Kit components**$^{(2)}$	
HSA-microcolloid	2.5 mg	HSA-microcolloid	1.0 mg
Stannous chloride·$2H_2O$	0.4 mg	Stannous chloride·$2H_2O$	0.17 mg
Monobasic sodium phosphate	0.458 mg	Human albumin	10 mg
Disodium phytate	0.25 mg	Disodium medronate	0.12 mg
Glucose anhydrous	15 mg	Sodium phosphate	10 mg
Poloxamer 238	2.5 mg	Poloxamer 188	1.1 mg

Preparation

The kits contain the sterile, lyophilized, preformed microcolloid in a multidose vial. Labeling with ^{99m}Tc-pertechnetate is carried out under aseptic conditions by adding 1–5 ml of ^{99m}Tc eluate (maximum of 3 GBq). The reaction is allowed to proceed at room temperature for 5 min. ^{99m}Tc-microaggregates form a sterile, pyrogen-free, milky-white solution suitable for intravenous injection. The pH is 4.0–7.0 on pH paper.

Description of the Kit

Preformed albumin microcolloid is easily labeled with reduced technetium at room temperature, provided the amount of stannous chloride is not less than 6 μg. No heating or pH adjustments are required. More than 90% of ^{99m}Tc-albumin microcolloids show a size distribution between 0.2 and 2.0 μm (Chia 1986; Honda et al. 1970; Taplin et al. 1964a; Yamada et al. 1969).

^{99m}Tc eluate used for labeling the preformed colloid should be obtained from a generator by daily elution in order to minimize ^{99}Tc carrier and should comply with specifications stated in the *European Pharmacopeia*. A flocculent precipitate is formed in the presence of Al^{3+} ion (>1 μg/ml) (Ponto et al. 1987).

Clinical Applications

Intravenous injection: Liver and spleen scintigraphy

Time of Examination. The time of examination should be 15–60 min after intravenous injection.

Recommended Activities for Indications

Liver and spleen scintigraphy:	40–150 MBq (1–4 mCi), injected intravenously 200 MBq (5.4 mCi) maximal activity for single-photon emission computer tomography (SPECT) ≤0.025 mg microaggregates/kg body weight The manufacturer's instructions should be followed.

Pediatric Dose. The amount of radioactivity for infants and children administered for liver/spleen scintigraphy is based on body weight, using the scaling factors given in Appendix 1, Table A1.2. The minimum value of the scaling factor is 0.1, since activities below 10% of the value of the adult radioactivity have resulted in poor image quality.

Additional Information

The use of ^{99m}Tc-albumin microcolloid in patients with a history of hypersensitivity to human albumin is contraindicated. ^{99m}Tc-albumin microcolloid should not be injected together with other drugs or components.

Albumin aggregates in the colloidal size range are obtained by heat denaturation of a 1% solution of human serum albumin (HSA) in alkaline medium (pH 10.0) (Honda et al. 1970). The size of the albumin colloid is affected by the pH, temperature, and the mode of agitation (Taplin et al. 1964 a).

Diffuse pulmonary accumulation of radiocolloids may result from elevated plasma levels of aluminum in patients with antacid therapy (Ponto et al. 1987). A number of clinical conditions associated with pulmonary uptake of ^{99m}Tc-labeled colloids have been reported (Hladik et al. 1987 a).

Hepatotoxic substances interfere with phagocytosis, causing transient changes, such as inhomogeneous or irregular distribution of radiocolloids in the liver/spleen or a shift of uptake from liver to spleen and/or bone marrow. Anesthetics (halothane) affect both phagocytic and catabolic reticuloendothelial system (RES) function, resulting in an increased uptake in the spleen (Hladik et al. 1987 b). The plasma expander dextran or protein deficiency may also reduce uptake of ^{99m}Tc-labeled colloids by the Kupffer cells (Hodges 1987).

Quality Control

Radiochemical Purity. ^{99m}Tc-albumin microcolloid is not described in the *European Pharmacopeia*. Thin-layer chromatography is recommended by the manufacturer, using 85% methanol as solvent. Free ^{99m}Tc-sodium pertechnetate is measured at R_f 0.7 and

^{99m}Tc-albumin microcolloid is identified at the origin. The radiochemical purity of ^{99m}Tc-albumin microcolloid should not be less than 95%.

The analysis of radiocolloids is based on the determination of free ^{99m}Tc-Na-pertechnetate, since colloidal activity remains at the start. Hydrolized ^{99m}Tc-activity cannot be distinguished from ^{99m}Tc-HSA colloid.

Recommended Methods

Thin-layer chromatography (USP 28)

Thin-layer chromatography (*USP*)		
Stationary phase:	Gelman ITLC-SG (fiberglass), 2×9.5 cm	
Solvent:	Methyl ethyl ketone (MEK)	
Developing time:	5 min	
R_f values:	^{99m}Tc-HSA microcolloid:	0.0–0.1
	^{99m}Tc reduced, hydrolized	0.0–0.1
	^{99m}Tc-pertechnetate:	0.9–1.0 (<5%)

Quantification of labeled components. Each chromatogram is measured (TLC linear analyzer or gamma counter), and the regional radioactivities are expressed as a percentage of the total recovered counts:

$$^{99m}\text{Tc-HSA microcolloid } (\%) = 100 - F$$
$$F\ (\%) = {}^{99m}\text{Tc-Na-pertechnetate (free).}$$

Results of analysis *(12 samples)*

Results were obtained using the analytical method described by the manufacturer.

Labeling and stability	15 min (%)	3 h (%)
^{99m}Tc-HSA microcolloid	95.9±0.26	94.7±0.32
^{99m}Tc-Na-pertechnetate	4.1±0.10	5.3±0.12

Pharmacokinetic Data

Intravenously injected colloids distribute within the body according to the phagocytic activity of the RES. Size has considerable influence on the biodistribution of colloidal particles. With a particle size of 0.3–0.6 μm, 80–90% of the radioactivity is extracted by the liver, 4–8% by the spleen, and <1% by the bone marrow (Nelp 1975). Larger colloidal particles show increased splenic uptake, whereas smaller particles (nanocolloids) localize in the bone marrow (10–15%) (Nelp and Bower 1969; Schuind et al. 1984).

In healthy persons, ^{99m}Tc-HSA microcolloid is rapidly removed from the blood by phagocytosis, approximately 94% in one passage (Shaldon et al. 1961). Fifteen minutes after the intravenous injection, only approximately 1% of the radioactivity is measured in blood (Chia 1986). The colloid disappearance rate has been used to estimate liver blood flow. Following the intravenous injection of tracer quantities (less than 0.25 mg/kg body weight), elimination half-times of 2–3 min were measured (Kitani and Taplin

1972). Using a smaller colloid (10–20 nm), a mean half-time of 2.6 min was reported (Iio et al. 1963). Values exceeding 3.0 min indicate reduced hepatic blood flow or a decreased liver function (cirrhosis), as observed in patients with various liver diseases (Iio et al. 1963; McAfee et al. 1975; Palmer et al. 1971; Taplin et al. 1964 b; Wagner et al. 1963). In patients with reduced hepatic uptake, the concentration of radioactivity within the spleen is increased.

^{99m}Tc-albumin microcolloid is metabolized in the Kupffer cells by proteolytic enzymes and eliminated from the liver (Reske et al. 1981; Taplin et al. 1964 a). In the case of ^{99m}Tc-albumin millimicrospheres, biological half-times of 10 min and 4.7 h were measured (Reske et al. 1981). In patients with overactive digestive function, RES ^{99m}Tc-labeled degradation products have been observed in the bile, accumulating in the gallbladder during the first 2–3 h post-intravenous injection (Kitani and Taplin 1972).

Repeated injections of high doses (5 mg/kg) of aggregated albumin (≤ 80 nm) over a 2-month period have been well tolerated in three subjects, showing no effect on the clearance rate when compared with control subjects (Iio et al. 1963). Subcutaneous injection of small doses (0.02 mg/kg) as well as a saturation dose of 4 mg/kg showed no evidence of hypersensitivity (Iio et al. 1963).

Radiation Dose

The liver, spleen, and red marrow are the most exposed organs. Calculations of the absorbed radiation dose resulting from liver and spleen scintigraphy are based on technetium-labeled colloids (International Commission on Radiological Protection 1987). The effective dose equivalent is 0.014 mSv/MBq. The effective whole-body dose in adults (70 kg) resulting from an intravenous injection of 185 MBq of ^{99m}Tc-albumin microcolloid is 2.6 mSv.

The effective dose equivalent per ICRP Publication 62 has been replaced by the quantity effective dose. Values per unit administered activity were published in Addendum 1 (International Commission on Radiological Protection 1991). Values calculated accordingly are slightly lower than the effective whole-body doses presented here.

Storage and Stability

Storage. The lyophilized kit is stored at 2–8 °C.

Stability. ^{99m}Tc-albumin microcolloid injection solution is stable for 3 h after preparation.

References

Chia HL (1986) Particulate radiopharmaceuticals in nuclear medicine. In: Cox PH (ed) Radiopharmacy and radiopharmacology yearbook II. Grune and Stratton, New York

Hladik WB III, Gregorio N, Braun TL, Stathis VJ, Ponto JA (1987 a) Radiopharmaceutical information and consultation services. In: Hladik WB, III, Saha GB, Study KT (eds) Essentials of nuclear medicine science. Williams & Wilkins, Baltimore, p 392

Hladik WB III, Ponto JA, Lentle BC, Laven DL (1987b) Iatrogenic alterations in the biodistribution of radiotracers as a result of drug therapy: reported instances. In: Hladik WB III, Saha GB, Study KT (eds) Essentials of nuclear medicine science. Williams & Wilkins, Baltimore, pp 189–219

Hodges R (1987) Iatrogenic alterations in the biodistribution of radiotracers as a result of drug therapy: theoretical considerations. In: Hladik WB III, Saha GB, Study KT (eds) Essentials of nuclear medicine science. Williams & Wilkins, Baltimore, pp 165–188

Honda T, Kazem I, Croll MN, Brady LW (1970) Instant labeling of macro- and microaggregated albumin with ^{99m}Tc. J Nucl Med 11:580–585

Iio M, Wagner HN Jr, Scheffel U, Jabbour B (1963) Studies of the reticuloendothelial system (RES). I. Measurement of the phagocytic capacity of the RES in man and dog. J Clin Invest 42:417–426

International Commission on Radiological Protection (1987) Technetium-labelled colloids. In: Annals of the ICRP, radiation dose to patients from radiopharmaceuticals, biokinetic models and data. ICRP publication 53, vol 18, no 1–4. Pergamon, Oxford, pp 179–183

International Commission on Radiological Protection (1991) ICRP Publication 62. In: Annals of the ICRP, radiological protection in biomedical research, vol 22, no 3. Pergamon, Oxford, pp 25–28

Kitani K, Taplin GV (1972) Biliary excretion of ^{99m}Tc-albumin microaggregate degradation products (a method for measuring Kupffer cell digestive function?). J Nucl Med 13:260–264

McAfee JG, Ause RG, Wagner HN Jr (1975) Diagnostic value of scintillation scanning of the liver. Arch Intern Med 116:95–110

Nelp WB (1975) An evaluation of colloids for RES function studies. In: Subramanian G, Rhodes BA, Cooper JF, Sodd VJ (eds) Radiopharmaceuticals. Society of Nuclear Medicine, New York, pp 349–355

Nelp WB, Bower RE (1969) The quantitative distribution of the erythron and the RE cell in the bone marrow organ of man. Blood 34:276–280

Palmer DI, Rifkind D, Brown DW (1971) ^{131}I-labeled colloidal serum albumin in the study of reticuloendothelial system function. III. Phagocytosis and catabolism compared in normal, leukemia, and immunosuppressed human subjects. J Infect Dis 123:465–469

Ponto JA, Swanson DP, Freitas JE (1987) Clinical manifestations of radio-pharmaceutical formulation problems. In: Hladik WB III, Saha GB, Study KT (eds) Essentials of nuclear medicine science. Williams & Wilkins, Baltimore, pp 268–289

Reske SN, Vyska K, Feinendegen LE (1981) In vivo assessment of phagocytic properties of Kupffer cells. J Nucl Med 22:405

Schuind F, Schoutens A, Verhas M, Verschaeren A (1984) Uptake of colloids by bone is dependent on bone blood flow. Eur J Nucl Med 9:461–463

Shaldon S, Chiandussi L, Guevara L, Caesar J, Sherlock S (1961) The estimation of hepatic blood flow and intrahepatic shunted blood flow by colloidal heat-denatured human serum albumin labeled with ^{131}I. J Clin Invest 40:1346–1354

Taplin GV, Dore EK, Johnson DE (1964) Hepatic blood flow and reticuloendothelial system studies with radiocolloids. In: Kniseley RM et al (eds) Dynamic clinical studies with radioisotopes. US Atomic Energy Commission, Division of Technical Information, TID 7678, pp 285–317

Taplin GV, Johnson DE, Dore EK, Kaplan HS (1964) Suspensions of radioalbumin aggregates for photoscanning the liver, spleen, lung and other organs. J Nucl Med 5:259–275

United States Pharmacopeial Convention (2005) United States Pharmacopeia (USP) 28 official monographs: Technetium Tc 99m albumin colloid injection. United States Pharmacopeial Convention, Rockville, Md, p 1851

Wagner HN Jr, Iio M, Hornick RB (1963) Studies of the reticuloendothelial system (RES). II. Changes in the phagocytic capacity of the RES in patients with certain infections. J Clin Invest 42:427–434

Yamada H, Johnson DE, Griswold ML et al (1969) Radioalbumin microaggregates for reticuloendothelial organ scanning and function assessment. J Nucl Med 10:453–454

12.3.1.4 ^{99m}Tc-Albumin Millimicrospheres (Size Range: 0.3–1.0 μm)

I. Zolle

Chemical name

Albumin millimicrospheres
HSA millimicrospheres
milli-HAM
^{99m}Tc-HSA millimicrospheres
^{99m}Tc-millimicrospheres
^{99m}Tc-milli-HAM

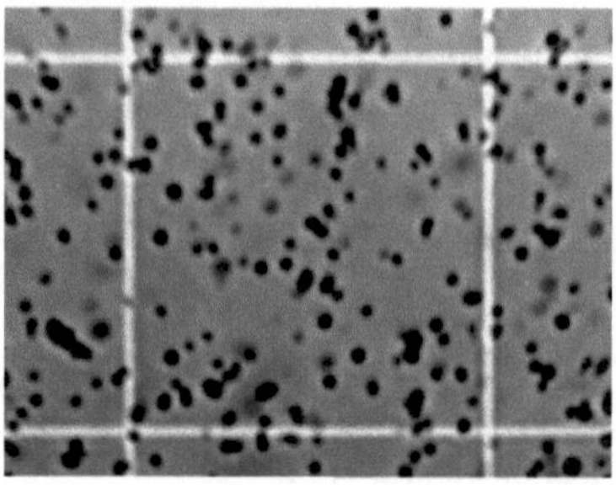

Albumin millimicrospheres (<1 μm)
(Small square 50×50 μm, 600-fold)

Kit components

HSA millimicrospheres	2.0 mg
Stannous chloride·2 H_2O	0.5 mg
Cysteine	0.2 mg
Pluronic F68	2.0 mg
Sodium chloride	18.0 mg

Listed trade names

Nanotec Sorin

Preparation

The kit contains sterile, lyophilized, preformed millimicrospheres in a multidose vial. Labeling with ^{99m}Tc-pertechnetate is carried out under aseptic conditions by adding 1–5 ml of ^{99m}Tc eluate (maximum of 3 GBq (80 mCi) to obtain a specific activity of 600 MBq/ml/0.4 mg). The reaction is allowed to proceed at room temperature for 15 min. ^{99m}Tc-millimicrospheres are a sterile, pyrogen-free, opalescent solution suitable for the intravenous injection. The pH is 6.0–7.5.

Description of the Kit

Preformed albumin millimicrospheres are easily labeled with reduced technetium at room temperature (Zolle et al. 1970, 1973). No heating or pH adjustments are required. Consistently high labeling yields were obtained with milli-HAM containing tin (II) salt (Villa et al. 1976), and when milli-HAM were labeled by electrolytic reduction using a tin anode and a platinum cathode (Angelberger et al. 1985; Köhn et al. 1985). More than 90% of ^{99m}Tc-albumin millimicrospheres show a size distribution between 0.3 and 0.8 μm (Köhn et al. 1985; Scheffel et al. 1972).

^{99m}Tc eluate used for labeling the preformed millimicrospheres should be obtained from a generator by daily elution in order to minimize ^{99m}Tc carrier and should comply with specifications stated in the *European Pharmacopeia.* The presence of Al^{3+} ion (>1 μg/ml) in the eluate may cause aggregation of the colloidal particles (Ponto et al. 1987).

For inhalation of ^{99m}Tc-albumin millimicrospheres, the vial (Nanotec) is transferred to an aerosol generator. Dry aerosols are obtained with ^{99m}Tc-milli-HAM suspended in ethanol (Angelberger et al. 1985; Köhn et al. 1985). These have advantages over moist aerosols, showing higher lung penetration between 20 and 30%, while in the case of aqueous nebulization, only 1–10% of the radioactivity are deposited in the lung (Santolicandro and Giuntini 1979).

Clinical Applications

Intravenous injection:	Liver and spleen scintigraphy
	Regional liver perfusion and RES function in cases of impaired liver function (cirrhosis, metal poisoning, transplantation)
	Bone marrow scintigraphy
Nebulization as aerosol:	Inhalation scintigraphy
	Assessment of pulmonary ventilation in patients with chronic obstructive lung disease, and for the differential diagnosis of acute pulmonary embolism in combination with lung perfusion scintigraphy
	Measurement of mucociliary function

Time of Examination

Liver and spleen scintigraphy:	10–60 min after the intravenous injection
Bone marrow imaging:	45–60 min after the intravenous injection
Inhalation scintigraphy:	Immediately after inhalation

Recommended Activities for Indications

Liver and spleen scintigraphy:	40–150 MBq (1–4 mCi), injected intravenously
	200 MBq (5.4 mCi) maximal activity for single-photon emission computed tomography (SPECT)
	≤0.025 mg millimicrospheres/kg body weight
Bone marrow scintigraphy:	185–370 MBq (5–10 mCi), injected intravenously
	400 MBq maximum recommended activity
Inhalation scintigraphy:	150 MBq (4 mCi) inhaled radioactivity
	The manufacturer's instructions should be followed.

Pediatric Dose. The amount of radioactivity for infants and children administered for liver/spleen and bone marrow scintigraphy is based on body weight, using the scaling factors given in Appendix 1, Table A1.2. The minimum value of the scaling factor is 0.1, since activities below 10% of the value of the adult radioactivity have resulted in poor image quality.

Additional Information

The use of ^{99m}Tc-albumin millimicrospheres in patients with a history of hypersensitivity to human albumin is contraindicated.

^{99m}Tc-albumin millimicrospheres should not be injected together with other drugs or components.

For intravenous injection (preferably not through a lying catheter), ^{99m}Tc-albumin millimicrospheres should be homogenously suspended to avoid in vivo aggregates. For this reason, the aspiration of blood into the syringe should be avoided.

Albumin millimicrospheres are obtained by heating a dispersed solution of human serum albumin (25%) in oil (Zolle et al. 1970, 1973). The size of the albumin particles depends mainly on the degree of homogenization, generally a size distribution between 0.3 and 3 μm is obtained (Zolle et al. 1970). For intravenous injection, particles larger than 1 μm have been removed by differential centrifugation (Angelberger et al. 1985; Köhn et al. 1985).

Hepatotoxic substances interfere with phagocytosis causing transient changes, such as inhomogeneous or irregular distribution of radiocolloids in the liver/spleen, or a shift of uptake from liver to spleen and/or bone marrow. Anesthetics (halothane) affect both phagocytic and catabolic RES function, resulting in an increased uptake in the spleen (Hladik et al. 1987).

Quality Control

Radiochemical purity. ^{99m}Tc-HSA-millimicrospheres are not described in the *European Pharmacopeia*. Thin-layer chromatography is recommended by the manufacturer using 85% methanol as solvent. Free ^{99m}Tc-pertechnetate is measured at an R_f of 0.6, and ^{99m}Tc-millimicrospheres are identified at the origin. The radiochemical purity of ^{99m}Tc-milli-HAM should not be less than 95%.

The analysis of radiocolloids is based on the determination of free ^{99m}Tc-Na-pertechnetate, since colloidal activity remains at the start. Hydrolized ^{99m}Tc-activity cannot be distinguished from ^{99m}Tc-milli-HAM.

Recommended Methods

Thin-layer chromatography		
Stationary phase:	Gelman ITLC-SG (fiberglass), 2×9.5 cm	
Solvent:	Methanol (85%)	
Developing time:	5 min	
R_f values:	^{99m}Tc-milli-HAM:	0.0–0.1
	^{99m}Tc reduced, hydrolized:	0.0–0.1
	^{99m}Tc-pertechnetate:	0.6–0.7 (<5%)

Quantification of labeled components. Each chromatogram is measured (TLC linear analyzer or gamma counter) and the regional radioactivities are expressed as a percentage of the total recovered counts:

^{99m}Tc-milli-HAM (%) = 100 − F

where F (%) = ^{99m}Tc-Na-pertechnetate (free).

Results of analysis *(12 samples)*
Results were obtained using the analytical method described by the manufacturer.

Labeling and stability	15 min (%)	3 h (%)
^{99m}Tc-HSA millimicrospheres	96.7 ± 0.57	96.3 ± 0.49
^{99m}Tc-Na-pertechnetate	3.2 ± 0.59	3.7 ± 0.48

^{99m}Tc-milli-HAM are labeled with a high radiochemical yield and show high stability of the radioactive label.

Pharmacokinetic Data

Trace amounts of intravenously injected ^{99m}Tc-albumin millimicrospheres (0.02 mg/kg body weight) are rapidly removed from the blood with a half-time of 1.4–2.0 min (Reske et al. 1981; Scheffel et al. 1972). From 83 to 86% is taken up in the liver by phagocytosis; the maximum activity in the liver is reached 5–10 min after the intravenous injection (Scheffel et al. 1972; Zolle et al. 1973). The plasma disappearance rate is related to the particle size of colloids. Using aggregated albumin (10–20 nm), a mean clearance half-time of 2.6 min was reported (Iio et al. 1963). In patients with reduced hepatic uptake, the spleen shows a high concentration of radioactivity (McAfee et al. 1975).

^{99m}Tc-albumin millimicrospheres are metabolized in the Kupffer cells. The elimination of ^{99m}Tc-milli-HAM from the liver was shown as a reliable indicator of specific Kupffer-cell function (Reske et al. 1981). A biphasic elimination was observed in man, with a fast ($T_{1/2}$ < 10 min) and a slowly degrading component ($T_{1/2}$ > 4 h) (Reske et al. 1981; Szabo et al. 1984). In tumor patients without detectable liver metastases or direct tumor invasion, elimination from the liver by phagocytosis was markedly delayed ($T_{1/2}$ of 8.55 h vs 4.28 h), while the clearance of ^{99m}Tc-millimicrospheres from the blood was increased ($T_{1/2}$ of 1.3 min vs 1.8 min) (Reske et al. 1981).

In the rat, ^{99m}Tc-milli-HAM have shown a monoexponential removal from the liver ($T_{1/2}$ of 8.8 h) with increasing intestinal activity 1–6 h after the intravenous injection (Villa et al. 1976). Biliary excretion of degradation products was observed with ^{99m}Tc-albumin microaggregates (Kitani and Taplin 1972; Wetterfors et al. 1960).

For the application of ^{99m}Tc-milli-HAM as an aerosol, preparations with an average diameter of approximately 0.5 μm have shown free diffusion into the lung periphery and high alveolar retention (Agnew et al. 1981; Köhn et al. 1985). Larger millimicrospheres (1–5 μm) sediment preferentially in the trachea and upper bronchial tree (Weiss et al. 1981).

^{99m}Tc-albumin millimicrospheres are well tolerated without complications. No acute reactions were seen when injecting 5–15 mg/kg body weight in 80 mice (Scheffel et al. 1972).

Hypersensitivity reactions are rare and caused mainly by agents used for suspension of ^{99m}Tc-milli-HAM. No cases of anaphylaxis have been reported after aerosol inhalation.

Radiation Dose

The liver, spleen, and red marrow are the most exposed organs. Calculations of the absorbed radiation dose resulting from liver and spleen scintigraphy are based on technetium-labeled colloids (ICRP Publication 53, International Commission on Radiological Protection 1987a). The effective dose equivalent is 0.014 mSv/MBq. The effective whole-body dose in adults (70 kg) resulting from an intravenous injection of 185 MBq (5 mCi) of ^{99m}Tc-millimicrospheres is 2.6 mSv.

Calculations of the absorbed radiation dose resulting from inhalation of ^{99m}Tc-millimicrospheres are based on technetium-labeled aerosols (ICRP Publication 53, International Commission on Radiological Protection 1987b). It is assumed that the label is released in the lung slowly, with a biological half-time of 24 h, and that the activity is excreted by the kidneys. The effective dose equivalent is 0.015 mSv/MBq. The dose to the bladder wall after inhalation of 150 MBq (4 mCi) is 1.95 mGy. The effective whole-body dose in adults (70 kg) resulting from inhalation of 150 MBq of ^{99m}Tc-millimicrospheres is 2.3 mSv.

The effective dose equivalent per ICRP 62 has been replaced by the quantity effective dose. Values per unit administered activity were published in Addendum 1 (International Commission on Radiological Protection 1991). Values calculated accordingly are slightly lower than the effective whole-body doses presented here.

Storage and Stability

Storage. The lyophilized kit is stored at 2–8 °C.

Stability. The ^{99m}Tc-milli-HAM injection is stable for 6 h after preparation.

References

Agnew JE, Bateman JR, Watts M, Paramananda V, Pavia D, Clarke SW (1981) The importance of aerosol penetration for lung mucociliary clearance studies. Chest 80(Suppl):843–846

Angelberger P, Zolle I, Strigl A, Köhn H, Mostbeck A, Fiedler W (1985) A dry aerosol of ^{99m}Tc-albumin-millimicrospheres for lung ventilation scintigraphy, In: Cox PH, Limouris G, Woldring MG (eds) Progress in radiopharmacology, vol 4. Martinus Nijhoff, The Hague, pp 73–86

Hladik WB, III, Ponto JA, Lentle BC, Laven DL (1987b) Iatrogenic alterations in the biodistribution of radiotracers as a result of drug therapy: reported instances. In: Hladik WB III, Saha GB, Study KT (eds) Essentials of nuclear medicine science. Williams & Wilkins, Baltimore, pp 189–219

Iio M, Wagner HN Jr, Scheffel U, Jabbour B (1963) Studies of the reticuloendothelial system (RES). I. Measurement of the phagocytic capacity of the RES in man and dog. J Clin Invest 42:417–426

International Commission on Radiological Protection (1987a) Technetium-labelled colloids (1987) In: Annals of the ICRP, radiation dose to patients from radiopharmaceuticals, biokinetic models and data. ICRP publication 53, vol 18, no 1–4, Pergamon, Oxford, pp 179–183

International Commission on Radiological Protection (1987b) Technetium-labelled aerosols (1987) In: Annals of the ICRP, radiation dose to patients from radiopharmaceuticals, biokinetic models and data. ICRP publication 53, vol 18, no 1–4. Pergamon, Oxford, pp 217–219

International Commission on Radiological Protection (1991) In: Annals of the ICRP, radiological protection in biomedical research, ICRP Publication 62, vol 22, no 3, Pergamon, Oxford, pp 25–28

Kitani K, Taplin GV (1972) Biliary excretion of 99mTc-albumin microaggregate degradation products (a method for measuring Kupffer cell digestive function?). J Nucl Med 13:260–264

Köhn H, Klech H, Angelberger P, Strigl A, Zolle I, Kummer F, Mostbeck A (1985) Dry aerosol of monodisperse millimicrospheres for ventilation imaging: production, delivery system, and clinical results in comparison with ^{81m}Kr and ^{127}Xe. Eur J Nucl Med 10:411–416

McAfee JG, Ause RG, Wagner HN Jr (1975) Diagnostic value of scintillation scanning of the liver. Arch Intern Med 116:95–110

Ponto JA, Swanson DP, Freitas JE (1987) Clinical manifestations of radio-pharmaceutical formulation problems. In: Hladik WB III, Saha GB, Study KT (eds) Essentials of nuclear medicine science. Williams & Wilkins, Baltimore, pp. 268–289

Reske SN, Vyska K, Feinendegen LE (1981) In vivo assessment of phagocytic properties of Kupffer cells. J Nucl Med 22:405

Santolicandro A, Giuntini C (1979) Patterns of deposition of labelled monodispersed aerosols in obstructive lung disease. J Nucl Med All Sci 23:115

Scheffel U, Rhodes BA, Natarajan TK, Wagner HN Jr (1972) Albumin microspheres for study of the reticuloendothelial system. J Nucl Med 13:498–503

Szabo Z, Vosberg H, Segall M, Feinendegen LE (1984) Messung der mittleren Retentionszeiten ^{99m}Tc-makierter HSA-Millimikrosphären in der Leber – Klinische Ergebnisse bei Patienten mit operiertem Mammakarzinom [in German]. Nucl Med 23:171–176

Villa M, Pretti O, Mosca R, Plassio G, Pasqualini R (1976) Preparation and evaluation of tin-albumin millimicrospheres for labelling with Tc-99m. J Nucl Biol Med 20:168–171

Weiss T, Dorow P, Felix R, Schmutzler H (1981) Pulmonales Aerosolverteilungsmuster und regionale mukoziliäre Clearance bei Patienten mit chronisch obstruktiver Atemwegserkrankung und small airways disease [in German]. Atemw Lungenkrkh 7:172–178

Wetterfors J, Gullberg R, Lilhedahl SO, Birke G, Olhage B (1960) Role of the stomach in albumin breakdown. Acta Med Scand 168:347

Zolle I, Bergmann H, Höfer R (1973) Millimicrospheres zur Funktionsprüfung des retikuloendothelialen Systems (RES) In: Teil, Fellinger K, Höfer R (eds) Radioaktive Isotope in Klinik und Forschung, Band 10, 2. Urban & Schwarzenberg, München-Berlin, 446–453

Zolle I, Hosain F, Rhodes BA, Wagner HN Jr (1970) Human serum albumin millimicrospheres for studies of the reticuloendothelial system. J Nucl Med 11:379

12.3.2 ^{99m}Tc-Labeled Nanocolloids

12.3.2.1 ^{99m}Tc-Rhenium Sulfide Nanocolloid (Size Range: 10–100 nm)

I. Zolle

Chemical name		Listed trade names	
Rhenium sulfide nanocolloid		NanoCis (TCK-17) (A+B)	CIS Bio
Tin(II)-sulfide nanocolloid		Lymphoscint (Solco)	GE Healthcare
^{99m}Tc-(Re)-sulfide nanocolloid			
^{99m}Tc-(Sn)-sulfide nanocolloid			
Kit components A		**Kit components B**	
Rhenium sulfide	0.48 mg	Sodium pyrophosphate · $10\,H_2O$	3.0 mg
Ascorbic acid	7.0 mg	Stannous chloride · $2\,H_2O$	0.5 mg
Gelatin	9.6 mg		
Water for injection	1.0 ml		
Hydrochloric acid (conc.)	37.4 µl		

Preparation

The NanoCis kit (TCK-17) consists of two vials, A and B. Vial A contains a sterile, pyrogen-free solution of ingredients. Vial B contains lyophilized sodium pyrophosphate and stannous chloride. Two milliliters of sterile water for injection are added to vial B to dissolve its content. Half a milliliter of solution B is transferred to vial A, and mixed well. Then, 1–2 ml of ^{99m}Tc eluate is added under aseptic conditions (not less than 370 MBq [10 mCi] to assure a specific activity of ≥100 MBq/ml/0.15 mg). The reaction vial is placed into a boiling water bath for 15–30 min. After cooling, the labeled colloid is ready for use. ^{99m}Tc-(Re)-sulfide nanocolloid is a sterile, pyrogen-free, brown solution, suitable for subcutaneous (interstitial) injection. The pH is between 4.0 and 7.0 on pH paper.

^{99m}Tc-Sn(II)-sulfide colloid (Lymphoscint) is prepared by adding 0.2–2 ml of sterile ^{99m}Tc eluate under aseptic conditions (not more than 1.5 GBq [40 mCi]). A volume of approximately 0.5 ml of the ^{99m}Tc eluate is recommended by the manufacturer to minimize ionic aluminum. The final volume may be adjusted with oxidant-free saline. The reaction vial is placed into a boiling water bath for 4 min. After cooling, the labeled colloid is ready for use. ^{99m}Tc-Sn(II)-sulfide nanocolloid is a sterile, pyrogen-free solution, suitable for subcutaneous (interstitial) injection. The pH is between 4.0 and 7.0 on pH paper.

Description of the Kit

^{99m}Tc-(Re)-sulfide colloid is formed by reduction with tin pyrophosphate in the boiling water bath. Rhenium sulfide is used as a carrier (Larson and Nelp 1966; Patton et al. 1966). The added ^{99m}Tc eluate should not exceed 2 ml, the ^{99m}Tc activity should not be less than 370 MBq (10 mCi). Gelatin is used as a stabilizer. The size distribution of ^{99m}Tc-(Re)-sulfide colloid is between 40 and 80 nm, comparable with the previously available ^{99m}Tc-(Sb)-sulfide colloid (5–15 nm).

^{99m}Tc-(Sn)-sulfide colloid is prepared by labeling a preformed colloid in the boiling water bath; gelatin is used for stabilization. ^{99m}Tc-tin(II)-sulfide colloid has a favorable particle size, namely ≤50 nm.

Factors causing low colloid formation (low specific activity) are primarily related to pH, incorrect order of mixing, low heating temperature, heating a large volume, inadequate boiling time, or a defect of kit formulation. ^{99m}Tc eluate used for colloid preparation should be obtained from a generator by daily elution in order to minimize ^{99}Tc carrier; preferably, a small volume is added in order to reduce the concentration of Al^{3+} ion (Ponto et al. 1987).

Clinical Applications

Subcutaneous (interstitial) injection:	Lymphoscintigraphy
	Visualization of lymphatic flow and regional lymph nodes in the extremities and the trunk
Subdermal or peritumoral injection:	Sentinel lymph node (SLN) scintigraphy
Oral application:	Gastroesophageal scintigraphy
	Esophageal motility disorders
	Gastroduodenal motor activity

Lymphoscintigraphy is performed for the diagnosis of peripheral edema caused by recurring erysipel or by affected lymph nodes as a result of metastatic infiltration, lymphogranuloma, lymphosarcoma, damage as a result of radiation treatment, or other causes. Other indications include secondary edema caused by blockade of the lymphatic flow, and thrombosis (Lofferer et al. 1974; Mostbeck et al. 1984). Lymphatic drainage from a primary tumor was studied in patients with breast cancer (Ege 1983). Several colloidal preparations have been evaluated (Kaplan et al. 1979, 1985; Nagai et al. 1982; Strand and Persson 1979).

SLN imaging can identify the first infected lymph node of a primary tumor before surgery. If the SLN is histologically tumor free, no other lymph node will contain metastatic disease (Alazraki et al. 1997).

Time of Examination

Lymphoscintigraphy:	Sequential scintigraphy 15 min after subcutaneous injection up to 1 h Visualization of lymph nodes between 2 and 6 h after injection
SLN imaging:	Sequential planar scintigraphy 5–10 min postinjection and 1–6 h postinjection (late images)
Gastroesophageal scintigraphy:	Sequential scintigraphy and static imaging may be performed

Recommended Activities for Indications

Lymphoscintigraphy of the extremities:	20–75 MBq (0.5–2 mCi) in a volume of 0.2–0.3 ml, subcutaneous injection into the interdigital spaces of the hands or feet. For a better resorption and to reduce the radiation burden to the injection site, the total dose (≤185 MBq [5 mCi]) may be divided between two and three injection sites. 40 MBq (1 mCi) maximum activity per injection site The volume of injection should not exceed 0.5 ml
Parasternal lymphatics:	20 MBq (0.5 mCi) in a volume of 0.3–0.5 ml, single subcostal injection at both sides
SLN:	50–80 MBq (1.3–2.2 mCi) in a volume of 0.1–0.5 ml, multiple subdermal injections (peritumoral) Guidelines for SLN detection should be followed (Cox et al. 1998)
Gastroesophageal reflux:	20 MBq (0.5 mCi) of ^{99m}Tc-(Re, Sn)-sulfide nanocolloid in a suitable liquid according to local practice 40 MBq (1 mCi) maximum recommended activity

Pediatric Dose. The amount of radioactivity for infants and children administered for lymphoscintigraphy is based on body weight, using the scaling factors given in Appendix 1 (Table A1.2). The minimum value of the scaling factor is 0.1, since activities below 10% of the value of the adult radioactivity have resulted in poor image quality.

Additional Information

^{99m}Tc-(Re, Sn)-sulfide colloid is not used for intravenous injection. A comparison with ^{99m}Tc-albumin nanocolloid showed considerable residual radioactivity in blood; approximately 11% were measured 3 h after the intravenous injection (De Schrijver et al. 1987).

The stability of colloids is affected by electrolytes, thus ^{99m}Tc-(Re, Sn)-sulfide colloid should not be injected together with other drugs or components.

Interstitial injection of a hypertonic colloidal solution may be associated with pain at the injection site. Therefore, ^{99m}Tc-(Re)-sulfide nanocolloid should be injected slowly.

^{99m}Tc-(Sn)-sulfide nanocolloid is supplied as an iso-osmolar solution, warranting a painless subcutaneous injection.

For interstitial injection ^{99m}Tc-(Re, Sn)-sulfide nanocolloid must be applied in a small volume not exceeding 0.5 ml. To ensure high specific activity, labeling should be performed with a small volume of ^{99m}Tc eluate (≤2 ml) and not less than 370 MBq (10 mCi).

The use of local anesthetic agents or hyaluronidase prior to administering the labeled preparation have been shown to negatively affect lymphatic uptake.

Iodinated contrast media used in lymphangiography may interfere with lymphatic imaging with ^{99m}Tc-(Re, Sn)-sulfide nanocolloid.

Quality Control

Radiochemical Purity. ^{99m}Tc-(Re, Sn)-sulfide nanocolloid is not described in the *European Pharmacopeia*. Paper chromatography on Whatman 1 paper is recommended by the manufacturer, using methyl ethyl ketone (MEK) as solvent. Free ^{99m}Tc-sodium pertechnetate is measured at an R_f of 1.0 and ^{99m}Tc-(Re, Sn)-sulfide nanocolloid is identified at the origin. The radiochemical purity of ^{99m}Tc-(Re, Sn)-sulfide nanocolloid should not be less than 95%.

The analysis of radiocolloids is based on the determination of free ^{99m}Tc-Na-pertechnetate, since colloidal activity remains at the start. Hydrolized ^{99m}Tc activity cannot be distinguished from the ^{99m}Tc-(Re, Sn)-sulfide nanocolloid.

Recommended Methods

Paper chromatography		
Stationary phase:	Whatman No. 1 paper, 2×9.5 cm	
Solvent:	Methyl ethyl ketone (MEK)	
Developing time:	5 min	
R_f values:	^{99m}Tc-(Re, Sn)-sulfide, nanocolloid:	0.0–0.1 (>95%)
	^{99m}Tc reduced, hydrolized:	0.0–0.1
	^{99m}Tc-pertechnetate:	0.9–1.0

Thin-layer chromatography on Gelman silica gel fiberglass sheets using acetone as solvent has been recommended for the identification of ^{99m}Tc-(Re)-sulfide microcolloid at the origin, and free ^{99m}Tc-pertechnetate at the solvent front (R_f = 1.0). This method is also recommended for the analysis of the nanocolloids.

Quantification of labeled components
Each chromatogram is measured (TLC linear analyzer or gamma counter) and the regional radioactivities are expressed as a percentage of the total recovered counts:

$$^{99m}\text{Tc-(Re, Sn)-sulfide nanocolloid (\%)} = 100 - F$$

where F (%) = ^{99m}Tc-Na-pertechnetate (free).

Results of analysis *(12 samples)*

Results were obtained using thin-layer chromatography and acetone as solvent.

Labeling and stability	15 min (%)	3 h (%)
^{99m}Tc-(Re, Sn)-sulfide nanocolloid	98.5 ± 0.67	96.3 ± 0.97
^{99m}Tc-Na-pertechnetate	1.4 ± 0.61	3.6 ± 1.07

Pharmacokinetic Data

After subcutaneous injection into the interdigital spaces of the hands or feet, ^{99m}Tc-(Re, Sn)-sulfide nanocolloid is transported with the interstitial liquid through the lymphatic capillaries into the lymph ducts, and is almost quantitatively retained by the regional lymph nodes (Mostbeck et al. 1984). Release of the colloid from the lymph nodes is slow and increasing with movement of the extremities. After 15 min of slow walking, 6–15% of the injected radioactivity has been measured in the regional lymph nodes (Mostbeck et al. 1984). Accumulation of the colloid in the liver is negligible.

Following the subcostal injection at both sides of the xyphoid (2–2.5 cm under the xyphoid) into the musculus rectus abdominalis, ^{99m}Tc-(Re, Sn)-sulfide nanocolloid is transported through the parasternal lymphatics into mediastinal and diaphragmal lymph nodes. The maximal accumulation is reached 3 h after injection (Ege 1983). Drainage from the interstitial injection site was observed between 1 and 35% in 24 h (Ege 1976).

Basic concepts for quantitative lymphoscintigraphy were derived from experimental studies in rabbits by correlating the uptake of different radiocolloids in the parasternal lymph nodes with measurements of the particle size (Strand and Persson 1979). For colloidal gold (Au-198) with a well defined particle diameter of approximately 5 nm, highest uptake in regional lymph nodes was recorded. ^{99m}Tc-(Sb)-sulfide colloid (Sb_2S_3), with colloidal particles between 5 and 15 nm, showed less uptake (approximately 60%), yet the highest uptake of all ^{99m}Tc-colloids that were evaluated. Larger particles cannot pass through the lymphatic capillary pores and are trapped in the interstitial fluid. Thus ^{99m}Tc-tin colloid, ^{99m}Tc-phytate, and ^{99m}Tc-sulfur colloid remain primarily at the injection site and show little lymphatic migration (Strand and Persson 1979).

^{99m}Tc-(Re, Sn)-sulfide nanocolloids show optimal physical characteristics (40–80 nm) for reliable visualization of anatomic lymph nodes.

Subdermal, peritumoral injection of ^{99m}Tc-(Re, Sn)-sulfide nanocolloid is used for visualization of the lymphatic drainage of a primary tumor and for identification of the SLN (Alazraki et al. 1997; Keshtgar et al. 1999; Nitz and Heidenreich 1999).

Radiation Dose

Calculations of the absorbed radiation dose after the subcutaneous administration of ^{99m}Tc-(Re)-sulfide nanocolloid into the extremities are based on the assumption that approximately 5–15% of the radioactivity are distributed over 10–20 lymph nodes (Mostbeck et al. 1984). The highest radiation dose is thus delivered at the injection site. If 37 MBq (1 mCi) are injected into each foot, the radiation absorbed dose at the injection site has been calculated as 400–700 mGy, and 22–27 mGy for each lymph node (Mostbeck et al. 1984). The weight of a lymph node was assumed to be 5 g.

Based on a model for the subcutaneous injection of ^{99m}Tc-(Sb)-sulfide colloid into the umbilical region, Bergquist et al. (1982) have calculated the radiation dose at the injection site assuming a tissue volume of 10 ml. Injection of 37 MBq (1 mCi) results in an average radiation dose of approximately 300 mGy at the injection site. The effective dose equivalent was determined as 0.005 mSv/MBq (Bergqvist et al. 1982).

Based on this information, the effective dose in adults (70 kg) resulting from the subcutaneous injection of 185 MBq (5 mCi) of ^{99m}Tc-(Re, Sn)-sulfide colloid is approximately 1 mSv.

Storage and Stability

Storage. The lyophilized kit is stored at 2–8 °C.

Stability. The ^{99m}Tc-(Re)-sulfide nanocolloid injection is stable for 4 h after preparation.

References

Alazraki N, Eshima D, Eshima LA, Herda SC, Murray DR, Vansant JP, Taylor AT (1997) Lymphoscintigraphy, the sentinel node concept, and the intraoperative gamma probe in melanoma, breast cancer and other potential cancers. Semin Nucl Med 27:55–67

Bergqvist L, Strand SE, Hafström L, Jönsson P-E (1984) Lymphoscintigraphy in patients with malignant melanoma: a quantitative and qualitative evaluation of its usefulness. Eur J Nucl Med 9:129–135

Cox CE, Pendas S, Cox JM, Joseph E, Shons AR, Yeatman T, Ku NN, Lyman GH, Berman C, Haddad F, Reintgen DS (1998) Guidelines for sentinel lymph node detection. Ann Surg 227:645–653

De Schrijver M, Streule K, Senekowitsch R, Fridrich R (1987) Scintigraphy of inflammation with nanometer-sized colloidal tracers. Nucl Med Commun 8:895–908

Ege GN (1976) Internal mammary lymphoscintigraphy – the rationale, technique, interpretation and clinical application: a review based on 848 cases. Radiology 118:101–107

Ege GN (1983) Lymphoscintigraphy – techniques and applications in the management of breast carcinoma. Semin Nucl Med 13:26–34

Kaplan WD, Davis MA, Rose CM (1979) A comparison of two technetium-99m-labeled radiopharmaceuticals for lymphoscintigraphy. J Nucl Med 20:933–937

Kaplan WD, Piez CW, Gelman RS, Laffin SM, Rosenbaum EM, Jennings CA, McCormick CA, Harris JR, Henderson IC, Atkins HL (1985) Clinical comparison of two radiocolloids for internal mammary lymphoscintigraphy. J Nucl Med 26:1382–1385

Keshtgar MRS, Waddington WA, Lakhani SR, Ell PJ (1999) The sentinel node surgical oncology. Springer, Berlin Heidelberg New York, and Eur J Nucl Med 26:57–67

Larson SM, Nelp WB (1966) Radiopharmacology of a simplified technetium-99m-colloid preparation for photoscanning. J Nucl Med 7:817–826

Lofferer O, Mostbeck A, Partsch H (1974) Lymphtransportstörung beim dicken Bein – Isotopenlymphographische Ergebnisse [in German]. Z. Hautkr. 49(14): 615–622

Mostbeck A, Kahn P, Partsch H (1984) Quantitative Lymphographie beim Lymphödem. In: Bollinger A, Partsch H (eds) Initiale Lymphstrombahn [in German]. Georg Thieme, Stuttgart, pp 116–122

Nagai K, Ito Y, Otsuka N, Muranaka A (1982) Deposition of small ^{99m}Tc-labeled colloids in bone marrow and lymph nodes. Eur J Nucl Med 7:66–70

Nitz DW P, Heidenreich P (eds) (1999) Sentinel-Lymphknoten In: Der Nuklearmediziner [in German]. Demeter, Munich

Patton DP, Garcia EN, Webber MM (1966) Simplified preparation of technetium-99m-sulfide colloid for liver scanning. Am J Roentgenol Radium Ther Nucl Med 97:880
Ponto JA, Swanson DP, Freitas JE (1987) Clinical manifestations of radio-pharmaceutical formulation problems. In: Hladik WB III, Saha GB, Study KT (eds) Essentials of nuclear medicine science. Williams & Wilkins, Baltimore, pp 268–289
Strand S-E, Persson BRR (1979) Quantitative lymphoscintigraphy I: basic concepts for optimal uptake of radiocolloids in the parasternal lymph nodes of rabbits. J Nucl Med 20:1038–1046

12.3.2.2 ^{99m}Tc-Albumin Nanocolloid (Size Range: 10–80 nm)

I. Zolle

Chemical names	**Listed trade names**	
Human albumin nanocolloid	Solco Nanocoll	GE Healthcare
Albumin nanoaggregates		
Aggregated albumin (AA)		
^{99m}Tc-HSA-nanocolloid		

Kit components	
Human albumin nanocolloid	0.5 mg
Stannous chloride · 2 H_2O	0.2 mg
Glucose anhydrous	
Dibasic sodium phosphate, anhydrous	
Sodium phytate, anhydrous	
Poloxamer 238	

Preparation

The kit contains the sterile, lyophilized, preformed colloid in a multidose vial. Labeling with ^{99m}Tc-pertechnetate is carried out under aseptic conditions by adding 1–5 ml of ^{99m}Tc eluate (maximum 5.5 GBq: approximately 150 mCi). The reaction is allowed to proceed at room temperature for 10 min. ^{99m}Tc-HSA nanocolloid is a sterile, pyrogen-free, opalescent solution suitable for the intravenous or subcutaneous injection. The pH is 4.0–7.0 on pH paper.

Description of the Kit

Preformed albumin nanocolloid is easily labeled with reduced technetium at room temperature. No heating or pH adjustment is required. Once formed, albumin nanocolloid is unaffected by time. More than 95% of the ^{99m}Tc-albumin nanocolloid show a size distribution between 10 and 80 nm (Chia 1986; Solco Nanocoll 1992).

^{99m}Tc eluate used for labeling the preformed colloid should be obtained from a generator by daily elution in order to minimize ^{99}Tc carrier and should preferably be in a small volume to avoid any interference of Al^{3+} ion (Ponto et al. 1987). The manufacturer recommends a minimum volume of ^{99m}Tc eluate, which is adjusted to the final volume with oxidant-free saline.

Labeling with a small volume (< 1 ml) of ^{99m}Tc-pertechnetate is also required if the colloid is used for lymphoscintigraphy.

Clinical Applications

Intravenous injection:	Bone marrow scintigraphy Scintigraphy of inflammation
Subcutaneous (interstitial) injection:	Lymphoscintigraphy Visualization of lymphatic flow and regional lymph nodes in the extremities and the trunk
Subdermal or peritumoral injection:	Sentinel lymph node (SLN) scintigraphy

Bone marrow scintigraphy with nanocolloids is based on a high extraction efficiency in bone marrow by phagocytosis (approximately 15% of the injection dose [ID]) (Hotze et al. 1984; McAfee et al. 1982; Munz 1984a; Nagai et al. 1982.). In areas where no phagocytes are present, no uptake is observed. This is the case in the fatty marrow, or where the phagocyte population has been replaced by other structures, such as metastatic growth. Enhanced phagocytic activity is causing increased colloid uptake. Defects in bone marrow distribution, nonuniform marrow distribution, or an expansion of active marrow into long bones has been diagnosed. Both, hot and cold lesions are visualized in the scintigram.

Biodegradable HSA nanocolloids offer considerable advantages over inorganic colloids. Large quantities labeled with I-131 have been used to study the phagocytic capacity of the reticuloendothelial system (RES) in healthy man and in patients with certain infections (Iio et al. 1963; Wagner et al. 1963).

Nanocolloids preferentially accumulate in inflammatory lesions associated with a number of conditions, i.e., osteomyelitis, osteitis, rheumatoid arthritis, arthrosis, joint prostheses, and wound healing, thus nanocolloid is clinically useful for detecting osteomyelitis and other bone or joint infections (De Schrijver et al. 1987; Froehlich 1985; Vorne et al. 1989).

Lymphoscintigraphy is performed for the diagnosis of peripheral edema caused by recurring erysipel or by affected lymph nodes as a result of metastatic infiltration, lymphogranuloma, lymphosarcoma, damage as a result of radiation treatment, or other causes. Other indications include secondary edema caused by blockade of the lymphatic flow, and thrombosis (Lofferer et al. 1974; Mostbeck et al. 1984). Lymphatic drainage from a primary tumor was studied in patients with breast cancer (Ege 1983). Several colloidal preparations have been evaluated for lymphoscintigraphy (Kaplan et al. 1979; Strand and Persson 1979; Kaplan et al. 1985).

SLN imaging can identify the first infected lymph node of a primary tumor before surgery. If the SLN is histologically tumor-free, no other lymph node will contain metastatic disease (Alazraki et al. 1997; Nitz and Heidenreich 1999).

Time of Examination

Bone marrow scintigraphy:	30 min after the intravenous injection
Inflammation imaging:	Static imaging 30–60 min after the intravenous injection
Lymphoscintigraphy:	Dynamic imaging immediately after the injection Static imaging 30–60 min after subcutaneous injection
SLN imaging:	Sequential planar scintigraphy 5–10 min postinjection and 1–6 h postinjection (late images)

Recommended Activities for Indications

Bone marrow scintigraphy:	185–370 MBq (5–10 mCi), injected intravenously 400 MBq maximum recommended activity 1.5–7 µg nanocolloid/kg body weight
Inflammation scintigraphy:	370–555 MBq (10-15 mCi), injected intravenously
Lymphoscintigraphy of the extremities:	20–75 MBq (0.5–2 mCi) in a volume of 0.2–0.3 ml, subcutaneous injection into the interdigital spaces of the hands or feet. For a better resorption and to reduce the radiation burden to the injection site, the total dose (≤185 MBq; 5 mCi) may be divided between two and three injection sites. 40 MBq (1 mCi) maximum activity per injection site The volume of injection should not exceed 0.5 ml
Parasternal lymphatics:	20 MBq (0.5 mCi) in a volume of 0.3–0.5 ml, single subcostal injection at both sides The manufacturer's instructions should be followed.
SLN:	50–80 MBq (1.3–2.2 mCi) in a volume of 0.1–0.5 ml, multiple subdermal injections (peritumoral) Guidelines for sentinel lymph node detection should be followed (Cox et al. 1998)

Pediatric Dose. The amount of radioactivity for infants and children administered for bone marrow and inflammation imaging or lymphoscintigraphy is based on body weight, using the scaling factors given in Appendix 1 (Table A1.2). The minimum value of the scaling factor is 0.1, since activities below 10% of the value of the adult radioactivity have resulted in poor image quality.

Additional Information

The use of ^{99m}Tc-albumin nanocolloid in patients with a history of hypersensitivity to human albumin is contraindicated.

The stability of colloids is affected by electrolytes; thus, ^{99m}Tc-albumin nanocolloid should not be injected together with other drugs or components. The use of local anesthetic agents or hyaluronidase prior to administering the labeled preparation has been shown to negatively affect lymphatic uptake.

Albumin aggregates in the size range ≤100 nm are obtained by heating a 1% solution of human serum albumin (HSA) in alkaline medium (pH 10.0) while shaking (Iio et al. 1963). Controlling the pH, temperature, agitation rate, and optical density can assure a high degree of uniformity in size.

Albumin nanocolloid is less affected by the concentration of Al^{3+} ion in the generator eluate than inorganic colloids. A reaction volume containing 1 µg Al^{3+} ion per milliliter had no effect; 2 µg/ml showed an increase in the particle diameter (80–100 nm) from 1–23% (Chia 1986; Haney 1971).

For interstitial injection ^{99m}Tc-albumin nanocolloid must be applied in a small volume not exceeding 0.5 ml. To ensure high specific activity, labeling should be performed with a small volume of ^{99m}Tc eluate (≤2 ml) and not less than 370 MBq (10 mCi).

Interstitial injection of a hypertonic colloidal solution may be associated with pain at the injection site. Therefore, ^{99m}Tc-albumin nanocolloid should be injected slowly.

Iodinated contrast media used in lymphangiography may interfere with lymphatic imaging with ^{99m}Tc-albumin nanocolloid.

Quality Control

Radiochemical Purity. ^{99m}Tc-albumin nanocolloid is not described in the *European Pharmacopeia*. Paper chromatography using Whatman 31 ET paper and saline as solvent is recommended by the manufacturer. Free ^{99m}Tc-sodium pertechnetate is measured at an R_f of 0.75 and ^{99m}Tc-albumin nanocolloid is identified at the origin. The radiochemical purity of ^{99m}Tc-albumin nanocolloid should not be less than 95%.

The analysis of radiocolloids is based on the determination of free ^{99m}Tc-Na pertechnetate, since colloidal activity remains at the start. Hydrolized ^{99m}Tc activity cannot be distinguished from the ^{99m}Tc-HSA nanocolloid.

Recommended Methods

Thin layer chromatography on Gelman silica gel fiberglass sheets using acetone as solvent has been in clinical use for the analysis of ^{99m}Tc-(Re)-sulfide microcolloid and is also recommended for ^{99m}Tc-nanocolloids.

Thin-layer chromatography		
Stationary phase:	Gelman ITLC-SG (fiberglass), 2×9.5 cm	
Solvent:	Acetone (or MEK)	
Developing time:	5 min	
R_f values:	^{99m}Tc-HSA nanocolloid:	0.0–0.1 (>95%)
	^{99m}Tc reduced, hydrolized:	0.0–0.1
	^{99m}Tc-pertechnetate:	0.9–1.0

Quantification of labeled components. Each chromatogram is measured (TLC linear analyzer or gamma counter), and the regional radioactivities are expressed as a percentage of the total recovered counts:

^{99m}Tc-HSA nanocolloid (%) = 100 – *F*

where *F* (%) = ^{99m}Tc-Na-pertechnetate (free).

Results of analysis *(12 samples)*
Results were obtained using thin-layer chromatography and acetone as solvent.

Labeling and stability	15 min (%)	3 h (%)
^{99m}Tc-HSA nanocolloid	96.7 ± 0.15	96.3 ± 0.16
^{99m}Tc-Na-pertechnetate	3.3 ± 0.50	3.7 ± 0.40

Pharmacokinetic Data

^{99m}Tc-albumin nanocolloid is removed from the circulation by phagocytosis. More than 95% of the intravenously injected nanocolloid is accumulated in the liver, spleen, and bone marrow 15 min after administration (Chia 1986; Iio et al. 1963). Nanocolloids show high accumulation in bone marrow (10–15%) (Hotze et al. 1984; McAfee et al. 1982).

Plasma clearance of colloidal particles is related to size; nanocolloids are normally cleared with a half-time of 2.6 min (Iio et al. 1963). Ten minutes after the intravenous injection, 10% of the injected dose was measured in blood (2.5% at 60 min), showing a constant decrease of radioactivity. In comparison, ^{99m}Tc-albumin microcolloid (0.2–1.0 μm) showed approximately 1.6% of the injected dose at 10 min (Chia 1986). The maximal rate of phagocytosis of aggregated albumin (AA) in man was determined as 1.07 mg per minute per kilogram of body weight (Iio et al. 1963).

An increase of extrahepatic accumulation of nanocolloid in the bone marrow is seen with hyperplastic bone marrow and with certain hematological disorders (polycytemia, leukemia) (Höfer et al. 1964). In patients with malignant disease, bone marrow scintigraphy may offer early detection of bone marrow infiltration (Hotze et al. 1984; Munz 1984b).

After subcutaneous injection into the interdigital spaces of the hands or feet, nanocolloid is transported with the interstitial liquid through the lymphatic capillaries into the lymph ducts and taken up almost quantitatively by the regional lymph nodes. Drainage from the interstitial injection site was observed between 1 and 35% in 24 h (Ege 1976).

Release of the colloid from the lymph nodes is slow and increasing with movement of the extremities. After 15 min of slow walking, 6–15% of the injected radioactivity has been measured in the regional lymph nodes (Mostbeck et al. 1984). Accumulation of the nanocolloid in the liver is negligible. Approximately 80% of the nanocolloid is transported by the lymphatic system (Saha 1987).

After the subcostal injection at both sides of the xyphoid (2–2.5 cm under the xyphoid) into the musculus rectus abdominalis, ^{99m}Tc-albumin nanocolloid is transported through the parasternal lymphatics into mediastinal and diaphragmal lymph nodes. The maximal accumulation is reached 3 h after injection (Ege 1983).

Subdermal, peritumoral injection of ^{99m}Tc-albumin nanocolloid is used for visualization of the lymphatic drainage of a primary tumor and for identification of the sentinel lymph node (Alazraki et al. 1997; Cox et al. 1998).

Data on the acute toxicity of intravenously or subcutaneously injected ^{99m}Tc-albumin nanocolloid have not been reported. Repeated injections of high doses of colloidal albumin aggregates (5 mg/kg) over a 2-month period have been well tolerated in three subjects, showing no effect on the clearance rate when compared with control subjects (Iio et al. 1963). Subcutaneous injection of small doses (0.02 mg/kg) as well as a saturation dose of 4 mg/kg showed no evidence of hypersensitivity.

Radiation Dose

After intravenous injection, the liver, spleen, and red bone marrow are the most exposed organs. The effective (whole body) dose equivalent is 0.014 mSv/MBq (International Commission on Radiological Protection 1987).

The effective dose in adults (70 kg) resulting from 370 MBq (10 mCi) of ^{99m}Tc-albumin nanocolloid is 5.2 mSv. The dose to liver and spleen after intravenous injection of 185 MBq (5 mCi) is 13.7 and 14.2 mGy, respectively. The dose to the bone marrow after intravenous injection of 370 MBq (10 mCi) is 5.5 mGy.

The effective dose equivalent (per ICRP 62) has been replaced by the quantity effective dose. Values per unit administered activity were published in Addendum 1 (International Commission on Radiological Protection 1991). Values calculated accordingly are slightly lower than the effective whole body dose presented here.

Calculations of the absorbed radiation dose following the subcutaneous administration of ^{99m}Tc-albumin nanocolloid into the extremities are based on the assumption that approximately 5–15% of the radioactivity is distributed over 10–20 lymph nodes (Mostbeck et al. 1984). The highest radiation dose is thus delivered at the injection site. If 37 MBq (1 mCi) are injected into each foot, the radiation absorbed dose at the injection site has been calculated as 400–700 mGy, and 22–27 mGy for each lymph node. The weight of a lymph node was assumed to be 5 g (Mostbeck et al. 1984).

Storage and Stability

Storage. The lyophilized kit is stored at 2–8 °C.

Stability. The ^{99m}Tc-albumin nanocolloid injection is stable for 6 h after preparation.

References

Alazraki N, Eshima D, Eshima LA, Herda SC, Murray DR, Vansant JP, Taylor AT (1997) Lymphoscintigraphy, the sentinel node concept, and the intraoperative gamma probe in melanoma, breast cancer and other potential cancers. Semin Nucl Med 27:55–67

Chia HL (1986) Particulate radiopharmaceuticals in nuclear medicine. In: Cox PH (ed) Radiopharmacy and Radiopharmacology yearbook II. Grune and Stratton, New York

Cox CE, Pendas S, Cox JM, Joseph E, Shons AR, Yeatman T, Ku NN, Lyman GH, Berman C, Haddad F, Reintgen DS (1998) Guidelines for sentinel lymph node detection. Ann Surg 227:645–653

De Schrijver M, Streule K, Senekowitsch R, Fridrich R (1987) Scintigraphy of inflammation with nanometer-sized colloidal tracers. Nucl Med Commun 8:895–908

Ege GN (1976) Internal mammary lymphoscintigraphy – the rationale, technique, interpretation and clinical application: a review based on 848 cases. Radiology 118:101–107

Ege GN (1983) Lymphoscintigraphy – techniques and applications in the management of breast carcinoma. Semin Nucl Med 13:26–34

Froehlich JW (ed) (1985) Nuclear medicine in inflammatory diseases. In: Nuclear medicine annual. Raven, New York, pp 23–72

Haney TA, Ascanio I, Gigliotti JA, Gusmano EA, Bruno GA (1971) Physical and biological properties of a ^{99m}Tc-sulfur colloid preparation containing disodium edetate. J Nucl Med 12:64–68

Höfer R, Ogris E, Pfeiffer G (1964) Kolloidspeicherung im Knochenmark bei Bluterkrankungen [in German]. Wien Z Inn Med 45:330–335

Hotze A, Mahlstedt J, Wolf F (1984) Knochenmarkszintigraphie: Methode, Indikationen, Ergebnisse [in German]. Giebeler, Darmstadt

Iio M, Wagner HN Jr, Scheffel U, Jabbour B (1963) Studies of the reticuloendothelial system (RES). I. Measurement of the phagocytic capacity of the RES in man and dog. J Clin Invest 42:417–426

International Commission on Radiological Protection (1987) Technetium-labeled colloids. In: Annals of the ICRP, radiation dose to patients from radiopharmaceuticals, biokinetic models and data. ICRP publication 53, vol. 18, no. 1–4. Pergamon, Oxford, pp 179–183

International Commission on Radiological Protection (1991) ICRP Publication 62. In: Annals of the ICRP, radiological protection in biomedical research, vol. 22, no. 3. Pergamon, Oxford, pp 25–28

Kaplan WD, Davis MA, Rose CM (1979) A comparison of two technetium-99m-labeled radiopharmaceuticals for lymphoscintigraphy. J Nucl Med 20:933–937

Kaplan WD, Piez CW, Gelman RS, Laffin SM, Rosenbaum EM, Jennings CA, McCormick CA, Harris JR, Henderson IC, Atkins HL (1985) Clinical comparison of two radiocolloids for internal mammary lymphoscintigraphy. J Nucl Med 26:1382–1385

Lofferer O, Mostbeck A, Partsch H (1974) Lymphtransportstörung beim dicken Bein – Isotopenlymphographische Ergebnisse. Z Hautkr 49:615–622

McAfee JG, Subramanian G, Aburano T, Thomas FD, Fernandes P, Gagne G, Lyons B, Zapf-Longo C (1982) A new formulation of Tc-99m minimicro-aggregated albumin for marrow imaging: comparison with other colloids, In-111 and Fe-59. J Nucl Med 23:21–28

Mostbeck A, Kahn P, Partsch H (1984) Quantitative Lymphographie beim Lymphödem. In: Bollinger A, Partsch H (eds) Initiale Lymphstrombahn [in German]. Georg Thieme, Stuttgart, pp 116–122

Munz DL (1984a) Knochenmarkszintigraphie: Grundlagen und klinische Ergebnisse [in German]. Der Nuklearmediziner 7:251–268

Munz DL (1984b) The scintigraphic bone marrow status in adult man: a new classification. In: Schmidt HAE, Adam WE (eds) Nuklearmedizin. Darstellung von Metabolismen und Organ-Funktionen. Schattauer, Stuttgart, p 640

Nagai K, Ito Y, Otsuka N, Muranaka A (1982) Deposition of small ^{99m}Tc-labeled colloids in bone marrow and lymph nodes. Eur J Nucl Med 7:66–70

Nitz DWP, Heidenreich P (1999) Sentinel-Lymphknoten. In: Der Nuklearmediziner. Demeter, Munich

Ponto JA, Swanson DP, Freitas JE (1987) Clinical manifestations of radio-pharmaceutical formulation problems. Hladik WB III, Saha GB, Study KT (eds) In: Essentials of nuclear medicine science. Williams & Wilkins, Baltimore, pp 268–289

Saha GB (1987) Normal biodistribution of diagnostic radiopharmaceuticals. In: Hladik WB III, Saha GB, Study KT (eds) Essentials of nuclear medicine science. Williams & Wilkins, Baltimore, pp 6–7

SolcoNanocoll (1992) Product monograph of the kit for the preparation of Tc-99m nanocolloid, issued by Sorin Biomedica, Italy

Strand S-E, Persson BRR (1979) Quantitative lymphoscintigraphy I: basic concepts for optimal uptake of radiocolloids in the parasternal lymph nodes of rabbits. J Nucl Med 20:1038–1046

Vorne M, Lantto T, Paakkinen S, Salo S, Soini I (1989) Clinical comparison of ^{99m}Tc-HM-PAO labeled leukocytes and ^{99m}Tc-nanocolloid in the detection of inflammation. Acta Radiol 30:633–637

Wagner HN Jr, Iio M, Hornick RB (1963) Studies of the reticuloendothelial system (RES). II. Changes in the phagocytic capacity of the RES in patients with certain infections. J Clin Invest 42:427–434

12.4 ^{99m}Tc-Labeled Myocardial Perfusion Agents

12.4.1 ^{99m}Tc-MIBI (Methoxyisobutyl Isonitrile)

F. Rakiás and I. Zolle

Chemical name	Chemical structure
2-Methoxy-isobutyl-isonitrile (MIBI) Tc(I)-Hexakis(2-methoxy-isobutyl-isonitrile) tetrafluroroborate Sestamibi (*Ph. Eur., USP*) Technetium Tc 99m sestamibi injection ^{99m}Tc-MIBI ^{99m}Tc-sestamibi injection	^{99m}Tc(I)-sestamibi complex

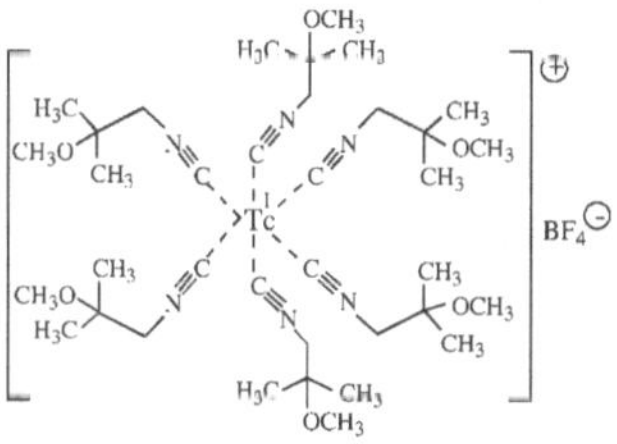

Kit components		Commercial products	
$Cu(MIBI)_4 \cdot BF_4$	1.0 mg	Cardiolite	Bristol-Myers Squibb (BMS)
Tin(II)-chloride dihydrate	0.075 mg	Miraluma	DuPont Merck/BMS
L-Cysteine hydrochloride hydrate	1.0 mg	Cardiospect	Rotop
Sodium citrate dihydrate	2.6 mg		
Mannitol	20 mg		

Preparation

The Cardiolite kit contains the lyophilized, sterile, pyrogen-free, inactive ingredients in a nitrogen atmosphere, ready for labeling with ^{99m}Tc-sodium pertechnetate (Council of Europe 2004). Labeling is performed according to the instructions given by the manufacturer (Bristol-Myers Squibb 2001).

A volume of 1–3 ml of ^{99m}Tc-pertechnetate (0.925–5.55 GBq, resp. 25–150 mCi) is added aseptically to the reaction vial. Before removing the syringe, an equal volume of headspace to normalize the pressure in the vial should be withdrawn. The shielded vial should be agitated vigorously to dissolve the lyophilized material, and then is placed in a boiling water bath, which should be shielded, for 10 min. After heating, the vial is placed into the lead shield and cooled at room temperature for approximately 15 min.

^{99m}Tc-sestamibi is a clear, colorless solution for intravenous injection, and the pH value is 5.3–5.9.

Another diagnostic kit for the preparation of ^{99m}Tc-sestamibi is Miraluma, approved for breast imaging (Bristol-Myers Squibb 2001).

Description of the Kit

The Cardiolite kit contains 2-methoxy-isobutyl-isonitrile (MIBI) as a preformed copper(I) complex, $Cu(MIBI)_4^+ \cdot BF_4^-$ [tetrakis (2-methoxy-isobutyl-isonitrile) Cu(I) tetrafluoroborate], which facilitates labeling by ligand exchange at elevated temperature.

An alternate method of heating by exposure of the reaction vial in a microwave oven has been reported (Hung et al. 1991). Using a reaction volume of 3 ml, the time for heating was reduced to 10 s, obtaining consistently high labeling yields. Separate studies using 3 ml saline have shown that heating for 8 s produced a temperature of 98.7±0.3 °C.

The Cardiolite kit contains no bacteriostatic agent. Only eluates from generators eluted within 24 h after the previous elution may be used for labeling, keeping the volume small (Van Duzee and Bugaj 1981).

Clinical Applications

^{99m}Tc(I)-sestamibi is used after intravenous injection:

- Myocardial perfusion studies
  - Diagnosis of ischemic heart disease
  - Diagnosis and localization of myocardial infarction
  - Assessment of global ventricular function (first pass technique for determination of ejection fraction and/or regional wall motion)
- Breast imaging in patients with an abnormal mammogram
- Parathyroid imaging in patients with hyperfunctioning adenoma

Myocardial perfusion imaging with ^{99m}Tc-sestamibi has been evaluated by multicenter clinical studies (Wackers et al. 1989). The advantages of ^{99m}Tc-sestamibi have been demonstrated in normal subjects (Marcass et al. 1990), in patients with ischemic heart disease by comparison with thallium-201 (Maisey et al. 1990; Sporn et al. 1988; Villanueva-Meyer et al. 1990); and validated by single-photon emission computer tomography (SPECT) (Berman et al. 1993).

Time of Examinations

Rest and stress scintigraphy:	1–1.5 h after the intravenous injection
Breast imaging:	Planar imaging is begun 5–10 min postinjection
Parathyroid imaging:	10–15 min postinjection

Recommended Activities for Indications. The activity-range for intravenous administration in patients (70 kg) is:

Diagnosis of reduced regional perfusion and myocardial infarction:	250–1,000 MBq (7–27 mCi)
Assessment of global ventricular function:	600–800 MBq (16–25 mCi), injected as a bolus
Miraluma breast imaging:	555–925 MBq (15–25 mCi)
Parathyroid imaging:	185–740 MBq (5–20 mCi), injected as a bolus

The performance of myocardial scintigraphy requires a strict protocol. Usually, the first examination is performed with exercise or pharmacological stress, with the intravenous injection of ^{99m}Tc-sestamibi administered at the highest heart rate (250–370 MBq, resp. 7–10 mCi). The stress scintigram is performed 1–1.5 h after the injection.

Three hours after the initial injection, a second dose of ^{99m}Tc-sestamibi (550–750 MBq) (15–20 mCi) is administered intravenously to examine the patient at rest. The rest scintigram is performed 1–1.5 h after the injection.

Two injections (stress and rest) are required in order to differentiate a transient (reversible) from a persistent perfusion defect, which is typical for ischemic heart disease (Borges-Neto et al. 1990; Büll et al. 1996). For verification of a scar after myocardial infarction (nonreversible perfusion defect), one injection at rest (185–300 MBq, resp. 5–8 mCi) may be sufficient. However, it is usually not possible to differentiate an acute myocardial infarction from ischemic defects, which are observed in patients with angina pectoris during chest pain (Tatum et al. 1997).

Based on diagnostic reference levels for radiopharmaceuticals adopted by the European member states, the total activity administered on a single day should not exceed 1,000 MBq (27 mCi) in the case of the combined rest–exercise protocol and 600 MBq (16.2 mCi) in the case of the one-day protocol (European Commission 1999). A rather large variation in the recommendations from country to country has been reported (Hesse et al. 2005).

Pediatric Dose. The amount of radioactivity for infants and children administered for myocardial scintigraphy is based on body weight, using the scaling factors given in Appendix 1, Table A1.2. The minimum value of the scaling factor is 0.1, since activities below 10% of the value of the adult radioactivity have resulted in poor image quality.

Additional Information

Cardiac imaging. Patients should fast for at least 4 h prior to the study. It is recommended that patients take a light fatty meal or drink a glass of milk after each injection prior to SPECT imaging. This will promote rapid hepatobiliary clearance of ^{99m}Tc-sestamibi, resulting in lower liver activity in the image.

The heart-to-background ratio will increase with time; thus, the imaging time is the best compromise between heart count rate and surrounding organ uptake. Scintigraphy is performed 1–2 h after rest and stress injections. There is no evidence for significant changes in myocardial tracer concentration or redistribution.

Either planar or tomographic imaging is performed for diagnosis of ischemic heart disease and myocardial infarction. Both may be performed electrocardiogram (ECG)-gated.

For assessment of global ventricular function the same standard techniques and projections are used, as established for ^{99m}Tc first-pass ejection studies.

Adequate hydration and frequent urination are necessary to reduce the radiation exposure of the bladder wall.

Breast imaging. Another diagnostic purpose is the investigation of patients suspected of breast cancer, particularly patients with an abnormal mammogram or a palpable breast mass (Imbriaco et al. 2001; Palmedo et al. 1996 and 1998).

Miraluma breast imaging is performed 10 min after the intravenous injection of 555 MBq (15 mCi) of ^{99m}Tc-sestamibi in the arm contralateral to the site of the breast abnormality. Then, planar images are obtained for 10 min in different positions according to the protocol. A malignant breast lesion is shown by increased uptake of the radiotracer. The detection of primary breast carcinoma may be carried out with either planar or SPECT acquisition.

Parathyroid imaging is performed for localization and identification of hyperfunctioning adenoma (Coakley 1991) by two scintigraphic procedures:

- Subtraction: Previously injected activity in thyroid gland is subtracted from the total neck image (^{99m}Tc-sestamibi + ^{123}I- or ^{99m}Tc-thyroid image) (Wei et al. 1992).
- Washout (planar or SPECT): Neck and thorax images are obtained at specified times (up to 4 h) after the injection of ^{99m}Tc-sestamibi (McBiles et al. 1995; Taillefer et al. 1992).

In order to assure high quality of the injection solution, the radiochemical purity is determined before administration of ^{99m}Tc(I)-sestamibi to patients (Hung 1991).

Quality Control

Radiochemical Purity. ^{99m}Tc-sestamibi is described in the *European Pharmacopeia* (Council of Europe 2005). Thin-layer chromatography using reverse-phase silica gel plates (octadecylsilyl-silica gel) and a mixture of acetonitrile, methanol, ammonium acetate, and tetrahydrofuran as solvent system is recommended for the separation of free ^{99m}Tc-sodium pertechnetate (R_f=0.9) and reduced, hydrolized ^{99m}Tc-activity (R_f=0–0.1). These impurities should not exceed 5% of the measured radioactivity.

The radiochemical purity is analyzed prior to administration of ^{99m}Tc-sestamibi. If the labeling yield is less than 90%, the preparation should be discarded.

Recommended Methods

The commonly used method is thin-layer chromatography on aluminum oxide-coated plastic TLC plates and absolute ethanol as solvent. Free ^{99m}Tc-sodium pertechnetate and reduced, hydrolized ^{99m}Tc-activity remain at the start. ^{99m}Tc-sestamibi is measured at an R_f of 0.9, moving with the solvent front. The radiochemical purity of ^{99m}Tc-sestamibi should not be less than 94%.

Thin-layer chromatography		
Stationary phase:	Baker-flex Alox 1B-F (precut to 2.5×7.5 cm)	
Solvent:	Ethanol (99.8%)	
Developing time:	10 min	
R_f values:	^{99m}Tc reduced, hydrolized:	0.0–0.1
	^{99m}Tc-pertechnetate:	0.0–0.1
	^{99m}Tc-MIBI complex:	0.9–1.0 (>94%)

Procedure:

- Pour enough ethanol into the TLC tank (beaker) to have a depth of 3–4 mm of solvent.
- Cover the tank (beaker) with parafilm and allow it to equilibrate for approximately 10 min.
- Dry the plates at 100 °C for 1 h and store in a desiccator. Remove predried plate from the desiccator just prior to use.

- Apply 1 drop of ethanol (95%), using a 1-ml syringe with a 22–26 gauge needle onto the aluminum oxide TLC plate, 1.5 cm from the bottom. Do not allow the spot to dry.
- Add two drops of ^{99m}Tc-MIBI injection solution side by side on top of the ethanol spot. Return the plate to a desiccator and allow the sample to dry (typically, 15 min).
- Develop the plate in the covered TLC tank in ethanol (99.8%) for a distance of 5.0 cm from the point of application.
- Cut the TLC plate 4.0 cm from the bottom and measure the ^{99m}Tc activity of each piece in the dose calibrator.

Calculate the percent radiochemical purity as:

$$^{99m}\text{Tc-sestamibi (\%)} = \frac{\text{Activity of upper piece}}{\text{Activity of both pieces}} \times 100$$

The time required to complete the entire procedure is approximately 35 min.

Results of analysis *(12 samples)*
Results were obtained using TLC and ethanol as solvent.

Labeling and stability	15 min (%)	6 h (%)
^{99m}Tc-Na-pertechnetate	5.4±0.3	5.6±0.3
^{99m}Tc-sestamibi	94.6±0.3	94.4±0.3

Paper chromatography
Another method, which is faster and offers a high separation of ^{99m}Tc-MIBI from impurities was introduced by Patel et al. (1995). Whatman 3MM paper strips are used and ethyl acetate as solvent. ^{99m}Tc-sestamibi moves with an R_f of 0.55–0.75, while ^{99m}Tc-pertechnetate and reduced, hydrolized ^{99m}Tc-activity remain at the start.

Stationary phase:	Whatman 3MM paper strips (cut to 0.5×6.0 cm)
Solvent:	Ethyl acetate
Developing time:	3 min
R_f values:	^{99m}Tc reduced, hydrolized: 0.0–0.1 ^{99m}Tc-Na-pertechnetate: 0.0–0.1 ^{99m}Tc-MIBI, 0.5–0.8 (>94%)

Quantification of labeled components. Each chromatogram is measured (TLC linear analyzer or gamma counter), and the regional radioactivities are expressed as a percentage of the total recovered counts:

$$^{99m}\text{Tc-sestamibi (\%)} = 100 - \%\,(F + H)$$

where % $(F + H) = {}^{99m}$Tc activity at origin.

Solvent extraction
A rapid and simple method to obtain information about the labeling yield of lipophilic ^{99m}Tc-sestamibi has found wide acceptance in nuclear medicine. Several solvents (e.g.,

ethyl acetate, chloroform, *n*-octanol) have been employed for extraction. Labeled impurities remain in the aqueous phase (saline).

Procedure:

- Add 0.1 ml of the ^{99m}Tc-sestamibi injection solution to a vial containing 3 ml of chloroform and 2.9 ml of saline.
- Close the vial, mix with a Vortex for 10 s, and let the phases separate for 1 min.
- Transfer the top layer (saline) to another vial and measure the activities in a dose calibrator. Lipophilic ^{99m}Tc-MIBI is in the chloroform fraction and the contaminants are in the saline layer.

The radiochemical purity is given by:

$$^{99m}\text{Tc-MIBI (\%)} = \frac{\text{Activity of chloroform fraction}}{\text{Activity of both fractions}} \times 100$$

Results of analysis. The activity in the chloroform fraction should not be less than 90% within 6 h after labeling.

Pharmacokinetic Data

Like thallous chloride [^{201}Tl$^+$], the cationic technetium complex accumulates in the viable myocardial tissue proportional to blood flow. Studies using cultures of myocardial cells have shown that uptake is not dependent on the functional capability of the sodium/potassium pump (Maublant et al. 1988). Cationic membrane transport inhibitors did not affect 1-min ^{99m}Tc-MIBI uptake kinetics when cells were preincubated for 1 min in solutions containing saturating concentrations of quabain (100 μM), a sodium/potassium ATPase inhibitor (Piwnica-Worms et al. 1990).

After the intravenous injection, ^{99m}Tc-sestamibi is distributed in the myocardium according to blood flow and diffusion. Myocardial extraction at rest is 65% (vs 85% Tl$^+$). One-hour-postinjection myocardial uptake at rest is 1.2%; during stress, it is 1.4% (Wackers et al. 1989).

The elimination from blood is fast: 3 min after the intravenous injection, 23% of the radioactivity is measured in blood, decreasing to 9% at 5 min, and 2.5% at 10 min. The effective half-life at rest of the fast early component is 2.18 min (2.13 min during exercise) (Wackers et al. 1989). Protein binding is low (less than 1%).

Elimination of ^{99m}Tc-sestamibi from the myocardium is 27% in 3 h; the biological half-time of elimination is approximately 6 h at rest and during stress. No redistribution of ^{99m}Tc-sestamibi is observed (Wackers et al. 1989).

The major metabolic pathway for clearance of ^{99m}Tc-sestamibi is the hepatobiliary tract. Activity from the gallbladder appears in the intestine within 1 h of injection. Twenty-nine percent of the injected dose is cleared unchanged through renal elimination in 24 h, and approximately 37% of the injected dose is cleared through the feces in 48 h (at rest) (Wackers et al. 1989).

The elimination during 3 h from the liver is 76%; from the spleen, 67%; and from the lung, 49% (Wackers et al. 1989).

The mechanism of localization in various types of breast tissue (benign, inflammatory, malignant, fibrous) has not been established. Malignant breast lesions show the highest uptake of ^{99m}Tc-sestamibi (sensitivity: 79–96%, specificity: 80–94%).

Preclinical safety. Acute intravenous toxicity studies were performed in mice, rats, and dogs. The lowest dose of the reconstituted Cardiolite kit that resulted in any deaths was 7 mg/kg (expressed as $Cu(MIBI)_4 \cdot BF_4$ content) in female rats. This corresponds to 500 times the maximal human dose (MHD) of 0.014 mg/kg for adults (70 kg). Neither rats nor dogs exhibited treatment related effects at reconstituted Cardiolite kit doses of 0.42 mg/kg (30 times MHD) and 0.07 mg/kg (5 times MHD), respectively, for 28 days.

Radiation Dose

The gallbladder wall, liver, spleen, and lung are the most exposed organs. The effective dose is 0.0085 mSv/MBq at rest and 0.0075 mSv/MBq after exercise (International Commission on Radiological Protection 1991). The values are calculated assuming a 3.5-h bladder voiding period.

When two separate injections of ^{99m}Tc-sestamibi are administered, namely 250 MBq (7 mCi) (exercise) and 750 MBq (20 mCi) (at rest) the effective whole-body dose in patients (70 kg) is 8.25 mSv. The effective dose resulting from an administered radioactivity of 250 MBq (7 mCi) at rest (verification of a scar) is 2.12 mSv.

The absorbed radiation dose to the gallbladder wall resulting from an intravenous injection of 750 MBq (20 mCi) of ^{99m}Tc-sestamibi corresponds to 29.3 mGy (at rest).

Storage and Stability

Storage. Cardiolite (Miraluma) kits should be stored at 15–25 °C.

Stability. ^{99m}Tc-sestamibi injection solution is stable for 6 h.

References

Berman DS, Kiat H, Friedman JD, Wang FP, Van Train K, Matzev L, Maddahi J, Germano G (1993) Separate acquisitions rest thallium-201/stress technetium-99m sestamibi dual-isotope myocardial perfusion single-photon emission computed tomography: a clinical validation study. J Am Coll Cardiol 22:1455–1464

Borges-Neto S, Coleman RE, Jones RH (1990) Perfusion and function at rest and treadmill exercise using technetium-99m-sestamibi: comparison of one- and two-day protocols in normal volunteers. J Nucl Med 31:1128–1132

Bristol-Myers Squibb (2001) Product Information. Cardiolite Kit for the preparation of technetium Tc-99m sestamibi for injection. Bristol-Myers Squibb Medical Imaging, Billerica, Mass.

Büll U, Kleinhans E, Reske SN (1996) Herz-Kreislauf-System. In: Büll U, Schicha H, Biersack H-J, Knapp WH, Reiners C, Schober O (eds) Nuklearmedizin [in German]. Georg Thieme, Stuttgart, pp 203–246

Coakley AJ (1991) Parathyroid localization – how and when? Eur J Nucl Med 18:151–152

European Commission (1999) Radiation protection 109, guidance on diagnostic reference levels (DRLs) for medical exposures. Office for Official Publications of the European Communities, Luxembourg

Council of Europe (2005) Technetium [^{99m}Tc] sestamibi injection solution, European pharmacopeia monograph 1926. Maisonneuve, Sainte-Ruffine

Hesse B, Tägli K, Cuocolo A et al (2005) EANM/ESC procedural guidelines for myocardial perfusion imaging in nuclear cardiology. Eur J Nucl Med Mol Imaging 32:855–897

Hung JC, Wilson ME, Brown ML, Gibbons RJ (1991) Rapid preparation and quality control method for technetium-99m-2-methoxy isobutyl isonitrile (technetium-99m-sestamibi). J Nucl Med 32:2162–2168

Imbriaco M, Del Vecchio S, Riccardi A, Pace L, Di Salle F, Di Gennaro F, Salvatore M, Sodano A (2001) Scintimammography with ^{99m}Tc-MIBI versus dynamic MRI for non-invasive characterization of breast masses. Eur J Nucl Med 28:56–63

International Commission on Radiological Protection (1991) Technetium-labelled MIBI. In: Annals of the ICRP, radiological protection in biomedical research. ICRP Publication 62, vol. 22, no. 3, Pergamon, pp 21–24

Maisey MN, Lowry A, Bischof-Delaloye A, Fridrich R, Inglese E, Khalil M, van der Schoot J (1990) European multicenter comparison of thallium-201 and technetium-99m-methoxyisobutyl isonitrile in ischemic heart disease. Eur J Nucl Med 16:869–872

Marcass C, Marzullo P, Oarodi O, Sambuceti G, L'Abbate A (1990) A new method of noninvasive quantitation of segmental myocardial wall thickening using technetium-99m 2-methoxy-isobutyl-isonitrile scintigraphy: results in normal subjects. J Nucl Med 31:173–177

Maublant JC, Gachon P, Moins N (1988) Hexakis (2-methoxy isobutylisonitrile) technetium-99m and thallium-201 chloride: uptake and release in cultured myocardial cells. J Nucl Med 29:48–54

McBiles M, Lambert AT, Cote MG, Kim SY (1995) Sestamibi parathyroid imaging. Sem Nucl Med 25:221–234

Palmedo H, Biersack HJ, Lastoria S, Maublant J, Prats E, Stegner HE, Bourgeois P, Hustinx R, Hilson AJW, Bischof-Delaloye A (1998) Scintimammography with technetium-99m methoxyisobutyl isonitrile: results of a prospective European multicenter trial. Eur J Nucl Med 25:375–385

Palmedo H, Grünwald F, Bender H, Schomburg A, Mallmann P, Krebs D, Biersack HJ (1996) Scintimammography with technetium-99m methoxyisobutylisonitrile: comparison with mammography and magnetic resonance imaging. Eur J Nucl Med 23:940–946

Patel M, Sadek S, Jahan S, Owunwanne A (1995) A miniaturized rapid paper chromatographic procedure for quality control of technetium-99m sestamibi. Eur J Nucl Med 22:1416–1419

Piwnica-Worms D, Kronauge JF, Delmon L, Holman BL, Marsh JD, Jones AG (1990) Effect of metabolic inhibition on technetium-99m-MIBI kinetics in cultured chick myocardial cells. J Nucl Med 31:464–472

Sporn V, Perez Balino N, Holman BL, Sosa Liprandi A, Masoli O, Mitta A, Camin LL, Castiglia S, McKusick KA (1988) Simultaneous measurement of ventricular function and myocardial perfusion using the technetium-99m isonitriles. Clin Nucl Med 13:77–81

Taillefer R, Boucher Y, Potvin C, Lambert R (1992) Detection and localization of parathyroid adenomas in patients with hyperparathyroidism using a single radionuclide imaging procedure with technetium-99m-sestamibi (double-phase study). J Nucl Med 33:1801–1807

Tatum JL, Jesse RL, Kontos MC, Nicholson CS, Schmidt KL, Roberts CS, Ornato JP (1997) Comprehensive strategy for the evaluation and triage of the chest pain patient. Ann Emerg Med 29: 116–125

United States Pharmacopeial Convention (2005) Official Monographs: USP 28, technetium Tc 99m sestamibi injection. United States Pharmacopeia, p 1863

Van Duzee BF, Bugaj JE (1981) The effect of total technetium concentration on the performance of a skeletal imaging agent. Clin Nucl Med 6(Suppl):P148

Villanueva-Meyer J, Mena I, Narahara KA (1990) Simultaneous assessment of left ventricular wall motion and myocardial perfusion with technetium-99m-methoxyisobutyl isonitrile at stress and rest in patients with angina: comparison with thallium-201 SPECT. J Nucl Med 31:457–463

Wackers FJTh, Berman DS, Maddahi J, Watson DD, Beller GA, Strauss HW, Boucher CA, Picard M, Holman BL, Fridrich R, Inglese E, Delaloye B, Bischof-Delaloye A, Camin L, McKusick K (1989) Technetium-99m hexakis 2-methoxyisobutyl isonitrile: Human biodistribution, dosimetry, safety, and preliminary comparison to thallium-201 for myocardial perfusion imaging. J Nucl Med 30:301–311

Wei JP, Burke GJ, Mansberger AR Jr (1992) Prospective evaluation of the efficacy of technetium-99m-sestamibi and iodine-123 radionuclide imaging of abnormal parathyroid glands. Surgery 112:1111–1117

12.4.2 ^{99m}Tc-Tetrofosmin

J. Imre and I. Zolle

Chemical name	**Chemical structure**
1,2-bis[bis(2-ethoxyethyl)-phosphino]-ethane Tetrofosmin (*USP*) Technetium Tc 99m tetrofosmin injection (*USP*) Tc(V) dioxo diphosphine complex ^{99m}Tc-tetrofosmin injection	^{99m}Tc(V)-tetrofosmin complex

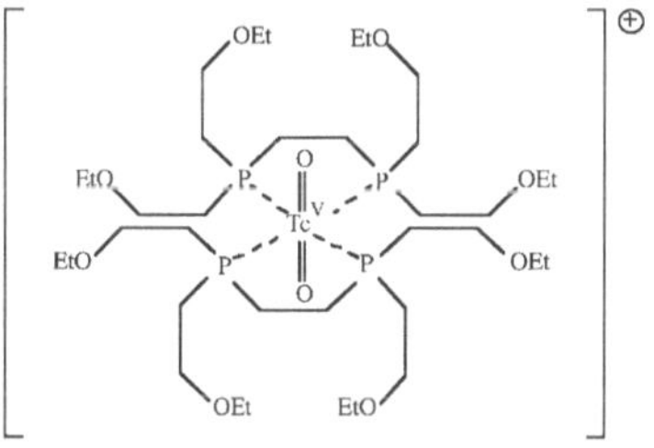

Kit components		**Commercial products**	
Tetrofosmin	0.23 mg	Myoview	GE Healthcare
Stannous chloride dihydrate	0.03 mg		
Disodium sulfosalicylate	0.32 mg		
Sodium D-gluconate	1.0 mg		
Sodium hydrogen carbonate	1.8 mg		

Preparation

The Myoview kit contains the sterile, lyophilized components in a nitrogen atmosphere. Labeling with ^{99m}Tc-sodium pertechnetate is performed by adding aseptically 4–8 ml of ^{99m}Tc activity to the vial, not exceeding 1.11 GBq/ml (30 mCi/ml). Before the syringe is removed from the vial, a volume of gas should be withdrawn from the space above the solution to normalize the pressure inside the vial. The shielded vial should be gently agitated to dissolve the lyophilized material, and allowed to react at room temperature for 15 min (Nycomed Amersham 1998).

^{99m}Tc(V)-tetrofosmin is a clear, colorless solution for intravenous injection, the pH value is 7.5–9.0.

Description of the Kit

The cationic Tc(V) dioxodiphosphine complex consists of a Tc *trans* oxo core having four phosphorus atoms of the bidentate diphosphine ligands arranged in a plane. A comparison with DMPE [1,2-bis(dimethylphosphino)ethane], originally introduced as the Tc(III) dichlorodiphosphine complex, may elucidate the effect of derivatization on complex structure and in vivo performance (Deutsch et al. 1981, 1989). The tetrofosmin ligand is characterized by four ether functional groups (Kelly et al. 1993).

Studies of the rate of formation of the $[^{99m}Tc(tetrofosmin)_2O_2]^+$ complex have indicated that complex formation depends on the ligand concentration and could also be enhanced by heating (Kelly et al. 1993). However, using the Myoview formulation, the Tc(V) dioxodiphosphine complex is formed rapidly at room temperature in high radiochemical purity at a ligand concentration lower than 30 μg/ml.

High chemical purity of the generator eluate is a prerequisite to obtain high labeling yields. To assure optimal reaction conditions, the eluate should be fresh (not older than 6 h), preferably obtained from a generator eluted regularly every 24 h. An interval of more than 72 h from the last elution excludes any generator eluate from being suitable for labeling Myoview.

The volume of ^{99m}Tc eluate should be at least 4 ml and not more than 8 ml. The activity concentration must not exceed 1.11 GBq/ml and should be diluted before addition, if necessary. The amount of ^{99m}Tc activity should not exceed 8.88 GBq (240 mCi) (Nycomed Amersham1998).

Clinical Applications

^{99m}Tc(V)-tetrofosmin is used for myocardial perfusion studies in patients with coronary artery disease:

- Diagnosis of ischemic heart disease
- Diagnosis of reduced regional perfusion and localization of myocardial infarction
- Detection of perfusion defects in myocardium at rest

^{99m}Tc-tetrofosmin has been evaluated as a radiotracer for myocardial perfusion imaging (Higley et al. 1993; Jain et al. 1993). Stress–rest imaging was performed using 1-day and 2-day protocols (Sridhara et al. 1994). Planar and tomographic imaging has been correlated with thallium-201 and coronary angiography (Heo et al. 1994). Pharmacological stress perfusion imaging with single-photon emission computer tomography (SPECT) was used as an alternative to dynamic exercise in patients with coronary artery disease (Cuocolo et al. 1996) and evaluated in a multicenter clinical trial (He et al. 1997). Left ventricular volumes and ejection fraction have been calculated from quantitative electrocardiographic-gated ^{99m}Tc-tetrofosmin myocardial SPECT (Yoshioka et al. 1999). ^{99m}Tc-tetrofosmin imaging at rest has been compared with rest redistribution of thallium-201 for predicting functional recovery after revascularization (Matsunari et al. 1997).

Time of Examinations. The rest and stress scintigraphy is performed 15–30 minutes after the intravenous injection (Higley et al. 1993).

Recommended Activities for Indications. The activity range for intravenous administration in patients (70 kg) is:

Diagnosis of reduced regional perfusion and myocardial infarction:	250–1,000 MBq (7–27 mCi)
First injection at peak exercise:	250–370 MBq (7–10 mCi)
Second injection at rest (4 h after the first injection):	550–750 MBq (15–20 mCi)
Single injection at rest:	185–300 MBq (5–8 mCi)

The performance of myocardial scintigraphy requires a strict protocol (He et al. 1997; Jain et al. 1993; Sridhara et al. 1994)

Usually, the first examination is performed with exercise or pharmacological stress, the intravenous injection of ^{99m}Tc-tetrofosmin is administered at the highest heart rate (250–350 MBq). The stress scintigram is performed 15 min after injection.

Four hours after the initial injection, a second intravenous injection of ^{99m}Tc-tetrofosmin (550–750 MBq) is administered to examine the patient at rest. The rest scintigram is performed 30 min after injection.

Two injections (stress and rest) are required in order to differentiate a transient (reversible) from a persistent perfusion defect, which is typical for ischemic heart disease (Jain et al. 1993; Sridhara et al. 1994). For verification of a scar after myocardial infarction (nonreversible perfusion defect), one injection at rest may be sufficient (Higley et al. 1993).

However, it is usually not possible to differentiate an acute myocardial infarction from ischemic defects, which are observed in patients with angina pectoris during chest pain (Tatum et al. 1997).

Based on diagnostic reference levels for radiopharmaceuticals adopted by the European member states, the total activity administered on a single day should not exceed 1,000 MBq in the case of the combined rest–exercise protocol and 600 MBq in the case of the 1-day protocol (European Commission 1999). A rather large variation in the recommendations from country to country has been reported (Hesse et al. 2005).

Pediatric Dose. Myoview is not recommended for use in infants and children since data are not available for these age groups.

Additional Information

Patients should fast overnight or have only a light breakfast prior to the study. Patients should drink sufficient water, and frequent bladder emptying should be encouraged in order to reduce the radiation exposure to the bladder wall.

Since no redistribution of ^{99m}Tc-tetrofosmin is observed, separate injections are required for stress and rest scintigraphy.

Planar or preferably SPECT imaging should begin no earlier than 15 min postinjection.

No significant changes in myocardial concentration or redistribution of ^{99m}Tc-tetrofosmin have been observed; therefore, images may be acquired up to at least 4 h postinjection (Higley et al. 1993; Jain et al. 1993).

Quality Control

Radiochemical Purity. ^{99m}Tc-Tetrofosmin is not described in the *European Pharmacopeia*. A rapid instant thin-layer chromatography (ITLC) procedure for determination of the radiochemical purity of ^{99m}Tc-tetrofosmin has been published (Van Hemert et al. 2001).

Thin-layer chromatography using Gelman silica gel strips and an organic solvent is described in the *United States Pharmacopeia* (United States Pharmacopeial Convention 2000) and is recommended by the manufacturer. Free ^{99m}Tc-pertechnetate moves with the solvent front. Reduced, hydrolized ^{99m}Tc activity and any hydrophilic complexes remain at the start. ^{99m}Tc-tetrofosmin is measured at an R_f of 0.4–0.7. The radiochemical purity of ^{99m}Tc-tetrofosmin should not be less than 90%.

The radiochemical purity is analyzed prior to administration of ^{99m}Tc-tetrofosmin. If the labeling yield is less than 90%, the preparation should be discarded.

Method Recommended by the Manufacturer

Thin-layer chromatography		
Stationary phase:	Gelman ITLC-SG strips (2×20 cm)	
Solvent:	Acetone-dichloromethane, 35:65 (v/v) Developing time: 20 min	
R_f values:	^{99m}Tc-reduced, hydrolized; hydrophilic complexes:	0.0–0.1
	^{99m}Tc-tetrofosmin complex (lipophilic):	0.4–0.7 (>90%)
	^{99m}Tc-pertechnetate:	0.9–1.0

Procedure:

- Pour enough solvent into the TLC tank (beaker) to have a depth of 1 cm.
- Cover the tank (beaker) with a lid and allow it to equilibrate for approximately 10 min.
- Mark the origin on the ITLC-SG strip with a pencil line 3 cm from the bottom.
- Apply 10–20 µl of sample at the origin of the silica gel strip. Do not allow the spot to dry.
- Place the strip into the chromatography tank and cover it immediately. Ensure that the strip is not adhering to the walls of the tank.
- Develop the plate in the covered TLC tank for a distance of 15 cm from the point of application.
- Cut the TLC strip at 3 cm and at 12 cm from the origin and measure the ^{99m}Tc activity of each piece in the dose calibrator.

Calculate the percent radiochemical purity as:

$$^{99m}\text{Tc-tetrofosmin}\ (\%) = \frac{\text{Activity of center piece}}{\text{Total activity of all three pieces}} \times 100$$

Quantification of labeled components. Each chromatogram is measured (TLC linear analyzer or gamma counter), and the regional radioactivities are expressed as a percentage of the total recovered counts:

$$^{99m}\text{Tc-tetrofosmin}\ (\%) = 100 - \%(F + H)$$

where F (%) = free ^{99m}Tc-pertechnetate, and H (%) = hydrolized ^{99m}Tc activity and hydrophilic complexes.

Results of analysis *(12 samples)*

Labeling and stability	15 min (%)	8 h (%)
^{99m}Tc-tetrofosmin complex	97.5 ± 0.40	96.6 ± 0.29
^{99m}Tc-hydrolized and hydrophilic complexes	2.1 ± 0.13	2.5 ± 0.19
^{99m}Tc-Na-pertechnetate	0.3 ± 0.08	0.8 ± 0.31

Pharmacokinetic Data

The lipophilic ^{99m}Tc-tetrofosmin complex accumulates in viable myocardial tissue proportional to blood flow. Uptake is due to diffusion; retention is based on viable myocytes (Platts et al. 1995; Takahashi et al. 1996). After the intravenous injection, the myocardial extraction of ^{99m}Tc-tetrofosmin at rest is 65% (Tl^+ 85%). Five minutes postinjection, myocardial uptake is 1.2%; during stress, 1.3%. The myocardial concentration of ^{99m}Tc-tetrofosmin remains unchanged up to at least 4 h postinjection (Higley et al. 1993; Jain et al. 1993).

The elimination from blood is fast; 10 min after the intravenous injection less than 5% of the radioactivity are measured in blood, corresponding to 3.5% in the plasma (Higley et al. 1993).

The major metabolic pathway for clearance of ^{99m}Tc-tetrofosmin is the hepatobiliary tract. Activity in the gallbladder reaches a maximum (10% of the injected dose) within 2 h of injection, falling to below 1% at 24 h (Higley et al. 1993).

Accumulation in the liver is 4.9–10.6% of the injected radioactivity; this value falls to 1.6% within 2 h, and after 8 h elimination is complete. Following exercise uptake is reduced to half the value with enhanced sequestration in skeletal muscle (Higley et al. 1993).

Initially, the lung show 0.7–3.0% uptake; elimination of radioactivity is fast, showing only traces of activity 4 h postinjection.

Urinary excretion is 39% in 48 h. Approximately 34% of the injected dose is cleared through the feces in 48 h (Higley et al. 1993).

Preclinical safety. Acute intravenous toxicity studies were performed in rats and rabbits, using up to 1,500 times the maximum single human dose. Repeat dose toxicity was assessed in rats and rabbits at 0, 10, 100, and 1,000 times the maximum human dose daily for 14 days.

No toxicologically significant findings were made on single-dose administration of 1,500 times the maximum equivalent human dose, or on 14-day repeat dose studies at a level of 100 times the maximum human dose. No mortalities have occurred (Kelly et al. 1993).

No significant mutagenic potential was seen in any of the tests used.

Radiation Dose

The gallbladder wall, intestinal tract, kidneys, bladder wall, the salivary glands, and the thyroid, are most exposed organs. Calculations of the effective dose were published by Higley et al. (1993), presenting effective dose values of 0.0089 mSv/MBq at rest and 0.0071 mSv/MBq after exercise, comparable with the effective dose values for technetium-MIBI (International Commission on Radiological Protection 1991). The values are calculated assuming a 3.5-h bladder voiding period.

When two separate injections of ^{99m}Tc-tetrofosmin are administered, namely 250 MBq (7 mCi) (exercise) and 750 MBq (20 mCi) (at rest), the effective whole-body dose in patients (70 kg) is 8.45 mSv. The effective dose resulting from an administered radioactivity of 250 MBq (7 mCi) at rest (verification of a scar) is 2.2 mSv.

The absorbed radiation dose to the gallbladder wall resulting from an intravenous injection of 750 MBq (20 mCi) of ^{99m}Tc-tetrofosmin is corresponding to 29.3 mGy (at rest).

Storage and Stability

Storage. The Myoview kit is stored at 2–8 °C.

Stability. ^{99m}Tc-tetrofosmin injection solution is stable for 8 h. It should be kept at 2–8 °C.

References

Cuocolo A, Nicolai E, Pace L, Nappi A, Sullo P, Cardei S, Argenziano L, Ell PJ, Salvatore M (1996) Technetium-99m-labeled tetrofosmin myocardial tomography in patients with coronary artery disease: comparison between adenosine and dynamic exercise stress testing. J Nucl Cardiol 3:194–203

Deutsch E, Glavan KA, Sodd VJ, Nishiyama H, Ferguson DL, Lukes SJ (1981) Cationic Tc-99m complexes as potential myocardial imaging agents. J Nucl Med 22:897–907

Deutsch E, Ketring AR, Libson K, Vanderheyden J-L, Hirth WJ (1989) The Noah's ark experiment: species dependent biodistributions of cationic ^{99m}Tc complexes. Int J Rad Appl Instrum 16:191–232

European Commission (1999) Radiation protection 109, guidance on diagnostic reference levels (DRLs) for medical exposures. Office for Official Publications of the European Communities, Luxembourg

He Z-X, Iskandrian AS, Gupta NC, Verani MS (1997) Assessing coronary artery disease with dipyridamole technetium-99m-tetrofosmin SPECT: a multicenter trial. J Nucl Med 38:44–48

Heo J, Cave V, Wasserleben V, Iskandrian AS (1994) Planar and tomographic imaging with technetium 99m-labeled tetrofosmin: correlation with thallium 201 and coronary angiography. J Nucl Cardiol 1:317–324

Hesse B, Tägli K, Cuocolo A et al (2005) EANM/ESC procedural guidelines for myocardial perfusion imaging in nuclear cardiology. Eur J Nucl Med Mol Imaging 32:855–897

Higley B, Smith FW, Smith T, Gemmell HG, Gupta PD, Gvozdanocic DV, Graham D, Hinge D, Davidson J, Lahiri A (1993) Technetium-99m-1,2-bis[bis(2-ethoxyethyl)phosphino]ethane: human biodistribution, dosimetry and safety of a new myocardial perfusion imaging agent. J Nucl Med 34:30–38

International Commission on Radiological Protection (1991) Technetium-labeled MIBI In: Annals of the ICRP, Radiological protection in biomedical research. ICRP publication 62, vol. 22, no. 3. Pergamon, Oxford, pp 21–24

Jain D, Wackers FJT, Mattera J, McMahon M, Sinusas AJ, Zaret BL (1993) Biokinetics of technetium-99m-tetrofosmin: myocardial perfusion imaging agent: implications for a one-day imaging protocol. J Nucl Med 34:1254–1259

Kelly JD, Forster AM, Higley B, Archer CM, Booker FS, Canning LR, Chiu KW, Edwards B, Gill HK, McPartlin M, Nagle KR, Latham IA, Pickett RD, Storey AE, Webbon PM (1993) Technetium-99m-tetrofosmin as a new radiopharmaceutical for myocardial perfusion imaging. J Nucl Med 34:222–227

Matsunari I, Fujino S, Taki J, Senma J, Aoyama T, Wakasugi T, Hirai J-I, Saga T, Yamamoto S, Tonami N (1997) Quantitative rest technetium-99m tetrofosmin imaging in predicting functional recovery after revascularization: comparison with rest-redistribution thallium-201. J. Am Coll Cardiol 29:1226–1233

Nycomed Amersham (1998) Product monograph for the Myoview kit for the preparation of technetium Tc-99m tetrofosmin for injection. Nycomed Amersham, Buckinghamshire

Platts EA, North TL, Pickett RD, Kelly JD (1995) Mechanism of uptake of technetium-tetrofosmin. I: uptake into isolated adult rat ventricular myocytes and subcellular localization. J Nucl Cardiol 2:317–326

Sridhara B, Sochor H, Rigo P, Braat SH, Itti R, Martinez-Duncker D, Cload P, Lahiri A (1994) Myocardial single-photon emission computed tomographic imaging with technetium-99m tetrofosmin: stress–rest imaging with same-day and separate-day rest imaging. J Nucl Cardiol 1:138–143
Takahashi N, Reinhardt CP, Marcel R, Leppo JA (1996) Myocardial uptake of ^{99m}Tc-tetrofosmin, sestamibi, and ^{201}Tl in a model of acute coronary reperfusion. Circulation 94:2605–2613
Tatum JL, Jesse RL, Kontos MC, Nicholson CS, Schmidt KL, Roberts CS, Ornato JP (1997) Comprehensive strategy for the evaluation and triage of the chest pain patient. Ann Emerg Med 29:116–125
United States Pharmacopeial Convention (2005) Official Monographs: USP 28, technetium Tc 99m tetrofosmin injection. United States Pharmacopeia (USP) 28–national formulary (NF) 23, p 1865
Van Hemert FJ, van Eck-Smit BLF, Schimmel KJM (2001) A rapid and stable ITLC procedure for the determination of the radiochemical purity of ^{99m}Tc-tetrofosmin. Nucl Med Commun 22:641–644
Yoshioka J, Hasegawa S, Yamaguchi H, Tokita N, Paul AK, Xiuli M (1999) Left ventricular volumes and ejection fraction calculated from quantitative electrocardiographic-gated ^{99m}Tc-tetrofosmin myocardial SPECT. J Nucl Med 40:1693–1698

12.5 ^{99m}Tc-Labeled Brain Perfusion Agents

12.5.1 ^{99m}Tc-HMPAO (Hexamethylpropylene Amine Oxime)

F. Rakiás and I. Zolle

Chemical name	**Chemical structure**
4,8-diaza-3,6,6,9-tetramethyl-undecane-2,10-dione-bisoxime (HMPAO) D,L-Hexamethylpropylene amine oxime (D,L-HMPAO) Exametazime (*Ph. Eur., USP*) Technetium ^{99m}Tc exametazime injection (*Ph. Eur., USP*) ^{99m}Tc(V)oxo-D,L-HMPAO complex	^{99m}Tc(V)-D,L-HMPAO complex

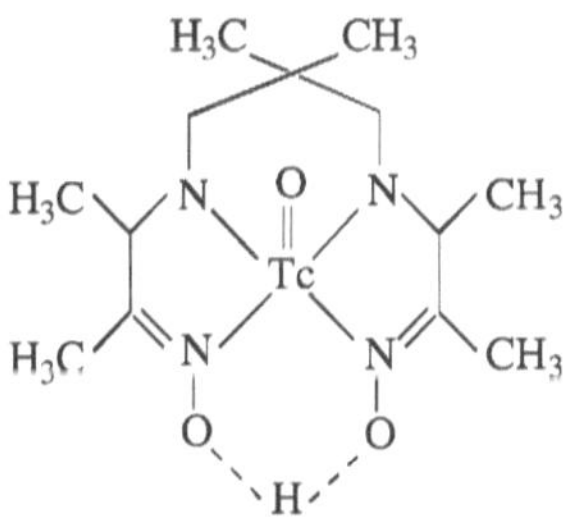

Kit components		**Commercial products**	
Exametazime	0.5 mg	Ceretec	GE Healthcare
Tin(II)-chloride dihydrate	7.6 µg	Neurospect	Rotop
Sodium chloride	4.5 mg		

Preparation

The Ceretec kit contains the lyophilized, sterile ingredients in a multidose vial. Labeling with ^{99m}Tc eluate is carried out under aseptic conditions by adding 5 ml of sterile sodium ^{99m}Tc-pertechnetate to the vial (0.37–1.11 GBq). The shielded vial should be gently inverted for 10 s to ensure complete dissolution of the lyophilisate. The reaction is allowed to proceed at room temperature for 5 min.

^{99m}Tc-hexamethylpropylene amine oxime (HMPAO) is a sterile, pyrogen-free, clear, colorless solution suitable for intravenous injection. The resulting pH is between 9.0 and 9.8.

In addition, each package consisting of five labeling units contains also five vials of methylene blue injection 1% USP (10 mg methylene blue USP in 1 ml water for injection), and five vials containing 0.003 *M* monobasic sodium phosphate USP and dibasic sodium phosphate USP in 4.5 ml of 0.9% sodium chloride injection USP. Each milliliter contains 0.276 mg monobasic sodium phosphate monohydrate, 0.142 mg dibasic sodium phosphate anhydrous, and 9 mg sodium chloride in water for injection for a sufficient amount. A mixture of methylene blue in phosphate buffer is used for stabilization of the ^{99m}Tc-HMPAO complex (Amersham Healthcare 1995).

Description of the Kit

^{99m}Tc-D,L-HMPAO complex is formed rapidly with reduced technetium at room temperature. The stannous tin content per vial should not decrease to less than 6 µg. Thus, high labeling is depending on maintaining tin in the reduced state. Any oxidant in the ^{99m}Tc eluate should be avoided. Only eluates from generators eluted regularly within 24 h after the previous elution may be used for labeling (Amersham Healthcare 1995). Isotopic dilution is observed with higher concentrations of ^{99}Tc in the first eluate of new generators or after weekends, reducing the labeling efficiency (Ponto et al. 1987). The formulation contains no antimicrobial preservative.

^{99m}Tc-exametazime injection solution should be used within 60 min after labeling. In order to maintain high stability of the ^{99m}Tc-HMPAO complex for up to 6 h after preparation, the producer provides methylene blue injection USP 1% or alternatively, cobalt(II)-chloride aqueous solution (250 µg/2.5 ml, *European Pharmacopeia* [*Ph. Eur.*]) for stabilization.

Stabilization of the ^{99m}Tc-HMPAO complex for intravenous use with methylene blue. Half a milliliter of methylene blue injection 1% USP should be withdrawn into a sterile syringe and injected into the 4.5-ml vial containing the buffer solution. Two milliliters of the methylene blue/phosphate buffer mixture should be gently swirled and withdrawn into a syringe for use as stabilizer. This mixture must be used within 30 min of preparation.

Within 2 min after labeling, add 2.0 ml of the methylene blue/phosphate buffer mixture to the Ceretec vial (Amersham Healthcare 1995). The volume of stabilized Ceretec injection solution is 7.0 ml, and the resulting pH between 6.5 and 7.5.

The ^{99m}Tc-HMPAO complex stabilized with methylene blue is stable for 4 h.

Stabilization of the ^{99m}Tc-HMPAO complex with cobalt(II)-chloride. Two milliliters of cobalt stabilizer solution (250 μg cobalt(II)-chloride·6 H_2O dissolved in 2.5 ml of water for injection) should be withdrawn with a 3-ml sterile syringe.

Within 1–5 min after labeling, 2.0 ml of the stabilizer solution should be injected into the Ceretec vial, and the vial agitated for 10 sec. The volume of stabilized Ceretec injection solution is 7.0 ml, and the resulting pH between 5.0 nd 8.0.

The ^{99m}Tc-HMPAO complex stabilized with cobalt(II)-chloride is stable for 6 h.

Clinical Applications

Technetium-99m exametazime injection is indicated for brain scintigraphy for the diagnosis of perfusion abnormalities of regional cerebral blood flow (rCBF):

- Detection of focal perfusion abnormalities
- Diagnosis of acute cerebral infarction (stroke) when computer tomography (CT) is negative
- Classification of defects of ischemic stroke
- Diagnosis of cerebrovascular disease and differentiation of focal abnormalities in CBF typical in multi-infarct dementia and degenerative dementia

The kinetic parameters of ^{99m}Tc-HMPAO retention in the human brain have been studied (Lassen et al. 1988), and measurements of CBF have been compared to ^{133}Xe (Andersen et al. 1988). CBF was quantified using dynamic single-photon emission computer tomography (SPECT) (Murase et al. 1992). The role of SPECT with ^{99m}Tc-HMPAO in ischemic stroke has been evaluated (Heiss 1983; Podreka et al. 1987). Reflow hyperemia in subacute stroke has been described as "luxury perfusion" (Lassen 1966) and was later explained by hyperfixation of ^{99m}Tc-HMPAO (Lassen and Sperling 1993). Increased regional blood flow in subacute stroke has been measured with ^{99m}Tc-HMPAO (Moretti et al. 1990). The effect of acetazolamide on CBF was reported in patients with severe internal carotid artery stenosis/occlusion (Asenbaum et al. 1995), and has been studied in primates (Dormehl et al. 1997). ^{99m}Tc-HMPAO SPECT was evaluated in patients with cerebrovascular disease by comparison with ^{18}F-fluoromethane positron emission tomography (PET) (Heiss et al. 1990). A pattern of focal abnormalities in CBF was derived in patients with degenerative dementia (Holman et al. 1992).

Time of Examination

Dynamic imaging:	Right after intravenous injection up to 10 min postinjection
Planar/tomographic imaging:	15 min up to 6 h postinjection

Recommended Activities for Indications

Cerebral perfusion scintigraphy:	370–740 MBq (10–20 mCi), injected intravenously 500 MBq (13.5 mCi) maximum recommended activity (Administration of Radioactive Substances Advisory Committee 1993) The manufacturer's instructions should be followed.

Pediatric Dose. The amount of radioactivity for infants and children administered for brain scintigraphy is based on body weight, using the scaling factors given in Appendix 1 (Table A1.4). The minimum value of the scaling factor is 0.1, since activities below 10% of the value of the adult radioactivity have resulted in poor image quality.

Additional Information

The ^{99m}Tc eluate used for labeling should be obtained from a ^{99m}Tc-generator, which is eluted daily at a 24-h interval. When using the stabilizing protocol, generator eluates older than 30 min should not be used. Highest radiochemical purity is obtained with ^{99m}Tc-pertechnetate right after elution.

The lipophilic ^{99m}Tc-HMPAO complex is transformed by hydrolysis (in vitro) to the "secondary" hydrophilic ^{99m}Tc-HMPAO complex, which cannot cross the blood–brain barrier (BBB) when injected. Since a high extraction fraction is essential for quantitation of CBF, high radiochemical purity of lipophilic ^{99m}Tc-HMPAO is required. The average value of lipophilic ^{99m}Tc-HMPAO has been reported as 83% (Lassen et al. 1988) and 80.8% (Murase et al. 1992), respectively. Care has been taken to minimize the time between the quality control measurements and administration of the ^{99m}Tc-HMPAO complex (Murase et al. 1992).

Stabilization of the ^{99m}Tc-HMPAO complex with methylene blue/phosphate buffer or with cobalt(II)-chloride will extend in vitro stability up to 6 h after labeling.

The patient should be asked to drink water frequently in order to stimulate excretion of radioactivity.

The ^{99m}Tc-exametazime complex may be used for radiolabeling of leukocytes. In this case, stabilization with methylene blue/phosphate buffer should not be used. A generator eluate more than 2-h-old should not be used for complex formation.

Quality Control

Radiochemical Purity. ^{99m}Tc-exametazime is included in the *Ph. Eur.* (Council of Europe 2005). Thin-layer chromatography on silica gel fiberglass plates using two solvent systems is described to determine the radiochemical purity of the injection solution. Free ^{99m}Tc-Na-pertechnetate (B) is determined in saline, the lipophilic ^{99m}Tc-HMPAO complex (D) is quantified indirectly by a subtraction method (Tab. 1). Previously, three solvent systems had been used, as recommended by the manufacturer.

The labeled product must be tested before application in patients. The radiochemical purity of lipophilic ^{99m}Tc-exametazime should not be less than 80% (*Ph. Eur.*).

Thin-layer chromatography. Aliquots are spotted and analyzed within 2 minutes after labelling, before stabilisation. The entire procedure of analysis takes approx. 15 minutes.

- System I: Gelman ITLC silica gel fiberglass plates and 2-butanone (methyl ethyl ketone [MEK]) as solvent: Reduced, hydrolized ^{99m}Tc-technetium (C) and the secondary ^{99m}Tc-HMPAO complex (A) remain at the start (A+C); the lipophilic ^{99m}Tc-HMPAO complex (D) and unbound ^{99m}Tc-Na-pertechnetate (B) are measured at the solvent front (B+D).

- System II: Gelman ITLC silica gel fiberglass plates and saline as solvent: Reduced, hydrolized ^{99m}Tc-technetium (C), the lipophilic and the secondary ^{99m}Tc-HMPAO complexes (A+D) remain at the start (A+C+D); unbound ^{99m}Tc-Na-pertechnetate (B) moves with the solvent front.

Table 1. Thin-layer chromatography on silica gel fiberglass plates using two solvent systems

System I (MEK)	Sum of reduced, hydrolized ^{99m}Tc activity and the secondary ^{99m}Tc-HMPAO complex at the start:	(A + C)
	Sum of lipophilic ^{99m}Tc-HMPAO complex and free ^{99m}Tc-pertechnetate at the solvent front:	(B + D)
System II (Saline)	Free ^{99m}Tc-pertechnetate at the solvent front:	(B)
	^{99m}Tc activity at origin corresponding to:	(A + C + D)
	A, B, and C represent labeled impurities, D is the lipophilic ^{99m}Tc-HMPAO complex	
	D (%) = 100 − % (A + C + B) (*Ph. Eur.*)	
	D (%) = % (A + C+D) − %(A+C) (*USP*)	

Impurities (A+C) measured in solvent I (MEK) and free ^{99m}Tc-pertechnetate (B), quantified in system II at the solvent front are subtracted from the total recovered activity, which is assumed as 100%. Free ^{99m}Tc-pertechnetate (B) should not exceed 10% of the total radioactivity, the lipophilic ^{99m}Tc-HMPAO complex should not be less than 80% (*Ph. Eur.*).

The difference between activity measured at the start (A+C+D) in solvent II (saline) minus activity measured at the start (A+C) in solvent I (MEK) corresponds to the percentage of the lipophilic ^{99m}Tc-HMPAO complex. Radioactivity measured at the solvent front (B) in system II indicates the amount of free ^{99m}Tc-pertechnetate (*USP*).

The percentage of the lipophilic ^{99m}Tc-HMPAO complex is calculated according to the specifications described in the pharmacopeias (Tab. 1).

Methods recommended by the manufacturer:
Thin-layer chromatography on Gelman silica gel sheets and paper chromatography on Whatman 1 strips is recommended by the manufacturer, using three solvent systems for the analysis of the lipophilic ^{99m}Tc-HMPAO complex (D), the secondary hydrophilic complex (A), unbound ^{99m}Tc-Na-pertechnetate (B), and reduced, hydrolized ^{99m}Tc activity (C) (Tab. 2). Since reduced, hydrolized ^{99m}Tc activity is determined separately, each impurity is quantified. The radiochemical purity of lipophilic ^{99m}Tc-exametazime should not be less than 80% (USP 28).

System I and II are identical with the methods recommended in the *Ph. Eur.*

System III is based on paper chromatography using Whatman 1 paper strips and acetonitrile-water (1 : 1) as solvent. Reduced, hydrolized ^{99m}Tc-technetium (C) remains at the start; both ^{99m}Tc-HMPAO complexes (A+D) and unbound ^{99m}Tc-Na-pertechnetate (B) are measured at the solvent front (A+B+D).

Table 2. Thin-layer chromatography and paper chromatography using three solvent systems (recommended for Ceretec)

System I	Sum of reduced, hydrolized ^{99m}Tc activity and the secondary ^{99m}Tc-HMPAO complex at the origin	(A + C)
System II	Free ^{99m}Tc-pertechnetate at the solvent front	(B)
System III	Reduced, hydrolized ^{99m}Tc-activity at the origin	(C)
	D (%) = 100 – % (A + B + C)	

The added impurities % (A + B + C) are subtracted from the total recovered activity (100%) according to the recommended methods (Tab. 2).

$$\text{Lipophilic } ^{99m}\text{Tc-HMPAO complex } (\%\ D) = 100 - \%(A + B + C)$$

A = Secondary ^{99m}Tc-HMPAO complex
B = ^{99m}Tc-pertechnetate
C = Reduced, hydrolized ^{99m}Tc activity
D = Lipophilic ^{99m}Tc-HMPAO complex

Results of analysis *(12 samples)*
Results were obtained using the analytical methods originally described for Ceretec (Amersham Healthcare 1995) outlined in Tab. 2.

Labeling and stability	15 min (%)	1 h (%)
^{99m}Tc-HMPAO lipophilic complex (D)	90.1±1.06	85.6±1.23
^{99m}Tc-HMPAO secondary complex (A)	4.5±0.90	6.5±1.79
^{99m}Tc-Na-pertechnetate (B)	2.9±0.51	3.9±0.62
^{99m}Tc-reduced, hydrolized (C)	2.4±0.47	4.0±1.42

Solvent extraction
A more rapid method to obtain information about the labelling yield (percentage of lipophilic ^{99m}Tc-exametazime) has found wide acceptance in nuclear medicine (Ballinger et al. 1988). Several solvents (ethyl acetate, chloroform, n-octanol) have been employed for the extraction of the lipophilic ^{99m}Tc-HMPAO complex. Labelled impurities remain in the aqueous phase.

Procedure: A sample of 0.1 ml of labeled Ceretec is added to a mixture of 3 ml of chloroform and 2.9 ml of saline. Close vial and mix well using a vortex mixer for 30 seconds. Let phases separate for 1 min, then transfer the top layer (saline) to another vial and measure both vials in a dose calibrator.

The radiochemical purity of the lipophilic ^{99m}Tc-HMPAO complex is expressed by:

$$^{99m}\text{Tc-HMPAO } (\%) = \frac{\text{Activity of chloroform fraction}}{\text{Sum of activities in both fractions}} \times 100$$

A comparison of results obtained with both methods (ITLC and solvent extraction) indicates excellent correlation between both separation methods:

ITLC separation ($n=12$)	87.0±2.0%
Solvent extraction ($n=12$)	85.5±3.7%

Pharmacokinetic Data

Neutral, lipophilic molecules with log *P*-values between 0.9 and 3.5 may cross the BBB by diffusion or active process, depending on the structural configuration (Holm et al. 1985; Troutner et al. 1984).

The lipophilic ^{99m}Tc-D,L-HMPAO complex (log $P=1.2$) can cross the BBB and is extracted with high efficiency ($E=0.8$) from blood at normal flow levels (Holmes et al. 1985; Lassen and Andersen 1988; Lassen et al. 1987). Cerebral extraction corresponds to approximately 5% of the injected radioactivity (Leonard et al. 1986; Neirinckx et al. 1987). The mesoform is not accumulated in the brain due to the stereospecificity of uptake (Sharp et al. 1986).

The lipophilic ^{99m}Tc-D,L-HMPAO complex is decomposed rapidly in vivo, both in the blood and in the brain. Due to this instability, the secondary ^{99m}Tc-D,L-HMPAO complex, a charged complex, is formed (Neirinckx et al. 1988). The secondary complex cannot pass the BBB and is trapped inside the brain and in blood cells, i.e., the ionized molecule is trapped.

After intravenous injection of a bolus of ^{99m}Tc-D,L-HMPAO complex, 50% of the radioactivity are eliminated from the circulation within 2–3 min; thus, when the lipophilic complex has been eliminated from the blood, no further exchange is observed (Lassen and Andersen 1988).

After approximately 5 min, when the lipophilic radiotracer has disappeared both in the blood and in brain, the distribution of radioactivity in the brain is a true image of the initial blood flow (Lassen et al. 1987). The radioactivity pattern remains constant for 24 h. Elimination from the brain is very slow, approximately 1% per hour (Ell et al. 1987). After 24 h, >70% of the tracer is still in the brain.

The kidneys excrete 41% of the injected radioactivity over the first 48 h. Between 8.5 and 13.0% pass through the liver, thereby showing the main bile ducts and the gall bladder. Lung uptake averages 9%, and about 2% of the injected radiotracer localize in the myocardium (Ell et al. 1987).

The lipophilic ^{99m}Tc-D,L-HMPAO complex accumulates in blood cells (80% in red cells, less in white cells and platelets); the blood shows approximately 12% radioactivity 1 h postinjection (Ell et al. 1987).

A relatively high concentration of activity was observed in lacrimal glands of volunteers (Meyer et al. 1990).

Radiation Dose

The excretory organs such as the kidneys and urinary bladder, along with the gallbladder wall, the upper large intestinal wall, the lower large intestinal wall, the small intestine, and the liver are the most exposed organs (ICRP 62). The effective (whole body) dose is 0.0093 mSv/MBq (International Commission on Radiological Protection 1991). The effective dose in adults (70 kg) resulting from 500 MBq (13.5 mCi) of intravenously injected ^{99m}Tc-HMPAO complex is 4.65 mSv.

Storage and Stability

Storage. The lyophilized kit should be stored at 15–25 °C.

Stability. The stabilized ^{99m}Tc-HMPAO complex may be used 4–6 h after preparation.

References

Administration of Radioactive Substances Advisory Committee (1993) Technetium-labeled exametazime (HMPAO) (1993) In: Notes for guidance on the administration of radioactive substances to persons for purposes of diagnosis, treatment or research, Appendix 1, Part A, pp 22–30; and Administration of Radioactive Substances Advisory Committee – ARSAC. 13(a) ibidem: Investigations – Children and young persons (1993) In: Notes for guidance on the administration of radioactive substances to persons for purposes of diagnosis, treatment or research, Appendix 1, Part D, pp 34–35

Amersham Healthcare (1995) Product information for the Ceretec kit for the preparation of technetium Tc-99m exametazime injection. Amersham Healthcare, UK

Andersen AR, Friberg HH, Schmidt JF, Hasselbalch SG (1988) Quantitative measurements of cerebral blood flow using SPECT and ^{99m}Tc-D,L-HMPAO compared to xenon-133. J Cereb Blood Flow Metab 8(Suppl 1):S69–S81

Asenbaum S, Reinprecht A, Brücke T, Wenger S, Podreka I, Deecke L (1985) A study of acetazolamide-induced changes in cerebral blood flow using ^{99m}Tc-HM-PAO SPECT in patients with cerebrovascular disease. Neuroradiology 27:509–516

Ballinger JR, Reid RH, Gulenchyn KY (1988) Radiochemical purity of [^{99m}Tc]HM-PAO. J Nucl Med 29:572–573 (letter)

Dormehl, IC, Oliver DW, Langen, K-J, Hugo N, Croft SA (1997) Technetium-99m-HMPAO, technetium-99m-ECD and iodine-123-IMP cerebral blood flow measurements with pharmacological interventions in primates. J Nucl Med 38:1897–1901

Ell PJ, DC Costa, ID Cullum, PH Jarritt, D Lui (1987) Ceretec rCBF atlas: the clinical application of rCBF imaging by SPECT: Amersham International, UK

Council of Europe (2005) Technetium ^{99m}Tc exametazime injection. In: European Pharmacopeia 5.0, monograph no 1925. Council of Europe, Maisonneuve, Sainte-Ruffine, p 1214

Heiss W-D (1983) Flow thresholds for functional and morphological damage of brain tissue. Stroke 14:329–331

Heiss W-D, Herholz K, Podreka I, Neubauer I, Pietrzyk U (1990) Comparison of ^{99m}Tc-HM-PAO SPECT with ^{18}F-fluoromethane PET in cerebrovascular disease. J Cereb Blood Flow Metab 10:687–697

Holm S, Andersen AR, Vorstrup S, Lassen NA, Paulson OB, Holmes RA (1985) Dynamic SPECT of the brain using a lipophilic technetium-99m complex, PnAO. J Nucl Med 26:1129–1134

Holman BL, Johnson KA, Gerada B, Carvalho PA, Satlin A (1992) The scintigraphic appearance of Alzheimer's disease: A prospective study using technetium-99m-HM-PAO SPECT. J Nucl Med 33:181–185

Holmes RA, Chaplin SB, Royston KG, Hoffmann TJ, Volkert WA (1985) Cerebral uptake and retention of ^{99m}Tc-hexamethyl-propyleneamine oxime (^{99m}Tc-HM-PAO). J Nucl Med Commun 6:443–447

International Commission on Radiological Protection (1991) Technetium-labelled HM-PAO (Ceretec) In: Annals of the ICRP, Radiological protection in biomedical research, ICRP publication 62, vol. 22., no. 3. Pergamon, Oxford, pp 11–13

Lassen NA (1966) The luxury-perfusion syndrome and its possible relation to acute metabolic acidosis localized within the brain. Lancet 2:1113–1115

Lassen NA, Andersen AR (1988) Technetium-99m compounds for measurement of cerebral blood flow [reply]. J Nucl Med 29:1664–1465
Lassen NA, Sperling B (1993) Hyperfixation of HM-PAO in subactute ischemic stroke leading to spuriously high estimates of cerebral blood flow by SPECT. Stroke 24:193–194
Lassen NA, Andersen AR, Friberg H, Neirinckx RD (1987) Technetium-99m-HM-PAO as a tracer of cerebral blood flow distribution: A kinetic analysis. J Cereb Blood Flow Metab 7 (Suppl 1):S535
Lassen NA, Andersen AR, Friberg L, Paulson OB (1988) The retention of ^{99m}Tc-D,L-HM-PAO in the human brain after intracarotid bolus injection: A kinetic analysis. J Cereb Blood Flow Metab 8(Suppl 1):S13–S22
Leonard JP, Nowotnik DP, Neirinckx RD (1986) Technetium-99m D,L-HM-PAO: a new radiopharmaceutical for imaging regional brain perfusion using SPECT – a comparison with iodine-123 HIPDM. J Nucl Med 27:1819–1823
Meyer JY, Thomson D, Mena I, Marcus CS (1990) Lacrimal gland dosimtery for the brain imaging agent ^{99m}Tc-HMPAO. J Nucl Med 31:1237–1239
Moretti JL, Defer G, Cinotti L, Cesaro P, Degos JD, Vigneron N, Ducassou D, Holman BL (1990) "Luxury perfusion" with ^{99m}Tc-HM-PAO and ^{123}I-IMP SPECT imaging during the subacute phase of stroke. Eur. J Nucl Med 16:17–22
Murase K, Tanada S, Fujita H, Sakaki S, Hamamoto K (1992) Kinetic behavior of Tc-99m HM-PAO in the human brain and quantification of cerebral blood flow using dynamic SPECT. J Nucl Med 33:135–143
Neirinckx RD, Canning LR, Piper IM, Nowotnik DP, Pickett RD, Holmes RA, Volkert WA, Forster AM, Weisner PS, Mariott JA, Chaplin SB (1987) Technetium-99m D,L-HM-PAO: a new radiopharmaceutical for SPECT imaging of regional cerebral blood perfusion. J Nucl Med 28:191–202
Neirinckx RD, Burke JF, Harrison RC, Forster AM, Andersen AR, Lassen NA (1988) The retention mechanism of technetium-99m-HMPAO: intracellular reaction with glutathione. J Cereb Blood Flow Metab 8(Suppl 1):S4–S12
Podreka I, Suess E, Goldenberg G, Brücke T, Müller C, Lang W, Neirinckx RD, Deecke L (1987) Initial experience with technetium-99m-HM-PAO brain SPECT. J Nucl Med 28:1657–1666
Ponto JA, Swanson DP, Freitas JE (1987) Clinical manifestations of radiopharmaceutical formulation problems. In: Hladik WB III, Saha GB, Study KT (eds) Essentials of nuclear medicine science. Williams & Wilkins, Baltimore, pp 268–270, 272–289
Sharp PF, Smith FW, Gemmel HG, Lyall D, Evans NTS, Gvozdanovic D, Davidson J, Tyrrell DA, Pickett RD, Neirinckx RD (1986) Technetium-99m HM-PAO stereoisomers as potential agents for imaging regional cerebral blood flow: Human volunteer studies. J Nucl Med 27:171–177
Troutner DE, Volkert WA, Hoffmann TJ, Holmes RA (1984) A neutral lipophilic complex of ^{99m}Tc with a multidentate amine oxime. Int J Appl Radiat Isot 35:467–470
United States Pharmacopeial Convention (2005) Official Monographs: USP 28, technetium Tc 99m exametazime injection. United States Pharmacopeia (USP) 28–national formulary (NF) 23, p 1855

12.5.2 ^{99m}Tc-ECD (Ethyl Cysteinate Dimer)

J. Imre and I. Zolle

Chemical name	**Chemical structure**
(*N,N′*-1,2-ethylenediyl-bis-L-cysteine) diethyl ester dihydrochloride (ECD) Ethyl-Cysteinate-Dimer (L,L-ECD) Bicisate (*USP*) Technetium Tc 99m bicisate injection (*USP*) Tc(V)oxo-*N,N′*-ethylene di-cysteinate diethyl ester ^{99m}Tc(V)oxo-L,L-ECD complex	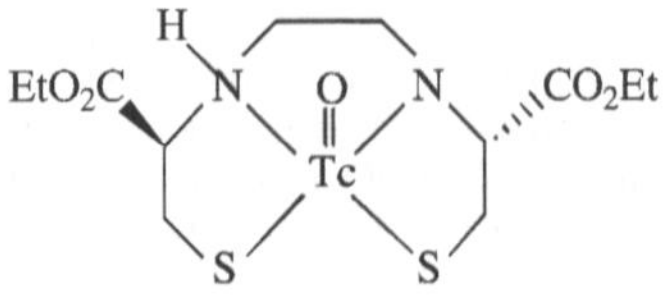 Tc(V)O-L,L-ECD
Kit components	**Listed trade names**
Vial A: Bicisate dihydrochloride 0.9 mg Stannous chloride, dihydrate 0.072 mg Edetate disodium, dihydrate 0.36 mg Mannitol 24 mg *Vial B:* Sodium phosphate dibasic Heptahydrate 4.1 mg Sodium phosphate monobasic monohydrate 0.46 mg Water for injection ad 1 ml	Neurolite Bristol-Myers Squibb

Preparation

The Neurolite kit consists of two nonradioactive vials, vial A and vial B.

Vial A contains the lyophilized active ingredients in a nitrogen atmosphere. A volume of 3 ml of saline is added to vial A, and the vial inverted to dissolve the kit content. Within 30 s, 1 ml from vial A should be withdrawn and injected into vial B.

Vial B contains 1 ml phosphate buffer, pH 7.6±0.4. It is placed into a lead shield. A volume of 2 ml of ^{99m}Tc-pertechnetate (925 MBq–3.7 GBq; 25–100 mCi) is added aseptically to vial B. Labeling is performed by adding aseptically 1 ml from vial A to vial B, and allowed to react for 30 min at room temperature (DuPont Merck Pharmaceutical 1995).

^{99m}Tc-ethyl cysteinate dimer (ECD) is a clear, colorless, sterile, solution suitable for intravenous injection. The pH value is between 6.6 and 7.8.

Description of the Kit

The kit formulation (vial A) contains bicisate dihydrochloride exclusively as the L,L-enantiomer (DuPont Merck Pharmaceutical 1995). After reconstitution with 3 ml saline, the pH of vial A is 2.7 ± 0.25. One third of bicisate (0.3 mg) is used for labeling. The rest is discarded. Storage of residual portions of Neurolite in a freezer and subsequent labeling within 4 weeks has been suggested (Verbeke et al. 1997).

Initially, reduced ^{99m}Tc activity and ethylenediaminetetraacetic acid (EDTA) form an intermediary complex, which is transformed slowly by ligand exchange to the ^{99m}Tc(V)-ECD complex, showing high in vitro stability. Originally, glucoheptonate had been used instead of EDTA (Cheesman et al. 1988).

The diaminodithiol (DADT)-derived ligand ECD forms a stable complex with pentavalent oxo-technetium (Cheesman et al. 1988). ^{99m}Tc-ECD is a neutral, lipophilic complex with the chemical formula $[TcO\text{-}L,L\text{-}ECD]^0$.

Clinical Applications

Technetium-99m bicisate injection is indicated for brain scintigraphy to delineate focal perfusion abnormalities:

- Diagnosis of acute cerebral infarction (stroke) when computer tomography (CT) is negative
- Detection of inflammatory conditions in the brain
- Detection of an abnormal focus in patients with head trauma after accidents
- Differentiation of focal abnormalities in cerebral blood flow typical in multi-infarct dementia and degenerative dementia

^{99m}Tc-bicisate uptake follows linearity, with blood flow values up to 20 ml/100 g/min; however, ECD underestimates higher flow rates (Tsuchida et al. 1992). A considerable disadvantage is 3–4% washout per hour, which is difficult to correct for. Another deficit of ECD is an overestimation of infarct size in patients with subactute stroke (Lassen and Sperling 1994). The role of single-photon emission computer tomography (SPECT) with ^{99m}Tc-bicisate in ischemic stroke has been evaluated (Brass et al. 1994; Moretti et al. 1990).

The kinetic parameters of ^{99m}Tc-ECD in the human brain have been compared with regional cerebral blood flow measurements with ^{133}Xe SPECT (Devous et al. 1993) and with positron emission tomography (PET) (Tsuchida et al. 1992). The response of ^{99m}Tc-ECD, ^{99m}Tc-examethylpropylene amine oxime HMPAO and ^{123}I-IMP (*N*-isopropyl-*p*-[^{123}I]-iodoamphetamine) to changes in cerebral blood flow was studied in primates (Dormehl et al. 1997).

Time of Examination. Images of the brain are obtained from 10 min up to 6 h after injection. Optimal images occur 30–60 min after injection.

Recommended Activities for Indications

Cerebral perfusion scintigraphy: 370–740 MBq (10–20 mCi), injected intravenously
500 MBq (13.5 mCi) maximum recommended activity (Administration of Radioactive Substances Advisory Committee 1993)
The manufacturer's instructions should be followed.

Pediatric Dose. The amount of radioactivity for infants and children administered for brain scintigraphy is based on body weight, using the scaling factors given in Appendix 1 (Table A1.2). The minimum value of the scaling factor is 0.1, since activities below 10% of the value of the adult radioactivity have resulted in poor image quality.

Additional Information

Patients should drink sufficient water before and after the study, and frequent bladder emptying should be encouraged in order to reduce the radiation exposure to the bladder wall.

Technetium ^{99m}Tc-bicisate is eliminated primarily by renal excretion. It should be used with caution in patients with renal or hepatic impairment.

The effect of acetazolamide on cerebral blood flow is underestimated with ^{99m}Tc-ECD (Dormehl et al. 1997).

Quality Control

Radiochemical Purity. ^{99m}Tc-bicisate is not described in the *European Pharmacopeia* (United States Pharmacopeial Convention 2005). Thin-layer chromatography (TLC) using Baker-flex silica gel strips and an organic solvent is recommended by the manufacturer. Free ^{99m}Tc-pertechnetate and reduced, hydrolized ^{99m}Tc activity remain at the start, ^{99m}Tc-bicisate is measured at an R_f of 0.9–1.0. The radiochemical purity of ^{99m}Tc-bicisate should not be less than 90%.

The radiochemical purity should be determined prior to administration of ^{99m}Tc-ECD complex. If the labeling yield is less than 90%, the preparation should be discarded.

Methods Recommended by the Manufacturer

Thin-layer chromatography		
Stationary phase:	Baker-flex SG 1B-F (precut to 2.5×7.5 cm)	
Solvent:	Ethyl acetate (high-performance liquid chromatography [HPLC] grade)	
Developing time:	15 min	
R_f values:	^{99m}Tc reduced, hydrolized:	0.0–0.1
	^{99m}Tc-pertechnetate:	0.0–0.1
	^{99m}Tc-ECD complex:	0.9–1.0 (>90%)

Procedure:

- Pour enough ethyl acetate into the TLC tank (beaker) to have a depth of 3–4 mm of solvent.
- Cover the tank (beaker) with parafilm and allow it to equilibrate for approximately 10 min.
- Dry the plates at 100 °C for 1 h and store in a desiccator. Remove predried plate from the desiccator just prior to use.
- With a pencil, draw a faint line across the TLC plate at heights of 2, 4.5, and 7 cm from the bottom of the TLC plate.
- Apply 5 µl of ^{99m}Tc-ECD injection solution at the center of the 2-cm mark. The diameter of the spot should not be greater than 10 mm.
- Allow the spot to dry for 5–10 min, no longer.
- Develop the plate in the covered TLC tank in fresh ethyl acetate to the 7.0-cm line (about 15 min). Remove the plate and dry in a ventilated hood.
- Cut the TLC plate at the 4.5-cm mark with scissors.
- Count the activity on each plate using a dose calibrator or a gamma counter.

The time required to complete the entire procedure is approximately 30 min.

The portion 4.5–7.0 cm contains the ^{99m}Tc-ECD complex and the bottom portion contains all labeled impurities.

Calculate the percent radiochemical purity as:

$$^{99m}\text{Tc-bicisate (\%)} = \frac{\text{Activity of upper piece}}{\text{Activity of both pieces}} \times 100$$

Results of analysis *(12 samples)*

Results were obtained by thin-layer chromatography described by the manufacturer.

Labeling and stability	15 min (%)	6 h (%)
^{99m}Tc-ECD complex	95.1 ± 0.41	93.2 ± 0.34
^{99m}Tc-Na-pertechnetate and ^{99m}Tc-reduced, hydrolized	4.8 ± 0.24	6.0 ± 0.17

Pharmacokinetic Data

After intravenous injection of a bolus of activity, ^{99m}Tc-L,L-ECD is distributed in the normal brain proportional to regional blood flow (Friberg et al. 1994). The lipophilic complex crosses the blood–brain barrier (BBB); cerebral uptake at 5 min postinjection is approximately 6.5% of the injected radioactivity (Holman et al. 1989; Leveille et al. 1989; Vallabhajosula et al. 1989). Within 4 h after injection, activity in the brain decreased to 3.7 ± 0.3% injection dose (ID).

The highest concentration of radioactivity in blood was measured at 1 min after intravenous injection, representing 19% of the injected dose (Vallabhajosula et al. 1989). The elimination from blood is rapid ($T_{1/2}$ = 0.8 min), falling to 10% of the injected radioactivity at 2 min. One hour after the intravenous injection, less than 5% of the radioactivity is present, mainly as the nonlipophilic complex. ^{99m}Tc-bicisate is rapidly metabolized in blood; the hydrophilic acid metabolites are excreted through the kidneys, resulting in high brain to soft tissue concentration ratios early after injection and increasing over several hours (Holman et al. 1989).

Brain retention of ^{99m}Tc-L,L-ECD is caused by intracellular stereospecific hydrolysis of the ester, producing the monoethyl ester (Walovitch et al. 1988). The D,D-isomer showed back-diffusion ($T_{1/2}$ 30 min). Once the metabolite is formed in the brain, it cannot cross the BBB in either direction (Walovitch et al. 1989). Effect of hypoxia on esterase function might interfere with reflow hyperemia in subacute stroke (Dormehl et al. 1997). Elimination from brain is expressed by two exponential functions, corresponding to half-times of 1.3 h (40%) and 42.3 h (60%, 3.8% ID), respectively (Vallabhajosula et al. 1989).

^{99m}Tc-L,L-ECD is rapidly metabolized in the liver to the monoacid, and also to the diacid, ^{99m}Tc-L,L-EC. At 5 min postinjection, only a third of the blood activity (2.4% ID) was identified as the neutral, parent complex. The remaining activity was a mixture of polar metabolites. The amount of ^{99m}Tc-bicisate is decreasing rapidly, 10 min after injection there remains little parent complex in the blood for extraction by the brain. Metabolites were identified in urine by HPLC as 19.5% monoacid and 37.8% diacid (^{99m}Tc-EC) and 26.3% monoacid and 48.9% diacid at 2 hr and at 6 hr after administration of ^{99m}Tc-L,L-ECD, respectively (Walovitch et al. 1991).

Excretion of ^{99m}Tc-L,L-ECD from the body is primarily by the kidneys, approximately 50% during the first 2 h, a total of 74% in 24 h. In feces 11.2 ± 6.2% were measured in 48 h. Total body retention of ^{99m}Tc activity was less then 30% at 4 hr (Vallabhajosula et al. 1989).

Radiation Dose

The excretory organs such as the kidneys and urinary bladder, along with the gallbladder wall, the upper large intestinal wall, the lower large intestinal wall, the small intestine, and the liver are the most exposed organs (International Commission on Radiological Protection 1991). The effective (whole body) dose for HMPAO is 0.0093 mSv/MBq (International Commission on Radiological Protection 1991). Based on this value, the effective dose in adults (70 kg) resulting from 500 MBq of intravenously injected ^{99m}Tc-ECD complex is 4.65 mSv.

Storage and Stability

Storage. The lyophilized kit should be stored at 15–25 °C. Vial A should be protected from light. ^{99m}Tc-ECD injection is kept at room temperature with adequate shielding.

Stability. The ^{99m}Tc-ECD injection is stable for 8 h after preparation.

References

Brass LM, Walovitch RC, Joseph JL, Leveille J, Marchand L, Hellman RS, Tikovsky RS, Masdeu JC, Hall KM, Van Heertum RL (1994) The role of single photon emission computed tomography brain imaging with ^{99m}Tc-bicisate in the localization and definition of mechanism of ischemic stroke. J Cereb Blood Flow Metab 14(Suppl 1):S91–S98

Cheesman EH, Blanchette MA, Ganey MV, Maheu LJ, Miller SJ, Watson AD (1988) Technetium-99m ECD: Ester-derivatized diaminedithiol Tc complexes for imaging brain perfusion. J Nucl Med (Abstr)29:788

Devous MD, Payne JK, Lowe JL, Leroy RF (1993) Comparison of technetium-99m-ECD to xenon-133 SPECT in normal controls and in patients with mild to moderate regional cerebral blood flow abnormalities. J Nucl Med 34:754–761

Dormehl, IC, Oliver DW, Langen, K-J, Hugo N, Croft SA (1997) Technetium-99m-HMPAO, technetium-99m-ECD and iodine-123-IMP cerebral blood flow measurements with pharmacological interventions in primates. J Nucl Med 38:1897–1901

DuPont Merck Pharmaceutical (1995) Product monograph Neurolite, issued by DuPont Merck Pharmaceutical, Wilmington, Del.

Friberg L, Andersen AR, Lassen NA, Holm S, Dam M, (1994) Retention of ^{99m}Tc-bicisate in the human brain after intracarotid injection. J Cereb Blood Flow Metab 14(Suppl 1):S19–S27

Holman BL, Hellman RS, Goldsmith SJ, Mean IG, Leveille J, Gherardi PG, Moretti JL, Bischof-Delaloye A, Hill TC, Rigo PM, Van Heertum RL, Ell PJ, Büll U, DeRoo MC, Morgan RA (1989) Biodistribution, dosimetry and clinical evaluation of Tc-99m ethyl cysteinate dimer (ECD) in normal subjects and in patients with chronic cerebral infarction. J Nucl Med 30:1018–1024

International Commission on Radiological Protection (1991) Technetium-labeled HM-PAO (Ceretec) In: Annals of the ICRP, radiological protection in biomedical research. ICRP publication 62, vol. 22, no. 3. Pergamon, Oxford, pp 11–13

Lassen NA, Sperling B (1994) ^{99m}Tc-Bicisate reliably images CBF in chronic brain diseases but fails to show reflow hyperemia in subacute stroke: report of a multicenter trial of 105 cases comparing 133-Xe and 99mTc-bicisate (ECD, Neurolite) measured by SPECT on the same day. J Cereb Blood Flow Metab 14(Suppl 1):S44–S48

Leveille J, Demonceau G, DeRoo MC, Rigo PM, Talliefer R, Morgan RA, Kupranick D, Walovitch RC (1989) Characterisation of technetium-99m-L,L-ECD for brain perfusion imaging. Part 2: biodistribution and brain imaging in humans. J Nucl Med 30:1902–1910

Moretti JL, Defer G, Cinotti L, Cesaro P, Degos JD, Vigneron N, Ducassou D, Holman BL (1990) "Luxury perfusion" with ^{99m}Tc-HM-PAO and ^{123}I-IMP SPECT imaging during the subacute phase of stroke. Eur. J Nucl Med 16:17–22

Tsuchida T, Yonekura Y, Sadato N, Iwasaki Y, Tamaki N, Konishi J, Fujita T, Matoba N, Nishizawa S, Magata Y (1992) Brain perfusion SPECT with [Tc-99m]-L,L-ethyl cysteinate dimer (ECD) in comparison with regional cerebral blood flow measured by PET: underestimation in the high flow range. J Nucl Med 33(Abstr):966

United States Pharmacopeial Convention (2005) USP 28 official monographs, technetium-Tc99m bicisate injection, United States pharmacopeia. United States Pharmacopeial Convention, p 1853

Vallabhajosula S, Zimmermann RE, Picard M, Stritzke P, Mena I, Hellman RS, Tikovsky RS, Stabin MG, Morgan RA, Goldsmith SJ (1989) Technetium-99m ECD: a new brain imaging agent: in vivo kinetics and biodistribution studies in normal human subjects. J Nucl Med 30:599–604

Verbeke K, Boonen C, Verbruggen AM (1997) Usefulness of residual fractions of L,L-ethylcysteinate dimer (Neurolite) for the preparation of ^{99m}Tc-L,L-ethylcysteinate dimer. Nucl Med Commun 18:535–539

Walovitch RC, Makuch J, Knapik G, Watson AD, Williams SJ (1988) Brain retention of ^{99m}Tc-ECD is related to in vivo metabolism. J Nucl Med 29(Abstr):747

Walovitch RC, Hill TC, Garrity ST, Cheesman EH, Burgess BA, O'Leary DH, Watson AD, Ganey MV, Morgan RA, Williams SJ (1989) Characterisation of ^{99m}Tc-L,L-ECD for brain perfusion imaging. Part 1: pharmacology of ^{99m}Tc-L,L-ECD in non-human primates. J Nucl Med 30:1892–1901

Walovitch RC, Franceschi M, Picard M, Cheesman EH, Hall KM, Makuch J, Watson MW, Zimmerman RE, Watson AD, Ganey MV, Williams SJ, Holman BL (1991) Metabolism of ^{99m}Tc-L,L-ethyl cysteinate dimer in healthy volunteers. Neuropharmacology 30:283–292

12.6 ^{99m}Tc-Labeled Leukocytes

I. Zolle and Gy. Jánoki

Chemical name	Kit components, Ceretec	
D,L-Hexamethylpropylene amine oxime	Exametazime	0.5 mg
D,L-HM-PAO	Tin(II)-chloride dihydrate	7.6 µg
Exametazime (*Ph. Eur., USP*)	Sodium chloride	4.5 mg
^{99m}Tc-D,L-HMPAO complex		

Preparation

The Ceretec kit contains the lyophilized, sterile ingredients in a multidose vial, sealed under nitrogen atmosphere. Labeling with ^{99m}Tc eluate is carried out under aseptic conditions by adding 5 ml of sterile sodium ^{99m}Tc-pertechnetate to the vial (0.37–1.11 GBq). The shielded vial should be inverted gently for 10 sec to ensure complete dissolution of the lyophilisate. The reaction is allowed to proceed at room temperature for 5 min. The resulting pH is between 9.0 and 9.8.

^{99m}Tc-hexamethylpropylene amine oxime (HMPAO) injection solution (0.1 mg/ml) is suitable for labeling leukocytes (GE Healthcare 2005).

Description of the Kit

^{99m}Tc-D,L-HMPAO complex is formed rapidly with reduced technetium at room temperature. A high labeling yield depends on maintaining tin in the reduced state. Any oxidant in the ^{99m}Tc eluate should be avoided. Only eluates from generators eluted regularly within 24 h after the previous elution may be used for labeling (GE Healthcare 2005). Isotopic dilution is observed with higher concentrations of ^{99}Tc in the first eluate of new generators or after weekends, reducing the labeling efficiency (Ponto et al. 1987).

The mean separation efficiency expressed as a percentage of the number of leukocytes present in the patients' blood is 40%. Only one fifth of the ^{99m}Tc-HMPAO injection solution (0.1 mg) is used for labeling. A maximum labeling efficiency of 80% during a 20-min incubation period was reported using 50 µg of D,L-HMPAO ligand in 1 ml of saline and 5 µg of tin(II)-salt, corresponding to an activity of 370–740 MBq (10–20 mCi) of ^{99m}Tc-D,L-HMPAO (Mortelmans et al. 1989).

Procedure for labeling autologous leukocytes. For cell labeling, aseptic techniques have to be used throughout (Danpure et al. 1988; Segall et al. 1994). The original labeling procedure described by the manufacturer has been adapted to using smaller amounts of blood.

1. Draw 2 ml of acid-citrate-dextrose (ACD) solution and 3 ml of a sedimentation agent into each of four 20-ml plastic nonheparinized syringes.

2. Withdraw 15 ml of patient's blood into each syringe (a total of 60 ml), and mix gently by inversion.
3. Allow tubes to stand for 30–40 min at room temperature for erythrocytes to sediment.
4. When red cells have sedimented to approximately half the original volume of the blood, carefully draw up the leukocyte-rich, platelet-rich plasma (LRPRP) into a sterile tube and centrifuge at 150×g for 10 min.
5. While centrifuging the tubes, reconstitute one vial of HMPAO with 1.5 ml of ^{99m}Tc eluate containing 700–750 MBq (19–20 mCi) of ^{99m}Tc-pertechnetate (400–500 MBq/ml, resp. 11–13 mCi/ml)). Gently invert the shielded vial for 10 sec to dissolve the lyophilisate. The generator eluate should not be more than 2 h old and the generator must have been eluted within the past 24 h.
6. Remove the supernatant PRP platelet-rich plasma (PRP) from the pellet of "mixed" leukocytes, leaving the pellet almost dry. Save 10–15 ml of PRP for step 8. Agitate the tube gently to loosen the cells, and then pool all the cells into one tube.
7. Add exactly 1 ml of ^{99m}Tc-HMPAO to the tube with pooled leukocytes (the radioactivity of ^{99m}Tc-HMPAO is 400–500 MBq (11–13 mCi)). Mix gently and incubate the cells for 10 min at room temperature.
8. While incubating the leukocytes, centrifuge the PRP (obtained in step 6) for 5 min at 2000×*g* to produce cell-free plasma (CFP).
9. After incubation, carefully add 3–5 ml of cell-free plasma (obtained in step 8) to the labeled cell suspension and mix.
10. Centrifuge at 150×*g* for 10 min.
11. Add 3–5 ml of CFP containing ACD to the pellet of leukocytes; gently swirl for mixing.
12. Measure the radioactivity in the cells and in the supernatant (from step 10) and calculate the labeling efficiency (defined as the radioactivity in the cells expressed as a percentage of the sum of activities measured in the cells and in the supernatant)
13. The labeled leukocytes are ready to be reinjected. This should be performed without delay.

Clinical Applications. Lipophilic ^{99m}Tc-exametazime has been shown to label leukocytes without affecting cell viability (Mortelmans et al. 1989; Peters et al. 1986; Roddie et al. 1988). HMPAO-labeled leukocytes have been used to locate site(s) of focal infection (e.g., abdominal abscess, abdominal sepsis) (Kelbaek et al. 1985); it is also indicated in conditions of fever of unknown origin, and in conditions not associated with infection such as inflammatory bowel disease (Arndt et al. 1993; Lantto et al. 1991). Labeled leukocytes have offered superior information when compared with bone scanning for the detection of osteomyelitis in children (Lantto et al. 1992). In a retrospective study in 116 patients with infection suspected to involve orthopedic implants, osteomyelitis, and septic arthritis, HMPAO-labeled leukocytes have been an effective tool in the diagnosis of chronic osteomyelitis and joint infection involving implants (sensitivity: >97%, specificity: >89%) (Devillers et al. 1995).

Time of Examination. Planar imaging is performed at 1, 2 or 24 h after injection of labeled leukocytes.

Recommended Activities for Indications. Abdominal scintigraphy: 185–370 MBq (5–10 mCi) by intravenous injection

Additional Information

Use only ^{99m}Tc eluate for labeling which has been obtained less than 2 h before use.

Do not use methylene blue for stabilization, if ^{99m}Tc-HMPAO injection is used for leukocyte labeling (GE Healthcare 2005).

^{99m}Tc-labeled leukocytes should be used within 1 h after preparation.

A prolonged and delayed lung transit time and/or an abnormally high liver uptake are indications that some damage of the labeled cells occurred (Bowring 1986).

Quality Control

Radiochemical Purity. ^{99m}Tc-exametazime is included in the *European Pharmacopeia* (Council of Europe 2002). Thin-layer chromatography on silica gel fiberglass plates, using two solvent systems is described to determine the radiochemical purity of the injection solution. In analogy, ^{99m}Tc-HMPAO-labeled leukocytes are analyzed under similar conditions. However, since cells do not migrate, free ^{99m}Tc-Na-pertechnetate and the unbound ^{99m}Tc-HMPAO complex are the major impurities to be quantified.

The labeled product is analyzed before application to the patient. The radiochemical purity of ^{99m}Tc-labeled leukocytes is generally 95%.

Thin-layer chromatography

Aliquots are taken after incubation (step 7 above) to determine the labeling yield and also of the resuspended labeled leukocytes (step 11 above).

Leukocyte labeling efficiency must be at least $52 \pm 4.1\%$. Higher labeling efficiencies have been reported (Mortelmans et al. 1989).

- System I: Gelman instant thin-layer chromatography (ITLC) silica gel fiberglass plates and 2-butanone (methyl ethyl ketone [MEK]) as solvent: Reduced, hydrolized ^{99m}Tc-technetium (C), the secondary ^{99m}Tc-HMPAO complex (A) and the ^{99m}Tc-HMPAO-labeled leukocytes (E) remain at the start (A+C+E); unbound ^{99m}Tc-HMPAO complex (D) and free ^{99m}Tc-Na-pertechnetate (B) are measured at the solvent front (B+D).
- System II: Gelman ITLC silica gel fiberglass plates and saline as solvent: Reduced, hydrolized ^{99m}Tc-technetium (C), the secondary ^{99m}Tc-HMPAO complexes (A), lipophilic ^{99m}Tc-HMPAO complex (D), and ^{99m}Tc-HMPAO-labeled leukocytes (E) remain at the start (A + C + D + E); unbound ^{99m}Tc-Na-pertechnetate (B) moves with the solvent front.

The difference between activity measured at the start (A + C + D + E) in solvent II (saline) minus activity measured at the start (A + C + E) in solvent I (MEK) corresponds to the percentage of unbound lipophilic ^{99m}Tc-HMPAO complex (D). Radioactivity measured at the solvent front in system II indicates the amount of free ^{99m}Tc-pertechnetate (B). Thus, the major radiochemical impurities (B and D) are used as an approximation of the purity of ^{99m}Tc-HMPAO-labeled leukocytes (E).

D (%) = %(A + C + D + E) – % (A + C + E)

B + D (%) = Unbound ^{99m}Tc-activity

^{99m}Tc-labeled leukocytes (%) = 100–%(B+D)

A = Secondary ^{99m}Tc-HMPAO complex
B = ^{99m}Tc-pertechnetate (free)
C = Reduced, hydrolized ^{99m}Tc activity
D = Lipophilic ^{99m}Tc-HMPAO complex
E = ^{99m}Tc-HMPAO-labeled leukocytes

Pharmacokinetic Data

Following reinjection of ^{99m}Tc-labeled leukocytes, normal distribution shows an initial transitory uptake in the lungs, and later, in the spleen, liver, and bone marrow. ^{99m}Tc-labeled leukocytes represent approximately 37% of the circulating pool 40 min after injection. The kidneys and the gall bladder may also be visualized. During the first hours (1 6 h) nonspecific bowel activity is seen, 24 h after injection activity in the colon predominates. ^{99m}Tc activity is slowly released from the cells, with an observed elution rate of 20% during the first 24 h (Mortelmans et al. 1989; Segall et al. 1994), excreted partly by the kidneys and partly by the liver into the gall bladder.

Damage to cells or clumping as a result of the labeling procedure will produce an abnormal distribution of ^{99m}Tc-labeled leukocytes (see Chap. 8).

Radiation Dose

Intravenously injected leukocytes are distributed in liver, spleen, bone marrow, and other tissues; 40% are assumed to circulate in the blood with a half-time of 7 h, after which they are taken up in the same organs and tissues and in the same proportions as for the early uptake. The total uptake in the liver is 20%, in the spleen 25%, in red bone marrow 30%, and 25% in other tissues. Whole-body elimination is assumed to be 50% in 70 days. The model is based on granulocytes, which form the majority of cells in a preparation of mixed leukocytes (International Commission on Radiological Protection 1987).

The effective (whole body) dose value is 0.011 mSv/MBq (International Commission on Radiological Protection 1991). The effective dose in adults (70 kg) resulting from an intravenous injection of 185 MBq of ^{99m}Tc-labeled leukocytes is 2.04 mSv. Good abscess images were reported with 100 MBq of ^{99m}Tc-labeled granulocytes, resulting in an effective dose to the patient of 1.2 mSv (Skretting et al. 1988).

Storage and Stability

Storage. Kits should be stored at room temperature, and not stored above 25 °C.

Stability. ^{99m}Tc-labeled leukocytes should be reinjected without delay after preparation. They should not be refrigerated or frozen.

References

Arndt JW, van der Sluys Veer A, Blok D, Griffoen G, Verspaget HW, Lamers CB, Pauwels EK (1993) Prospective comparative study of technetium-99m-WBCs and indium-111-granulocytes for the examination of patients with inflammatory bowel disease. J Nucl Med 34:1052–1057

Bowring CS (1986) Imaging and quantitative scanning. In: Lewis SM, Baily RJ (eds) Radionuclides in hematology. Churchill Livingstone, Edinburgh, pp 151–172

Council of Europe (2005) Technetium ^{99m}Tc exametazime injection. In: European pharmacopeia 5.0, monograph no 1925. Council of Europe, Maisonneuve, Sainte-Ruffine, p 1214

Danpure HJ, Osman S, Carroll MJ (1988) The development of a clinical protocol for the radiolabelling of mixed leukocytes with ^{99m}Tc-hexamethylpropyleneamine oxime. Nucl Med Commun 9:465–475

Devillers A, Moisan A, Jean S, Arvieux C, Bourguet P (1995) Technetium-99m hexamethylpropylene amine oxime leukocyte scintigraphy for the diagnosis of bone and joint infections: a retrospective study in 116 patients. Eur J Nucl Med 22:302–307

GE Healthcare (2005) Product information for the Ceretec kit for the preparation of technetium Tc-99m exametazime injection. GE Healthcare, UK

International Commission on Radiological Protection (1987) Technetium-labelled white blood cells (leukocytes). In: Annals of the ICRP, radiation dose to patients from radiopharmaceuticals, biokinetic models and data. ICRP publication 53, vol. 18, no. 1–4. Pergamon, Oxford, pp 231–232

International Commission on Radiological Protection (1991) Technetium-labelled white blood cells (leukocytes). In: Annals of the ICRP, radiological protection in biomedical research. ICRP publication 62, vol. 22, no. 3. Pergamon, Oxford, pp 25–28

Kelbaek H, Fogh J, Gjorup T, Bülow K, Vestergaard B (1985) Scintigraphic demonstration of subcutaneous abscesses with ^{99m}Tc-labelled leukocytes. Eur J Nucl Med 10:302–303

Lantto EH, Lantto TJ, and Vorne M (1991) Fast diagnosis of abdominal infections and inflammations with technetium-99m-HMPAO labeled leukocytes. J. Nucl Med 32:2029–2034

Lantto T, Kaukonen J-P, Kokkola A, Laitinen R, Vorne M (1992) Tc-99m HMPAO labeled leukocytes superior to bone scan in the detection of osteomyelitis in children. Clin Nucl Med 17:7–10

Mortelmans L, Malbrain S, Stuyck J, De Backker C, Heynen MJ, Boogaerts M, De Roo M, Verbruggen A (1989) In vitro and in vivo evaluation of granulocyte labelling with (^{99m}Tc)D,L-HMPAO. J Nucl Med 30:2022–2028

Peters AM, Danpure HJ, Osman S, Hawker RJ, Henderson BL, Hodgson HJ, Kelly JD, Neirinckx RD, Lavender JP (1986) Clinical experience with ^{99m}Tc-HMPAO for labeling leukocytes and imaging inflammation. Lancet 25:946–949

Ponto JA, Swanson DP, Freitas JE (1987) Clinical manifestations of radiopharmaceutical formulation problems. In: Hladik WB III, Saha GB, Study KT (eds) Essentials of nuclear medicine science. Williams & Wilkins, Baltimore, pp 268–289

Roddie ME, Peters AM, Danpure HJ, Osman S, Henderson BL, Lavender PJ, CarrolMJ, Neirinckx RD, Kelly JD (1988) Inflammation: imaging with Tc-99m HMPAO-labelled leukocytes. Radiology 166:767–772

Segall GM, Lang H V, Chaovapong W (1994) In vitro evaluation of white blood cell labelling with ^{99m}Tc-radiopharmaceuticals. Nucl Med Commun 15:845–849

Skretting A, Benestad HB, Sundrehagen H (1988) Whole-body distribution of ^{99m}Tc labelled autologous human granulocytes and radiation dose to cells and organs. Eur J Nucl Med 14:1–7

United States Pharmacopeial Convention (2005) Official Monographs: USP 28, technetium Tc 99m exametazime injection. United States Pharmacopeia (USP) 28–national formulary (NF) 23, p 1855

12.7 ^{99m}Tc-Labeled Bone Imaging Agents

12.7.1 ^{99m}Tc-Pyrophosphate (PYP)

S. Kladnik and I. Zolle

Chemical name	Chemical structure
Sodium pyrophosphate·10 H_2O (PYP) Tin(II) diphosphate Technetium ^{99m}Tc tin pyrophosphate injection (*Ph. Eur.*) Technetium Tc 99m (pyrophosphate injection (*USP*) ^{99m}Tc-(Sn)-pyrophosphate injection ^{99m}Tc-PYP	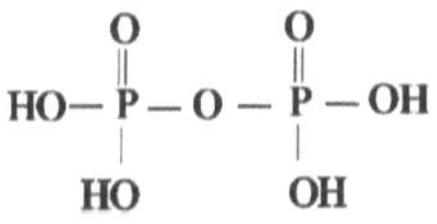 Diphosphoric acid

Kit components	
TechneScan PYP	
PYP	11.9 mg
Tin(II)-chloride-dihydrate	3.4 mg
AngioCis	
PYP	100 mg
Tin(II)-chloride dihydrate	1.6 mg
HematoCis	
PYP	0.67 mg
Tin(II)-chloride dihydrate	0.01 mg

Commercial products	
TechneScan PYP	Mallinckrodt/Tyco
Pyroscint	Bristol-Meyers Squibb
AngioCis (TCK-7)	CIS Bio
HematoCis (TCK-11)	CIS Bio

Preparation

Each vial contains the freeze-dried, sterile components under nitrogen atmosphere in a multidose vial. Depending on the intended use, the vial is reconstituted with sterile saline or with sterile sodium ^{99m}Tc-pertechnetate injection solution. All transfers and vial stopper entries must be done using aseptic techniques. Following reconstitution, the vial is agitated to dissolve the lyophilized material.

Labeling with ^{99m}Tc-pertechnetate is performed by adding 1–5 ml of sterile sodium ^{99m}Tc-pertechnetate solution to obtain a suitable radioactivity concentration for intravenous application (Cis International 1985 a).

^{99m}Tc-(Sn)-pyrophosphate (^{99m}Tc-PYP) injection solution is clear and free of particulate matter, the pH value is 5.–7.0.

Pretreatment of red blood cells (RBC) with PYP for in vivo labeling with ^{99m}Tc-Na-pertechnetate. Stannous pyrophosphate cold kits (except HematoCis) are used for in vivo labeling of erythrocytes with ^{99m}Tc-pertechnetate. In this case, the vial is reconstituted with sterile, nonpyrogenic saline or water, containing no preservatives. A volume of 3 ml resp. 10 ml is recommended by the manufacturer. The vial is agitated to dissolve the lyophilized material. After two resp. 5 min at room temperature, the preparation is ready for injection. A volume corresponding to 3.0–4.0 mg of stannous pyrophosphate is injected intravenously; 30 min later, ^{99m}Tc-Na-pertechnetate (555–740 MBq) (15–20 mCi) is injected, also intravenously, for in vivo labeling of pretreated erythrocytes. The amount of PYP should not exceed 0.5 mg/kg body weight (Mallinckrodt Medical 1993). Factors to be considered for in vivo labeling have been presented (Zimmer et al. 1979).

In vitro labeling of RBC with ^{99m}Tc-Na-pertechnetate. HematoCis (TCK-11) consists of two vials. Vial A contains a lyophilized, sterile formulation of 0.67 mg sodium pyrophosphate decahydrate and 0.01 mg stannous chloride dihydrate under nitrogen atmosphere. Vial B contains 10 ml of sterile saline under nitrogen atmosphere. Three milliliters of saline from vial B should be injected into vial A (this reducing solution must be used within 1 h). It should be stirred with a vortex for complete dissolution of freeze-dried material (Cis International 1985b).

One and a half milliliters of the reducing solution A should be withdrawn and added to 2 ml of blood. After 5 min's incubation at room temperature and gentle agitation, the preparation is centrifuged and the plasma is removed. From vial B, 1–2 ml of saline should be added, the vial centrifuged, and then the supernatant withdrawn.

For labeling, 0.8–1.4 ml of ^{99m}Tc-Na-pertechnetate (74–740 MBq) (2–20 mCi) is to be added to the RBC and incubated at room temperature for 5 min. The vial contents should be mixed well and then centrifuged to remove the supernatant. Measurements of both vials are then taken, and the labeling yield calculated. Generally, the labeling yield is approximately 97%.

Labeled RBC may be resuspended with 0.5–1 ml of saline from vial B or with patient's plasma. The preparation is ready for intravenous injection of ^{99m}Tc-RBC.

After heat alteration at 49.5 °C for 15 min, in vitro labeled RBC may be used for spleen scintigraphy. For this procedure, ^{99m}Tc-RBC should not be resuspended in plasma.

Advantages of in vitro labeling:

- High specific activity of ^{99m}Tc-RBC
- Controlled unbound ^{99m}Tc activity
- Bolus injection
- Heat alteration for spleen scintigraphy

For different kit applications, the manufacturer's instructions should be followed.

Description of the Kit

Kits containing tin-pyrophosphate as a sterile, nonpyrogenic formulation are either reconstituted with sterile saline or with sterile sodium ^{99m}Tc-pertechnetate solution. ^{99m}Tc-PYP has been used for imaging myocardial infarction; the cold kits serve as stannous agent for in vivo labeling of RBC. No bacteriostatic preservative is present in kits.

Stannous pyrophosphate and ^{99m}Tc-(Sn)-pyrophosphate are sterile, pyrogen-free, clear, colorless solutions suitable for intravenous injection.

Two stable complexes of technetium pyrophosphate have been identified at carrier levels by polarography, namely Tc(III) and Tc(IV) (Russell and Cash 1979). At pH below 6.0, ^{99m}Tc-(Sn)-pyrophosphate is described as a stable complex with Tc(IV).

The active ingredient is sodium pyrophosphate ($Na_4P_2O_7 \cdot 10\ H_2O$). For reduction of ^{99m}Tc-pertechnetate to lower oxidation states, tin(II) chloride or tin(II) fluoride is used. An optimal ratio of reducing agent/pyrophosphate must be maintained to prevent ^{99m}Tc-Sn-colloid formation (Srivastava et al. 1977). Technical problems with kit production have been reported (Kowalsky and Dalton 1981).

Clinical Applications

^{99m}Tc(tin)-pyrophosphate has been introduced for skeletal imaging (Subramanian et al. 1972) and was evaluated in patients with bone disease (Cohen et al. 1972; Fletcher et al. 1973; Rampon et al. 1974). The diagnostic value of bone scintigraphy has been evaluated retrospectively (Kuntz et al. 1975) and compared with compounds showing higher in vivo stability, namely diphosphonate derivatives (Henne et al. 1975; Rudd et al. 1977). The advantages of ^{99m}Tc-PYP for imaging damaged myocardial tissue have been demonstrated (Bonte et al. 1974; Buja et al. 1977; Cowley et al. 1977; Davis et al. 1976; Kelly et al. 1979; Willerson et al. 1975, 1977; Zaret et al. 1976). Application of the cold kit PYP as an agent for labeling RBC with technetium has gained general acceptance (Hegge et al. 1978; Pavel et al. 1977; Thrall et al. 1978).

^{99m}Tc-RBC for radionuclide angiography
- Regional imaging of blood pools (deep vein visualization)
- Electrocardiogram (ECG)-triggered cardiac radionuclide ventriculography (ejection fraction, wall motion)
- Detection of gastro intestinal hemorrhage, blood loss
- Determination of RBC mass or blood volume
- Spleen scintigraphy (heat-treated labeled RBC)

Time of Examination

Acute myocardial necrosis:	90–120 min after intravenous injection of ^{99m}Tc(tin)-pyrophosphate Patients are imaged 24 h up to 6 days following the acute episode.
Blood pool scintigraphy:	10 min after intravenous injection of sodium ^{99m}Tc-pertechnetate
Spleen scintigraphy:	60 min after intravenous injection of ^{99m}Tc-RBC (denatured)

Recommended Activities for Indications

Myocardial infarct imaging:	555–740 MBq (15–20 mCi) ^{99m}Tc(tin)-pyrophosphate: 0.5 mg/kg body weight (BW)
Blood pool imaging:	555–740 MBq (15–20 mCi) ^{99m}Tc-pertechnetate
Spleen scintigraphy:	37–75 MBq (1–2 mCi) ^{99m}Tc-RBC

Pediatric Dose. The amount of radioactivity for infants and children is based on BW, using the scaling factors given in Appendix 1, Table A1.2. The minimum value of the scaling factor is 0.1, since activities below 10% of the value of the adult radioactivity have resulted in poor image quality.

Additional Information

Patients should drink sufficient water and frequent bladder emptying should be encouraged, to reduce the radiation exposure to the bladder wall.

Pyrophosphate in aqueous solutions slightly hydrolyzes to monophosphates, causing the formation of free pertechnetate and hydrolized colloidal technetium.

Alteration of body distribution of ^{99m}Tc-pyrophosphate has been observed under various conditions (Crawford and Gumerman 1978; Hladik et al. 1982, 1987).

Early studies in experimental models suggested that hydroxyapatite deposition is not detectable until 12–24 h after infarction (Bonte et al. 1974).

Patients with cardiotoxic lesions resulting from an excess treatment with Adriamycin (doxorubicin) showed a diffuse accumulation of ^{99m}Tc-PYP in the heart, which otherwise is known only as localized uptake in focal lesions of myocardial infarction. Similar images have been observed in patients after defibrillation and reanimation treatment (Chacko et al. 1977).

Excess aluminum (in patients taking drugs for peptic ulcer, containing aluminum hydroxide) could interfere with the biodistribution of ^{99m}Tc-PYP; almost all the radioactivity was found in the liver and spleen, possibly due to formation of a colloidal ^{99m}Tc-species (Hladik et al. 1987).

A reduction in the labeling yield of RBC has been reported for heparin, tin-overload, aluminum, prazocin, methyldopa, hydralazine, digoxin, quinidine, β-adrenergic blockers, (e.g., propanolol), calcium channel blockers (e.g., verapamil), nitrates (e.g., nitroglycerin), doxorubicin, iodinated contrast agents, and Teflon tubing (catheter) (Hladik et al. 1987).

It is recommended to perform in vivo labeling of RBC prior to the administration of iodinated contrast media (Tatum et al. 1983).

Diminished cardiac activity and increased renal activity has been observed when a heparinized catheter was used for in vivo red cell labeling with pyrophosphate. This was interpreted as formation of a ^{99m}Tc-heparin complex that localizes avidly in the kidneys (Hegge et al. 1978). Injection of Sn-pyrophosphate and ^{99m}Tc-pertechnetate through a heparin lock should be avoided.

Legal Aspects. Quality requirements of ^{99m}Tc-PYP are stated in the official monographs of the *United States Pharmacopeia* (*USP*) and the *European Pharmacopeia* (*Ph. Eur.*):
Technetium ^{99m}Tc-tin pyrophosphate injection (Ph. Eur.) (Council of Europe 2005)
Technetium Tc-99m pyrophosphate injection (USP) (United States Pharmacopeial Convention 2005)

Quality Control

Radiochemical Purity. The *Ph. Eur.* requires thin-layer chromatography (TLC) on instant (I)TLC-silica gel (SG) fiberglass sheets for the identification of free ^{99m}Tc-Na-pertechnetate, using methyl ethyl ketone (MEK) as solvent. Unbound ^{99m}Tc-pertechnetate is measured at an R_f of 1.0. ^{99m}Tc-PYP and reduced, hydrolized ^{99m}Tc activity is identified at the origin.

Reduced, hydrolized ^{99m}Tc activity is determined separately using sodium acetate (13.6%) as solvent. Hydrolized ^{99m}Tc-oxide is measured at the origin, and ^{99m}Tc-PYP and ^{99m}Tc-pertechnetate move with the solvent front.

The sum of the measured radioactivity due to impurities obtained in two-solvent systems should not exceed 10% of the total radioactivity, according to the limits stated in the *Ph. Eur.* Analysis of various Tc-99m-labeled bone scanning agents has been reported (Krogsgaard 1976).

The radiochemical purity of ^{99m}Tc-(Sn)-pyrophosphate should not be less than 90%.

Conditions for thin-layer chromatography are shown below (Table 1 and 2) as are typical results of ^{99m}Tc-PYP analyses at different times after preparation.

Table 1. Thin-layer chromatography on silica gel plates using two solvent systems

	System I	System II
Stationary phase	ITLC-SG (fiberglass)	ITLC-SG (fiberglass)
Solvent:	MEK	Sodium acetate 13.6%
Developing time:	10 min	10 min

Table 2. Relative migration of labeled components and quantification of impurities

System I	^{99m}Tc-reduced, hydrolized and ^{99m}Tc-PYP at the start	
	Free ^{99m}Tc-pertechnetate at the solvent front	(A)
System II	Reduced, hydrolized ^{99m}Tc activity at the start	(B)
	^{99m}Tc-pertechnetate and ^{99m}Tc-PYP at the solvent front	
	^{99m}Tc-PYP (%) = 100 – %(A+B)	

Results of analysis *(12 samples)*
Results were obtained using the analystical methods described in the *Ph. Eur.*

Labeling and stability	15 min (%)	6 h (%)
^{99m}Tc-PYP complex	96.8	93.7
^{99m}Tc-Na-pertechnetate	0.8	1.4
^{99m}Tc reduced, hydrolized	2.3	4.9

Pharmacokinetic Data

After intravenous injection in patients, ^{99m}Tc-(Sn)-pyrophosphate accumulates in regions of active osteogenesis, and also in injured myocardium, mainly in necrotic tissue (Buja et al. 1977). Uptake in infarcted myocardium in experimental model (dog) was reported as 4.4% injected dose (ID)/g, compared with 0.31% in normal myocardial tissue (Bevan et al. (1980).

Due to in vivo instability, ^{99m}Tc-PYP is no longer used for bone scintigraphy.

Uptake of ^{99m}Tc-PYP in acute myocardial infarction was detected with higher sensitivity than were the known ^{99m}Tc-diphosphonate complexes (Kelly et al. 1979). Using a rat model, the percentage of injected activity per gram of infarct was approximately 2.4 times higher than with ^{99m}Tc-MDP (Davis et al. 1976). Visualization of the infarct can be effected from 24 h–7 days after the onset of symptoms and with maximum sensitivity (96%) between 48 and 72 h (Kelly et al. 1979).

Pyrophosphate is subject to enzymatic hydrolysis by pyrophosphatase, which has been demonstrated in bone, the kidneys, and other tissues (Kornberg 1962; Russell et al. 1970). Several ^{99m}Tc-phosphate species are formed, showing a different biological distribution and causing poor image quality of the bone scintigram (Eckelman and Volkert 1982).

Two hours after the intravenous injection of ^{99m}Tc-PYP, 10–30% of the injected radioactivity is taken up by bone structures; approximately 10% remain in the vascular space, declining to within 2–3% 24 h postinjection. (Krishnamurthy et al. 1975; Subramanian et al. 1975). The average urinary excretion is 60% of the administered dose in 24 h.

The slow clearance of ^{99m}Tc-PYP is due to high binding to plasma proteins, 42% at 2 h postinjection (Saha and Boyd 1978).

Stannous pyrophosphate has an affinity for RBC. It has been reported that stannous ion binds to the β-chain of hemoglobin. When ^{99m}Tc-Na-pertechnetate is injected 20–30 min after the intravenous injection of the reducing solution, high labeling of the preloaded erythrocytes is obtained. Normally, ^{99m}Tc-Na-pertechnetate will diffuse freely in and out of the RBC; however, in the presence of stannous ion, it is reduced intracellularly and reacts with hemoglobin. Intravenous injection of 10–20 µg/kg BW of stannous pyrophosphate results in efficient labeling of the blood pool. At 10 min postinjection, 77 ± 15% of the injected radioactivity remains in the circulation, and at 100 min postinjection, 71 ± 14%. The radioactivity is decreasing slowly, approximately 6% in 2 h (Mallinckrodt Medical 1993).

The effect of tin is long lasting. Preloaded RBC show labeling with ^{99m}Tc-Na-pertechnetate for up to 8 days after the injection of stannous pyrophosphate (Ancri et al. 1977). With doses of 0.02 mg of Sn(II)/kg BW, no negative effects were observed (Mallinckrodt Medical 1993). In vivo RBC labeling should not be repeated within 3 months.

Radiation Dose

^{99m}Tc-PYP is used for myocardial scintigraphy after intravenous injection. ^{99m}Tc-labeled erythrocytes are used for blood pool imaging, heat-denatured RBC for imaging the spleen. The most exposed organs are bone, bone marrow, bladder, kidneys, heart, and spleen. The radiation absorbed dose values previously obtained using the medical internal radiation dose (MIRD) scheme (Weber et al. 1989) correlate well with the effective (whole body) dose equivalent given in the ICRP 53 (International Commission on Radiological Protection (1987 a). A voiding period of 2 h was assumed.

The effective (whole body) dose equivalent for ^{99m}Tc(Sn)-pyrophosphate is 0.008 mSv/MBq (ICRP 53, International Commission on Radiological Protection 1987 a). The effective dose in adults (70 kg) resulting from 740 MBq (20 mCi) of intravenously injected ^{99m}Tc(Sn)-pyrophosphate for cardiac imaging is approximately 5.9 mSv.

The effective (whole body) dose equivalent for ^{99m}Tc-labeled erythrocytes is 0.0085 mSv/MBq (International Commission on Radiological Protection 1987 b). The effective dose in adults (70 kg) resulting from 740 MBq (20 mCi) of intravenously injected ^{99m}Tc-pertechnetate for angioscintigraphy corresponds to approximately 6.3 mSv. The absorbed radiation dose to the heart resulting from ^{99m}Tc-labeled RBC is 17.0 mGy and to the kidneys, 7.4 mGy. Approximately 15% of the activity is excreted in the urine during the first day (Porter et al. 1983). Activity is assumed to be distributed in the blood, being removed with a half-time of 60 h by renal excretion.

The effective dose value for the intravenous injection of denatured RBC is 0.019 mSv/MBq (International Commission on Radiological Protection 1991). The effective dose in adults (70 kg) resulting from 75 MBq (2 mCi) of ^{99m}Tc-RBC (denatured) for spleen scintigraphy is approximately 1.4 mSv. The dose to the spleen after intravenous injection of 75 MBq (2 mCi) of denatured ^{99m}Tc-RBC is 42 mGy.

Storage and Stability

Storage. The lyophilized kit is to be stored in the refrigerator at 2–8 °C.

Stability. ^{99m}Tc-pyrophosphate injection solution is stable for 4 h at room temperature.

References

Ancri D, Lonchampt M, Basset J (1977) The effect of tin on the tissue distribution of Tc-99m sodium pertechnetate. Radiology 124:445–450

Bevan JA, Tofe AJ, Benedict JJ, Francis MD, Barnett BL (1980) Tc-99m HMDP (hydroxymethylene diphosphonate): a radiopharmaceutical for skeletal and acute myocardial infarct imaging. II. Comparison of Tc-99m hydroxymethylene diphosphonate (HMDP) with other technetium-labeled bone imaging agents in a canine model. J Nucl Med 21:967–970

Bonte FJ, Parkey RW, Graham KD, Moore J, Stokely EM (1974) A new method for radionuclide imaging of myocardial infarcts. Radiology 110:473–474

Buja LM, Tofe AJ, Kulkarni PV et al (1977) Sites and mechanisms of localization of technetium-99m phosphorus radiopharmaceuticals in acute myocardial infarcts and other tissues. J Clin Invest 60:724–740

Chacko AK, Gordon DH, Bennett JM et al (1977) Myocardial imaging with Tc-99m pyrophosphate in patients on Adriamycin treatment for neoplasia. J Nucl Med 18:680–683

Cis International (1985a) Product information for the TCK-7 kit (AngioCis) for the preparation of technetium Tc-99m (Sn) pyrophosphate (PYP). Cis International, France

Cis International (1985b) Product information for the TCK-11 kit (HematoCis) for the preparation of technetium Tc-99m red blood cells (RBC). Cis International, France

Cohen Y, Perez R, Henry R, Panneciere C (1972) Use of technetium 99m-labelled sodium pyrophosphate in skeletal scintigraphy. C R Acad Sci Hebd Seances Acad Sci D 275:1719–1721

Council of Europe (2005) Technetium ^{99m}Tc tin pyrophosphate injection. In: European pharmacopeia, monograph 129. Council of Europe, Maisonneuve, Sainte-Ruffine, p 1230

Cowley MJ, Mantle JA, Rogers WJ et al (1977) Technetium stannous pyrophosphate myocardial scintigraphy: reliability and limitations in assessment of acute myocardial infarction. Circulation 56:192–198

Crawford JA, Gumerman LW (1978) Alteration of body distribution of ^{99m}Tc-pyrophosphate by radiographic contrast material. Clin Nucl Med 3:305–307

Davis MA, Holman BL, Carmel AN (1976) Evaluation of radiopharmaceuticals sequestered by acutely damaged myocardium. J Nucl Med 17:911–917

Eckelman WC, Volkert WA (1982) In vivo chemistry of ^{99m}Tc-chelates. Int J Appl Radiat Isot 33:945–951

Fletcher JW, Solaric Georges E, Henry RE, Donato RM (1973) Evaluation of ^{99m}Tc-pyrophosphate as a bone imaging agent. Radiology 467–469

Hegge FN, Hamilton GW, Larson SM et al (1978) Cardiac chamber imaging: a comparison of red blood cells labelled with Tc-99m in vitro and in vivo. J Nucl Med 19:129–134

Henne W, Pixberg H-U, Pfannenstiel P (1975) Technetiumpolyphosphat und Technetiumdiphosphonat – Eine vergleichende Untersuchung [in German]. Nucl Med 14:83–90

Hladik WB, Nigg KK, Rhodes BA (1982) Drug-induced changes in the biologic distribution of radiopharmaceuticals. Sem Nucl Med 12:184–218

Hladik WB, Ponto JA, Lentle BC, Laven DL (1987) Iatrogenic alterations in the biodistribution of radiotracers as a result of drug therapy: reported instances. In: Hladik WB III, Saha GB, Study KT (eds) Essentials of nuclear medicine science. Williams & Wilkins, Baltimore, pp 193–202

International Commission on Radiological Protection (1987a) Technetium-labelled phosphates and phosphonates. In: Annals of the ICRP, radiation dose to patients from radiopharmaceuticals, biokinetic models and data. ICRP publication 53, vol 18, no 1–4. Pergamon, Oxford, pp 213–215

International Commission on Radiological Protection (1987b) Technetium-labelled erythrocytes. In: Annals of the ICRP, radiation dose to patients from radiopharmaceuticals, biokinetic models and data. ICRP publication 53, vol 18, no 1–4. Pergamon, Oxford, pp 209–210

International Commission on Radiological Protection (1991) Technetium-labelled denatured erythrocytes. In: Annals of the ICRP, radiological protection in biomedical research. ICRP publication 62, vol 22, no 3. Pergamon, Oxford, pp 25–28

Kelly RJ, Chilton HM, Hackshaw BT, Ball JD, Watson NE, Kahl FR, Cowan RJ (1979) Comparison of Tc-99m pyrophosphate and Tc-99m methylene diphosphonate in acute myocardial infarction: concise communication. J Nucl Med 20:402–406

Kornberg A (1962) On the metabolic significance of phosphorolytic and pyrophosphorolytic reactions. In: Kasha M and Pullman B (eds) Horizons in biochemistry. Academic, New York, pp 251–264

Kowalsky RJ, Dalton DR (1981) Technical problems associated with the production of technetium [^{99m}Tc] tin(II) pyrophophate kits. Am J Hosp Pharm 38:1722–1726

Krishnamurthy GT, Huebotter RJ, Walsh CF, Taylor JR, Kehr MD, Tubis M, Blahd WH (1975) Kinetics of ^{99m}Tc-labeled pyrophosphate and polyphosphate in man. J Nucl Med 16:109–115

Krogsgaard OW (1976) Radiochemical purity of various Tc-99m-labelled bone scanning agents. Eur J Nucl Med 1:15–17

Kuntz D, Rain JD, Lemaire V, Sainte-Croix A, Ryckewaert A (1975) Diagnostic value of bone scintigraphy with technetium pyrophosphate. Study of 250 patients. Rev Rhum Mal Osteoartic 42:19–24

Mallinckrodt Medical (1993) Product information for TechneScan PYP for the preparation of (Sn) pyrophosphate (PYP) injection solution. Mallinckrodt Medical, The Netherlands

Pavel DG, Zimmer AM, Patterson VN (1977) In vivo labeling of red blood cells with ^{99m}Tc: a new approach to blood pool visualization. J Nucl Med 18:305–308

Porter WC, Dees SM, Freitas JE, Dworkin HJ (1983) Acid-citrate-dextrose compared with heparin in the preparation of in vivo/in vitro technetium-99m red blood cells. J Nucl Med 24:383–387

Rampon S, Bussiere JL, Prin P, Sauvezie B, Leroy V, Missioux D et al (1974) 250 studies of bone radioisotope scanning by tin pyrophosphate labeled with technetium 99m. Analytical and clinical study. Rev Rhum Mal Osteoartic 4:745–751

Rudd TG, Allen DR, Hartnett DE (1977) Tc-99m methylene diphosphonate versus Tc-99m pyrophosphate: biologic and clinical comparison. J Nucl Med 18:872–876

Russell CD, and Cash AG (1979) Complexes of technetium with pyrophosphate, etidronate, and medronate. J Nucl Med 20:532–537

Russell RGG, Mühlbauer RC, Bisaz S et al (1970) The influence of pyrophosphate, condensed phosphates, phosphonates and other phosphate compounds on the dissolution of hydroxyapatite in vitro and on bone resorption induced by parathyroid hormone in tissue culture and in thyroparathyroidectomized rats. Calcif Tissue Res 6:183–196

Saha GB, Boyd CM (1978) Plasma protein-binding of ^{99m}Tc-pyrophosphate. Int. J Nucl Med Biol 5:236–239

Srivastava SC, Meinken G, Smith TD, Richards P (1977) Problems associated with stannous ^{99m}Tc-radiopharmaceuticals. Int J Appl Radiat Isot 28:83–95

Subramanian G, McAfee JG, Bell EG, Blair RJ, Mara RE, Relston PH (1972) ^{99m}Tc-labeled polyphosphates as skeletal imaging agent. Radiology 102:701–704

Subramanian G, McAfee JG, Blair RJ, Kallfelz FA, Thomas FD (1975) Technetium-99m methylene diphosphonate – a superior agent for skeletal imaging: comparison with other technetium complexes. J Nucl Med 16:744–755

Tatum JL, Burke TS, Hirsch JI et al (1983) Pitfall to modified in vivo method of technetium-99m red blood cell labeling – iodinated contrast media. Clin J Nucl Med 8:585–587

Thrall JH, Freitas JE, Swanson D, Rogers WL, Clare JM, Brown ML, Pitt B (1978) Clinical comparison of cardiac blood pool visualization with technetium 99m red blood cells labeled in vivo and with technetium 99m human serum albumin. J Nucl Med 19:796–803

United States Pharmacopeial Convention (2005) Official Monographs: USP 28, technetium Tc 99m pyrophosphate injection. United States Pharmacopeia (USP) 28–national formulary (NF) 23, p 1862

Weber DA, Makler PT Jr, Watson EE, Coffey JL, Thomas SR, London J (1989) Radiation absorbed dose from technetium-99m-labeled bone imaging agents: MIRD dose estimate report no 13. J Nucl Med 30:1117–1122

Willerson JT, Parkey RW, Bonte FJ, et al. (1975) Technetium-99m stannous pyrophosphate myocardial scintigrams in patients with chest pain of varying etiology. Circulation 51:1046–1052

Willerson JT, Parkey RW, Buja LM et al (1977) Are ^{99m}Tc-stannous pyrophosphate myocardial scintigrams clinically useful? Clin Nucl Med 2:137–145

Zaret BL, Di Cola VC, Donabedian RK, Puri S, Wolfson S, Freedman GS, Cohen LS (1976) Dual radionuclide study of myocardial infarction: relationships between myocardial uptake of potassium-43, technetium-99m stannous pyrophosphate, regional myocardial blood flow and creatine phosphokinase depletion. Circulation 53:422–428

Zimmer AM, Pavel DG, Karesh SM (1979) Technical parameters of in vivo red blood cell labelling with technetium-99m. Nuklearmedizin 18:241–245

12.7.2 ^{99m}Tc-Diphosphonates

I. Zolle and S. Kladnik

^{99m}Tc-DPD (Dicarboxypropane diphosphonate)

Chemical name	**Chemical structure**
3,3-Diphosphono-1,2-propane-dicarboxy acid, tetrasodium salt (DPD) 1,2-dicarboxypropane diphosphonate	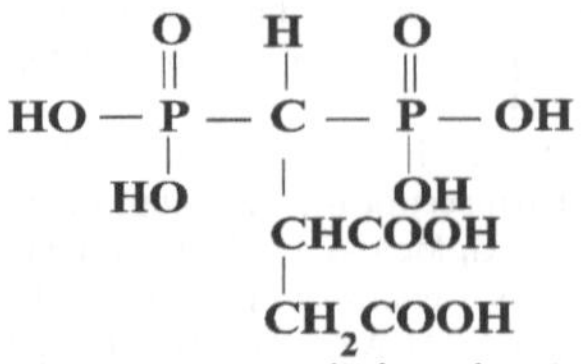 Dicarboxypropane diphosphonic acid
Kit components	**Commercial products**
DPD 13.0 mg Tin(II)-chloride dihydrate 0.23 mg *N*-(4-aminobenzoyl)-L-glutamic acid, monosodium salt 1.0 mg	Teceos CIS bio international

^{99m}Tc-HDP (Hydroxymethylene diphosphonate)

Chemical name	**Chemical structure**
Hydroxymethylene diphosphonic acid, disodium salt (HMDP, HDP) Hydroxymethylene diphosphonate Oxidronate (*USP*) Technetium Tc 99m oxidronate injection (*USP*)	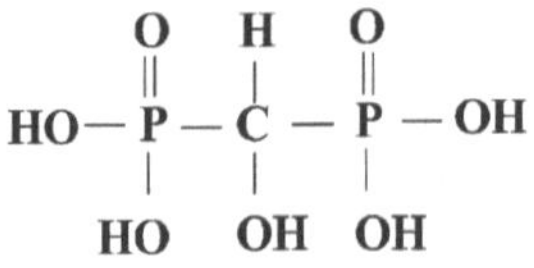 Hydroxymethylene diphosphonic acid
Kit components	**Commercial products**
Oxidronate 3.0 mg Tin(II)-chloride dihydrate 0.24–0.45 mg Gentisic acid resp. 0.84 mg Ascorbic acid 0.75 mg Sodium chloride 10 mg	TechneScan HDP Mallinckrodt Medical OsteoCis (TCK-21) CIS bio international

^{99m}Tc-HEDSPA (Hydroxyethylidene diphosphonate)

Chemical name	**Chemical structure**
1-Hydroxyethylidene-1,1-diphosphonic acid, disodium salt (HEDSPA) Ethylidene-1-hydroxy-1,1-disodium phosphonate (EHDP) Hydroxyethylidene diphosphonate Etidronate (*USP*) Technetium Tc-99m etidronate injection (*USP*)	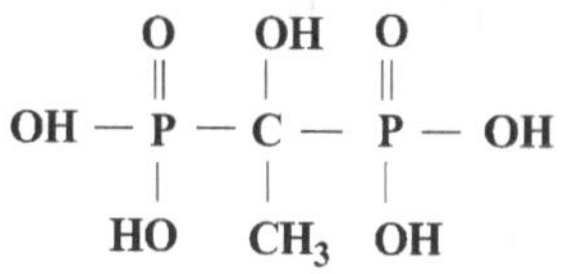 Hydroxethylidene diphosphonic acid

Kit components		**Commercial products**	
Etidronate disodium salt	0.75 mg	HEDSPA	Union Carbide*
Stannous tartrate	0.18 mg	Osteoscan	Proctor and Gamble
Hydrochloric acid q.s.	pH=4	No longer commercially available	

* HEDSPA Unit dose kit: Etidronate disodium tin kit (80 μg stannous ion)

^{99m}Tc-MDP (Methylene diphosphonate)

Chemical name	**Chemical structure**
Methylene diphosphonic acid, disodium salt (MDP) Methylene diphosphonate Medronate (*Ph. Eur.; USP*) Technetium Tc 99m medronate injection (*Ph. Eur.; USP*)	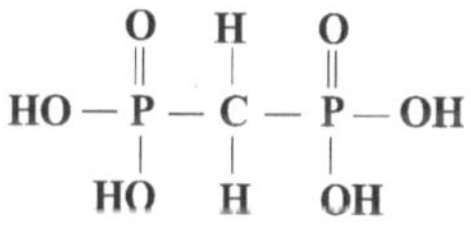 Methylene diphosphonic acid

Kit components		**Commercial products**	
MedroCis:		MedroCis (TCK-14)	CIS bio international
Methylene diphosphonate	10.0 mg	Lenoscint	Bristol-Myers Squibb
Tin(II)-chloride dihydrate	1.0 mg	Amerscan MDP	GE Healthcare
Ascorbic acid	1.8 mg	Amerscan Stannous Agent	GE Healthcare

Preparation

Kits for the preparation of diphosphonate complexes contain the lyophilized ingredients in a multidose vial. Labeling with ^{99m}Tc-pertechnetate is performed by adding 2–10 ml of sterile ^{99m}Tc eluate by aseptic techniques. When calculating the amount of radioactivity to be added, the labeling efficiency, number of patients, time of injection and radioactive decay must be taken into account. Activities corresponding to 6.6–18.5 GBq

(180–500 mCi) are recommended by the manufacturer. The vial is agitated to dissolve the lyophilized material. After 5–20 min at room temperature, the preparation is ready for injection.

^{99m}Tc-diphosphonate injection solutions are clear and free of particulate matter; the pH value is 3.5–7.5 (3,3-diphosphono-1,2-propane-dicarboxy acid [DPD] 6.5–7.5).

A special formulation of medronate (Amerscan Stannous Agent) is used for in vivo loading of red blood cells with stannous ion, preparatory to labeling with sodium ^{99m}Tc-pertechnetate.

Amerscan Stannous Agent contains a freeze-dried, sterile formulation of 6.8 mg sodium medronate, corresponding to 5.4 mg medronic acid as disodium salt, and 4.0 mg stannous fluoride. Upon reconstitution with 6 ml saline, the recommended amount for adults (70 kg) by intravenous injection is 2 ml, corresponding to 1.8 mg medronic acid and 1.3 mg stannous fluoride.

Description of the Kit

Each vial contains the freeze-dried, sterile components under nitrogen atmosphere in a multidose vial. ^{99m}Tc-Na-pertechnetate used for labeling must be free from any oxidizing agent, using a solution of 0.9% sodium chloride (saline) as diluent. No bacteriostatic agents are present in kits. In order to prevent oxidative reactions, kits are purged with nitrogen gas, the amount of tin reducing agent might be increased, and/or antioxidants might be added (Saha 1998).

The active ingredients are diphosphonic acids – dicarboxypropane diphosphonic acid (DPD), hydroxymethylene diphosphonic acid (HDP or HMDP), hydroxyethylidene diphosphonic acid (EHDP), methylene diphosphonic acid (MDP) – or their sodium salts. As stabilizers are used *N*-(4-aminobenzoyl)-L-glutaminic acid, gentisic acid (2,5-dihydroxybenzoic acid), ascorbic acid, or carbamide. For reduction of ^{99m}Tc-pertechnetate to lower oxidation states, tin(II) chloride and tin(II) fluoride are employed.

Commercial kit formulations contain various quantities of active ingredients; some formulations use gentisic acid as a stabilizer of stannous ion (Tofe and Francis 1976; Tofe et al. 1980). Reduced, hydrolized ^{99m}Tc activity has been a common impurity. Colloid formation is also caused by aluminum in the ^{99m}Tc eluate (Ponto et al. 1987). An optimal ratio of reducing agent/diphosphonate must be maintained to prevent ^{99m}Tc-Sn-colloid formation (Srivastava et al. 1977). Optimal labeling conditions require carrier-free ^{99m}Tc eluate obtained from a generator that is eluted daily (Van Duzee and Bugaj 1981).

Generally, diphopsphonate kits are used for several patient doses, keeping the injected mass of the ^{99m}Tc-diphosphonate complex below 1.0 mg (Castronovo and Callahan 1972). Thus, also in cases, when a single dose is prepared, the standard labeling conditions are applied.

Pretreatment of red blood cells (RBC) with MDP for in vivo labeling with ^{99m}Tc-Na-pertechnetate. Stannous medronate freeze-dried formulation is reconstituted with sterile, nonpyrogenic saline. A volume of 6 ml is added (product information for Amerscan Stannous Agent). The vial is agitated to dissolve the lyophilized material. After 5 min at room temperature, the preparation is ready for injection (2 ml for one patient). The injection solution should be clear and free of particulate matter with a pH value between

5.5 and 7.5. Thirty minutes later, ^{99m}Tc-Na-pertechnetate (555–740 MBq) (15–20 mCi) is injected, also intravenously, for in vivo labeling of pretreated erythrocytes.

Factors to be considered for in vivo labeling have been presented (Zimmer et al. 1979).

For different kit applications, the manufacturer's instructions should be followed.

Historical background. HEDSPA was the first Tc-99m-diphosphonate complex introduced to clinical practice by Castronovo and Callahan (1972). The use of EHDP was proposed by several groups (Pendergrass et al. 1973; Tofe and Francis 1972; Subramanian et al. 1972; Yano et al. 1973). Large numbers of patients were studied and compared with polyphosphate and pyrophosphate (Ackerhalt et al. 1974; Dunson et al. 1973; Genant et al. 1974; Hughes et al. 1975; Krishnamurthy et al. 1974); however, not all reports were favorable.

MDP was introduced as a superior agent (Subramanian et al. 1975) and served as a reference for later developments (Russel and Cash 1979).

HMDP, or HDP, showed favorable characteristics in early studies (Bevan et al. 1980; Domstad et al. 1980; Francis et al. 1980; Fogelman et al. 1981); after critical evaluation it was concluded that there was no significant difference between HMDP and MDP.

DPD was developed at Höchst, Germany (Schwarz and Kloss 1981; Schwarz et al. 1991), and compared favorably with MDP (Buell et al. 1982; Godart et al. 1986; Hale et al. 1981; Pauwels et al. 1983; Schroth et al. 1984). DPD is another efficient phosphonate complex that has been widely used as Teceos. The merits of available diphosphonate bone scanning agents have been discussed by Fogelman 1982.

Clinical Applications

Skeletal scintigraphy with ^{99m}Tc-diphosphonate complexes.

- Diagnosis of bone disease, e.g., primary bone tumors, metastatic bone tumors
- Metabolic bone disease, osteomyelitis
- Localization of fractures (stress and hairline fractures, fractures of the small bones of the hands and feet)
- Evaluation of painful arthroplasties of the knee and hip
- Evaluation of bone pain in patients with negative x-ray

Radionuclide angiography with ^{99m}Tc-RBC

- Regional imaging of blood pools (deep vein visualization)
- Electrocardiogram (ECG)-triggered cardiac radionuclide ventriculography (ejection fraction, wall motion)

An overview on the clinical application of radionuclide imaging in skeletal disease provides methodological detail (McAfee 1987).

Amerscan Stannous Agent facilitates in vivo labeling of RBC; therefore, ^{99m}Tc-medronate is also used for blood pool scintigraphy (Gray et al. 1979).

Time of Examination

Bone imaging: 2 h after intravenous injection

Blood pool scintigraphy: 10 min after intravenous injection of sodium ^{99m}Tc-pertechnetate

Recommended Activities for Indications

Bone scintigraphy:	370–740 MBq (10–20 mCi)
Blood pool imaging:	555–740 MBq (15–20 mCi) (^{99m}Tc-pertechnetate)

Pediatric Dose. The amount of radioactivity for infants and children administered for bone scintigraphy is based on body weight, using the scaling factors given in Appendix 1, Table A1.2. The minimum value of the scaling factor is 0.1, since activities below 10% of the value of the adult radioactivity have resulted in poor image quality.

Additional Information

Patients should be encouraged to drink sufficient water and to empty the bladder before scintigraphy is started. Frequent bladder emptying is recommended to reduce the radiation exposure to the bladder wall.

When kits containing large amounts of stannous ion were used for bone imaging, ^{99m}Tc-labeling of RBC for up to 2 weeks after administration of ^{99m}Tc-PYP was observed (Ancri et al. 1977).

Alteration of body distribution of ^{99m}Tc-phosphonates has been observed under various conditions (Hladik et al. 1982, 1987).

Presence of aluminum in the generator eluate and treatment with aluminum-containing drugs may result in colloid formation, with liver and spleen uptake and trapping of particles in lung capillaries (Zimmer and Pavel 1978). Colloid formation is also observed at elevated pH above 8.5 (Chaudhuri 1976).

The influence of iron and iron-containing drugs on the retention of diphosphonates has been observed frequently (Hladik 1987). A localized muscular accumulation of iron-dextran at the injection site was reported (Byun et al. 1976).

After intravenous infusion of iron containing drugs or in cases of chronic iron overload in chronic diseases, a change in the bone-to-kidney ratio of diphosphonate complexes has been reported, namely, a decrease in bone uptake and an increase in accumulation in renal parenchyma (McRae et al. 1976). Dissociation of the ^{99m}Tc-diphosphonate complex and a conversion into the renaltropic gluconate was proved in the presence of ionic iron(II) and calcium. In vivo alteration of the complex was concluded from an increase in renal uptake even after infusion of dextrose.

Long-term treatment with glucocorticosteroids results in a decrease in bone uptake. This phenomenon may not be explained as interference between a pharmaceutical and a radiopharmaceutical but by the drug-induced osteoporosis (Conklin et al. 1983).

Another case report described the diffuse accumulation of ^{99m}Tc-MDP in the liver of a patient with methotrexate-induced hepatotoxicity. The authors postulated that inhibition of protein synthesis by methotrexate caused a disruption of the cell membranes, resulting in the inflow of calcium ions. The high affinity of MDP for calcium leads to diffuse uptake of tracer in the liver (Hladik et al. 1982).

Since the stability of ^{99m}Tc-diphosphonate complexes may be affected, ^{99m}Tc-diphosphonates should not be mixed with other drugs or components nor injected simultaneously.

Legal Aspects. Quality requirements of ^{99m}Tc-diphosphonates are stated in the official monographs of the *European Pharmacopeia* (*Ph. Eur.*) and the *United States Pharmacopeia* (*USP*):

^{99m}Tc-MDP	Technetium Tc-99m medronate injection in *USP* 28 (United States Pharmacopeial Convention 2005)
	Technetium ^{99m}Tc medronate injection in *Eur. Ph.* 5.0 (Council of Europe 2005)
^{99m}Tc-HDP	Technetium Tc-99m oxidronate injection in *USP* 28
^{99m}Tc-EHDP	Technetium Tc-99m etidronate injection in *USP* 28

Quality Control

Radiochemical Purity. ^{99m}Tc-medronate is included in the pharmacopoeias and may serve as an example for the analysis of other ^{99m}Tc-diphosphonates. The *Ph. Eur.* requires thin-layer chromatography (TLC) on instant (I)TLC-silica gel (SG) fiberglass sheets, using two separate solvent systems.

- System I: Separation of ^{99m}Tc-pertechnetate in organic solvent methyl ethyl ketone (MEK): ^{99m}Tc-diphosphonates and hydrolized ^{99m}Tc activity remain at the start; free ^{99m}Tc-pertechnetate is measured at the solvent front.
- System II: Separation of reduced, hydrolized ^{99m}Tc activity using sodium acetate (13.6%): ^{99m}Tc-diphosphonates and free ^{99m}Tc-pertechnetate move with the solvent front; colloidal activity is measured at the start.

Impurities. Free ^{99m}Tc-pertechnetate and colloidal activity are expressed as a percentage of the sum of the radioactivity measured in systems A and B, not exceeding 5%. ^{99m}Tc-pertechnetate does not exceed 2%.

The radiochemical purity of ^{99m}Tc-medronate is not less than 95% (*Ph. Eur.*).

Thin-layer chromatography

Analysis of various Tc-99m-labelled bone scanning agents has been reported (Krogsgaard 1976).

Conditions for thin-layer chromatography are shown below (Table 1 and 2), as are results of ^{99m}Tc-MDP analyses at different times after preparation. Three different solvents were evaluated in each TLC system.

Table 1. Thin-layer chromatography using silica gel and three solvents in each system

	System I	System II
Stationary phase:	ITLC-SG (fiberglass)	ITLC-SG (fiberglass)
Solvent: (a)	MEK	Sodium acetate 13.6%
(b)	Methanol–acetone, 1:1	Saline
(c)	Acetone	Saline
Developing time:	10 min	10 min

Table 2. Relative migration of labeled components and quantification of impurities

System I	^{99m}Tc-reduced, hydrolized and ^{99m}Tc-MDP at the start Free ^{99m}Tc-pertechnetate at the solvent front	(A)
System II	Reduced, hydrolized ^{99m}Tc activity at the start ^{99m}Tc-pertechnetate and ^{99m}Tc-MDP at the solvent front	(B)
	^{99m}Tc-MDP (%) = 100–%(A+B)	

Results of analysis *(12 samples)*

Shown are results of ^{99m}Tc-MDP analyses obtained with three different solvents, as described in Table 1. The consistent values of free ^{99m}Tc-pertechnetate and hydrolized ^{99m}Tc activity measured at different times after preparation (*a*), favor the use of MEK and sodium acetate (13.6%), as stated in the Ph. Eur. However, saline may be used instead.

Labeling and stability	15 min (%)			6 h (%)		
Solvent no.	a	b	c	a	b	c
^{99m}Tc-MDP complex	98.0	96.7	96.4	97.4	95.2	95.0
^{99m}Tc-Na-pertechnetate	1.4	2.8	3.2	1.9	4.3	4.5
^{99m}Tc-reduced, hydrolized	0.6	0.4	0.4	0.6	0.5	0.5

Biological Test (*Ph. Eur.*). The biodistribution of ^{99m}Tc-MDP should be determined in three rats, according to the procedure in the *Ph. Eur.* to demonstrate that not less than 2.5% of the radioactivity is found in the femur and not more than 1.0% is found in the liver. Calculate the radioactivity per unit mass in the femurs (A_1), muscle (A_2) and blood (A_3). The ratio of radioactivity per unit mass in the femur and muscle (A_1/A_2) is not less than 100, and in the femur and blood (A_1/A_3) is not less than 40.

Pharmacokinetic Data

Following intravenous injection, 45–50% of ^{99m}Tc-diphosphonates (MDP, HDP, DPD) accumulate in the skeleton, while most of the rest is excreted in the urine. Maximum bone accumulation occurs 1 h after injection and remains constant for 72 h. The determining factors for bone uptake of ^{99m}Tc-phosphonate complexes are an increased blood flow to the skeleton and reactive bone formation, causing avid extraction by the bone mineral matrix (Jones et al. 1976; Sahni et al. 1993). Symmetric areas of increased activity concentration are seen in the metaphyseal zones in the growing skeleton. Ankles, knees, elbows, wrists, shoulder joints, pelvic bones, and vertebrae show increased uptake in the normal anterior bone scintigram (Saha 1987).

The disappearance of ^{99m}Tc-phosphonate complexes from blood is affected by skeletal fixation and urinary excretion. Blood clearance proceeds with three half-times of elimination, a rapid phase ($T_{1/2}$ = 3.5 min), an intermediate phase ($T_{1/2}$ = 27 min), and a slow phase ($T_{1/2}$ = 144 min). Transfer to the bone matrix corresponds to the intermediate phase. The total radioactivity in blood at 5 min, 2, 4, and 24 h after the intravenous injection was measured as 40, 10, 5.8, and 2.3%, respectively (Subramanian et al. 1975 a).

Besides an increase in the osteogenic activity, bone uptake of ^{99m}Tc-diphosphonates is closely related to chemical structure, as demonstrated by the skeletal uptake of

EHDP, MDP, and HDP, differing by the substituents at the methylene bridge. Whole-body retention (24 h) in volunteers showed values of 18.4 (EHDP), 30.3 (MDP), and 36.6% (HDP), indicating a greater uptake of oxidronate (HDP) (Fogelman et al. 1981). These results agree with the conformational requirements derived from structure–activity relationship studies performed with mono- and disubstituted diphosphonates (Wang et al 1980).

For normal renal function, elimination by glomerular filtration of the nonfixed complex is approximately 32% in 1 h, 47% in 2 h, and 60% in 6 h. The percentage of cumulative activity excreted in the urine during 24 h is 79.2% (EHDP) and 76.5% (MDP), indicating 75–80% 24-h urinary excretion of bone scanning agents (Subramanian et al. 1975b). The relatively lower urinary excretion of HDP in dogs, namely, approximately 59%, is primarily resulting from high skeletal uptake, indicated by a femur-to-muscle ratio of 35 (Bevan et al. 1980).

Protein binding of MDP has been reported as 22% at 2 h; in the case of EHDP, approximately 30% 3 h after injection (Saha and Boyd 1979; Subramanian et al. 1975a).

^{99m}Tc-DPD shows lower binding to plasma protein and is excreted to a smaller extent in the urine (Schwarz et al. 1991). Whole-body retention at 24 h was 40.6% for DPD and 27.0% in the case of MDP (Buell et al. 1982). The bone-to-soft tissue ratio, which is important for skeletal scintigraphy, showed consistently higher values for ^{99m}Tc-DPD, namely, an increase of 11.4% in normal volunteers and 7.3% in patients 2 h postinjection (Buell et al. 1982).

The detection of bone metastases is based on increased regional uptake of ^{99m}Tc-diphosphonates; in a comparative study the diagnostic efficacy was determined for ^{99m}Tc-MDP, ^{99m}Tc-HDP, and ^{99m}Tc-DPD, demonstrating equal detection rates for the three agents (Pauwels et al. 1983).

Diphosphonates form stable Tc(IV) complexes. Contrary to inorganic phosphorus compounds, which are degraded in vivo, the organically bound diphosphonate complexes show high in vivo stability (Davis and Jones 1976).

Stannous medronate has an affinity for RBC. It has been shown that stannous ion binds to the β-chain of hemoglobin. When ^{99m}Tc-Na-pertechnetate is injected 20–30 min after the intravenous injection of the reducing solution, high labeling of the preloaded erythrocytes is obtained (Callahan et al. 1982). Approximately 15% of the injected activity is excreted in the urine during the first day (Porter et al. 1983).

Radiation Dose

^{99m}Tc-diphosphonate complexes are injected intravenously for bone scintigraphy. ^{99m}Tc-labeled erythrocytes are used for blood pool imaging. The most exposed organs are bone, bone marrow, bladder, kidneys, and the heart. The radiation absorbed dose values previously obtained using the medical internal radiation dose (MIRD) scheme (Weber et al. 1989) correlate well with the effective (whole body) dose equivalent given in the ICRP 53 (International Commission on Radiological Protection 1987a). A voiding period of 3 h was assumed.

The effective (whole body) dose value for ^{99m}Tc-diphosphonates is 0.0057 mSv/MBq (International Commission on Radiological Protection 1998). The effective dose in adults (70 kg) resulting from 740 MBq (20 mCi) of intravenously injected ^{99m}Tc-diphosphonate complexes for bone scintigraphy is approximately 4.2 mSv.

The effective (whole body) dose value for ^{99m}Tc-labeled erythrocytes is 0.0066 mSv/MBq (International Commission on Radiological Protection 1991). The effective dose in adults (70 kg) resulting from 740 MBq (20 mCi) of intravenously injected ^{99m}Tc-pertechnetate for angioscintigraphy corresponds to approximately 4.9 mSv. The absorbed radiation dose to the heart resulting from ^{99m}Tc-labeled RBC is 17.0 mGy, and to the kidneys, 7.4 mGy. Approximately 15% of the activity is excreted in the urine during the first day (Porter et al. 1983). Activity is assumed to be distributed in the blood, being removed with a half-time of 60 h, by renal excretion (International Commission on Radiological Protection 1987b).

Storage and Stability

Storage. The lyophilized kits are to be stored in the refrigerator at 2–8 °C.

Stability. ^{99m}Tc–diphosphonate injection solution is stable for 6 h at room temperature.

References

Ackerhalt RE, Blau M, Bakshi S, Sondel JA (1974) A comparative study of three ^{99m}Tc-labeled phosphorus compounds and ^{18}F-fluoride for skeletal imaging. J Nucl Med 15:1153–1157

Ancri D, Lonchampt M, Basset J (1977) The effect of tin on the tissue distribution of Tc-99m sodium pertechnetate. Radiology 124:445–450

Bevan JA, Tofe AJ, Benedict JJ, Francis MD, Barnett BL (1980) Tc-99m HMDP (hydroxymethylene diphosphonate): a radiopharmaceutical for skeletal and acute myocardial infarct imaging. I. Synthesis and distribution in animals. J Nucl Med 21:961–966

Buell U, Kleinhans E, Zorn-Bopp E, Reuschel W, Muenzing W, Moser EA, Seiderer M (1982) A comparison of bone imaging with Tc-99m DPD and Tc-99m MDP: concise communication. J Nucl Med 23:214–217

Byun HH, Rodman SG, Chung KE (1976) Soft tissue concentration of Tc-99m phosphates associated with injection of iron dextran complex. J Nucl Med 17:374–375

Callahan RJ, Froelich JW, McKusick KA, Leppo J, Strauss WH (1982) A modified method for the in vivo labeling of red blood cells with Tc-99m: concise communication. J Nucl Med 23:315–318

Castronovo FP, Callahan RJ (1972) New bone scanning agent: ^{99m}Tc-labeled 1-hydroxyethylidene-L,L-disodium phosphonate. J Nucl Med 13:823–827

Chaudhuri TK (1976) The effect of aluminium and pH on altered body distribution of Tc-99m EHDP. Int J Nucl Med Biol 3:37

Conklin JJ, Alderson PO, Zizic TM et al (1983) Comparison of bone scan and radiograph sensitivity in the detection of steroid-induced ischemic necrosis of bones. Radiology 147:221–226

Council of Europe (2005) Technetium ^{99m}Tc medronate injection. In: European Pharmacopeia 5.0, monograph no 641. Council of Europe, Maisonneuve, Sainte-Ruffine, p 1219

Davis MA, Jones AG (1976) Comparison of ^{99m}Tc-labeled phosphate and phosphonate agents for skeletal imaging. Semin Nucl Med 6:19–31

Domstad PA, Coupal JJ, Kim EE, Blake JS, DeLand FH (1980) ^{99m}Tc-hydroxymethane diphosphonate: a new bone imaging agent with a low tin content. Radiology 136:209–211

Dunson GL, Stevenson JS, Cole CM, Mellor MK, Hosain F (1973) Preparation and comparison of technetium-99m diphosphonate, polyphosphate and pyrophosphate in nuclear bone imaging radiopharmaceuticals. Drug Intell Clin Pharm 7:470–474

Fogelman I (1982) Diphosphonate bone scanning agents – current concepts. Eur J Nucl Med 7:506–509

Fogelman I, Pearson DW, Bessent RG, Tofe AJ, Francis MD (1981) A comparison of skeletal uptakes of three diphosphonates by whole-body retention: concise communication. J Nucl Med 22:880–883

Francis MD, Ferguson DL, Tofe AJ, Bevan JA, Michaels SE (1980) Comparative evaluation of three diphosphonates: in vitro adsorption (C14-labeled) and in vivo osteogenic uptake (Tc-99m complexed). J Nucl Med 21:1185–1189

Genant HK, Bautovich GJ, Singh M (1974) Bone seeking radionuclides: An in-vivo study of factors affecting skeletal uptake. Radiology 113:373–382

Godart G, Durez M, Bevilaqua M, Abramovici J, Robience Y (1986) Technetium-99m MDP vs technetium-99m dicarboxypropane diphosphonate. A clinical comparison in various pathologic conditions. Clin Nucl Med 11:92–97

Gray WR, Hickey D, Parkey RW, Bonte FJ, Roan P, Willerson JT (1979) In: Parkey RW, Bonte FJ, Buja LM, Willerson JT (eds) Qualitative blood pool imaging in clinical nuclear cardiology. Appleton-Century-Crofts, NY, pp 297–308

Hale TI, Jucker A, Vgenopoulos K, Sauter B, Wacheck W, Bors L (1981) Clinical experience with a new bone seeking ^{99m}Tc-radiopharmaceutical. Nucl Compact 12:54–55

Hladik WB, Nigg KK, Rhodes BA (1982) Drug-induced changes in the biologic distribution of radiopharmaceuticals. Semin Nucl Med 12:184–218

Hladik WB, Ponto JA, Lentle BC, Laven DL (1987) Iatrogenic alterations in the biodistribution of radiotracers as a result of drug therapy: reported instances. In: Hladik WB III, Saha GB, Study KT (eds) Essentials of nuclear medicine science. Williams & Wilkins, Baltimore, pp 193–202

Hughes SPF, Jeyasingh K, Lavender PJ (1975) Phosphate compounds in bone scanning. J Bone Joint Surg (Br) 57:214–216

International Commission on Radiological Protection (1987a) Technetium-labelled phosphates and phosphonates. In: Annals of the ICRP, radiation dose to patients from radiopharmaceuticals, biokinetic models and data. ICRP publication 53, vol 18, no 1–4. Pergamon, Oxford, pp 213–215

International Commission on Radiological Protection (1987b) Technetium-labelled erythrocytes. In: Annals of the ICRP, radiation dose to patients from radiopharmaceuticals, biokinetic models and data. ICRP publication 53, vol 18, no 1–4. Pergamon, Oxford, pp 209–210

International Commission on Radiological Protection (1991) Technetium-labelled erythrocytes. In: Annals of the ICRP, radiological protection in biomedical research. ICRP publication 62, vol 22, no 3. Pergamon, Oxford, pp 25–28

International Commission on Radiological Protection (1998) Technetium-labelled phosphates and phosphonates (1998) In: Annals of the ICRP, radiation dose to patients from radiopharmaceuticals, addendum to ICRP 53, publication 80, vol 28, no 3. Pergamon, Oxford, p 75

Jones AG, Francis MD, Davis MA (1976) Bone scanning: radionuclide reaction mechanism. Semin Nucl Med 6:3–18

Krishnamurthy GT, Tubis M, Endow JS, Singhi V, Walsh CF, Blahd WH (1974) Clinical comparison of the kinetics of ^{99m}Tc-labeled polyphosphate and diphosphonate. J Nucl Med 15:848–855

Krogsgaard OW (1976) Radiochemical purity of various Tc-99m-labeled bone scanning agents. Eur J Nucl Med 1:15–17

McAfee JG (1987) Radionuclide imaging in metabolic and systemic skeletal diseases. Semin Nucl Med 17:334–349

McRae J, Hambright P, Valk P, Bearden AJ (1976) Chemistry of Tc-99m tracers. II. In vitro conversion of tagged HEDP and pyrophosphate (bone seekers) into gluconate (renal agent). Effects of Ca and Fe(II) on in vivo distribution. J Nucl Med 17:208–211

Pauwels EKJ, Blom J, Camps JAJ, Hermans J, Rijke AM (1983) A comparison between the efficacy of ^{99m}Tc-MDP, ^{99m}Tc-DPD, ^{99m}Tc-HDP for the detection of bone metastases. Eur J Nucl Med 8:118–122

Pendergrass HP, Postsaid MS, Castronovo FP (1973) The clinical use of Tc-99m diphosphonate (HEDSPA). Radiol 107(3): 557–562

Ponto JA, Swanson DP, Freitas JE (1987) Clinical manifestations of radio pharmaceutical formulation problems. In: Hladik WB III, Saha GB, Study KT (eds) Essentials of nuclear medicine science. Williams & Wilkins, Baltimore, pp 271–274

Porter WC, Dees SM, Freitas JE, Dworkin HJ (1983) Acid-citrate-dextrose compared with heparin in the preparation of in vivo/in vitro technetium-99m red blood cells. J Nucl Med 24:383–387

Russell CD, Cash AG (1979) Complexes of technetium with pyrophosphate, etidronate, and medronate. J Nucl Med 20:532–537

Saha GB (1987) Normal biodistribution of diagnostic radiopharmaceuticals. In: Hladik WB III, Saha GB, Study KT (eds) Essentials of nuclear medicine science. Williams & Wilkins, Baltimore, pp 3–19

Saha GB (1998) Fundamentals of radiopharmacy, 4th edn. Springer, Berlin Heidelberg New York
Saha GB, Boyd CM (1979) A study of protein-binding of ^{99m}Tc-methylene diphosphonate in plasma. Int J Nucl Med Biol 6:201–206
Sahni M, Guenther HL, Fleish H, Collin P, Martin J (1993) Biphosphonates act on rat bone resorption through the mediation of osteoblasts. J Clin Invest 91:2004–2011
Schroth H-J, Hausinger F, Garth H, Oberhausen E (1984) Comparison of the kinetics of methylene diphosphonate (MDP) and dicarboxypropane diphosphonic acid (DPD), two radio-diagnostics for bone scintigraphy. Eur J Nucl Med 9:529–532
Schwarz A, Kloss G (1981) Technetium-99m DPD – a new skeletal imaging agent. J Nucl Med 22(Abstr):7
Schwarz A, Oberhausen E, Schroth HJ, Hale TJ (1991) Teceos. Scientific information on the labeling unit technetium-99m-3,3-diphosphono-1,2-propanedicarboxylic acid. Behring Diagnostika, Marburg, Germany
Srivastava SC, Meinken G, Smith TD, Richards P (1977) Problems associated with stannous ^{99m}Tc-radiopharmaceuticals. Int J Appl Radiat Isot 28:83–95
Subramanian G, McAfee JG, Blair RJ, Mehter A, Connor T (1972) ^{99m}Tc-EHDP: a potential radiopharmaceutical for skeletal imaging. J Nucl Med 13:947–950
Subramanian G, McAfee JG, Blair RJ, Kallfelz FA, Thomas FD (1975a) Technetium-99m methylene diphosphonate – a superior agent for skeletal imaging: comparison with other technetium complexes. J Nucl Med 16:744–755
Subramanian G, McAfee JG, Blair RJ (1975b) An evaluation of ^{99m}Tc-labeled phosphate compounds as bone imaging agents. In: Subramanian G, Rhodes BA, Cooper JF, Sodd VJ (eds) Radiopharmaceuticals. Society of Nuclear Medicine, New York, pp 319–328
Tofe AJ, Francis MD (1972) In vitro optimization and organ distribution studies in animals with the bone scanning agent ^{99m}Tc-Sn-EHDP. J Nucl Med 13(Abstr):472
Tofe AJ, Francis MD (1976) In vitro stabilization of a low tin bone imaging kit. J Nucl Med 16:414–422
Tofe AJ, Bevan JA, Fawzi MB, Francis MD, Silberstein EB, Alexander GA, Gunderson DE, Blair K (1980) Gentisic acid: a new stabilizer for low tin skeletal imaging agents: concise communication. J Nucl Med 21:366–370
United States Pharmacopeial Convention (2005) United States pharmacopeia USP 28 official monographs: technetium Tc-99m medronate injection, technetium Tc-99m oxidronate injection, technetium Tc-99m etidronate injection. United States Pharmacopeial Convention, Rockville, Md.
Van Duzee BF, Bugaj JE (1981) The effect of total technetium concentration on the performance of a skeletal imaging agent. Clin Nucl Med 6(Suppl):148
Wang TST, Fawwaz RA, Johnson LJ, Mojdehi GE, Johnson PM (1980) Bone-seeking properties of Tc-99m carbonyl diphosphonic acid, dihydroxy-methylene diphosphonic acid, and monohydroxy-methylene diphosphonic acid: concise communication. J Nucl Med 21:767–770
Weber DA, Makler PT Jr, Watson EE, Coffey JL, Thomas SR, London J (1989) Radiation absorbed dose from technetium-99m-labeled bone imaging agents: MIRD dose estimate report no 13. J Nucl Med 30:1117–1122
Yano Y, McRae J, Van Dyke DC, Anger HO (1973) Technetium-99m-labeled stannous ethane-1-hydroxy-1-diphosphonate: a new bone scanning agent. J Nucl Med 14:73–78
Zimmer AM, Pavel DG (1978) Experimental investigations of the possible cause of liver appearance during bone scanning. Radiology 126:813–816
Zimmer AM, Pavel DG, Karesh SM (1979) Technical parameters of in vivo red blood cell labeling with Technetium-99m. Nuklearmedizin 18:241–245

12.8 ^{99m}Tc-Labeled Renal Imaging Agents

12.8.1 ^{99m}Tc-DMSA (Dimercaptosuccinic Acid)

J. Környei and I. Zolle

Chemical name	Chemical structure
Dimercaptosuccinic acid (DMSA) Succimer (*Ph. Eur.; USP*) Technetium Tc 99m succimer injection (*Ph. Eur.; USP*) ^{99m}Tc(III)-dimercaptosuccinate (DMS) ^{99m}Tc(V)-DMSA	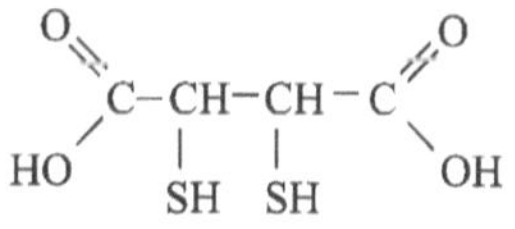 Dimercaptosuccinic acid

Kit components		Listed trade names	
Amerscan DMSA:		Amerscan DMSA	GE Healthcare
Dimercaptosuccinic acid	1 mg	RenoCis (TCK-12)	CIS Bio
Tin(II)-chloride dihydrate	0.42 mg	TechneScan DMSA	Mallinckrodt/Tyco
Inositol	50.0 mg	Nephroscint	Bristol-Myers Squibb
Ascorbic acid	0.7 mg		
Sodium chloride	2.9 mg	DMSA	Rotop

Preparation

The kit contains the lyophilized ingredients in a multidose vial. Labeling is performed by adding 1–6 ml of ^{99m}Tc eluate under aseptic conditions. Sterile ^{99m}Tc-sodium pertechnetate (370–925 MBq, maximum 3.7 GBq) (10–25 mCi) is injected into the vial, and then mixed well to dissolve the lyophilisate. The reaction is allowed to proceed at room temperature for 15 min.

^{99m}Tc-dimercaptosuccinic acid (DMSA) complex is a sterile, pyrogen-free, clear colorless solution suitable for intravenous injection. The pH should be between 2.3 and 3.5 (*Ph. Eur.*).

Description of the Kit

The kit contains an isomeric mixture of DMSA as the mesoisomer (>90%) and the D,L-isomers (<10%). Upon addition of sterile sodium ^{99m}Tc-pertechnetate injection to the sterile, nonpyrogenic formulation, the renal ^{99m}Tc(III)-DMSA complex is formed.

The lyophilized formulation is under nitrogen atmosphere to avoid oxidative processes; the introduction of air into the vial must be avoided. A breather needle should not be used. It is recommended to use a separate kit preparation for each patient. If multiple doses are obtained from the same vial, care should be taken to avoid introducing air into the reaction vial, and the doses should be drawn as close together as possible and administered immediately (Taylor et al. 1980).

The antioxidant inositol is used for stabilization of the ^{99m}Tc(III)-DMSA complex.

^{99m}Tc eluate used for labeling should be obtained from a generator that is eluted daily (Van Duzee and Bugaj 1981).

Initially, two complexes are formed under acidic conditions (pH 2.5), which yield the ^{99m}Tc(III)-DMSA complex at the end of the reaction time (15 min). An excess amount of stannous ion as well as a high molar ratio of Sn(II) to Sn(IV) have been postulated to yield Tc(III)-dimercaptosuccinate (DMS) (Ikeda et al. 1976).

At least four ^{99m}Tc-DMSA complexes have been identified. It has been shown that pH, stannous ion concentration, and the concentration and purity of ^{99m}Tc-pertechnetate added for on-site preparation affect the formation of the different complexes (Ikeda et al. 1977 a, b).

The formation of pentavalent ^{99m}Tc-DMSA complex occurs at elevated pH; therefore, 0.2 ml of a sterile solution of 3.5% sodium bicarbonate is added to the lyophilized DMSA kit before adding ^{99m}Tc-sodium pertechnetate (Ramamoorthy et al. 1987). The optimal pH for ^{99m}Tc(V)-DMSA formation is pH 7.5–8.0; the resulting concentration of stannous ion produced a complex with high stability (6 h). It was observed that addition of ^{99m}Tc-pertechnetate should follow immediately after mixing the powder with the alkaline sodium bicarbonate solution (Johannsen et al. 1979; Ramamoorthy et al. 1987; Westera et al. 1985).

The negatively charged Tc(V)-oxo complex with DMSA consists of a TcO core having four sulfur atoms of the bidentate DMSA ligands arranged in a plane. The chemical formula is $[TcO\,(DMSA)_2]^{-1}$ (Saha 1997).

Clinical Applications

Renal scintigraphy:	Static imaging (planar or tomographic) of the functional renal cortex Morphological studies of the renal cortex indicating space-occupying lesions and areas of reduced function kidney localization, delineating size and shape of the kidneys as well as defects
Scintigraphy of medullary carcinoma of the thyroid (MCT):	Detection of metastatic lesions

For static imaging of the kidneys, the radiotracer is retained in the renal parenchyma by tubular fixation. The scintigram shows solely functional parenchyma. Necrotic tissue and inflammatory processes are not imaged; a tumor, cyst or abscess appears as cold area.

Pentavalent ^{99m}Tc(V)-DMSA has been used in the diagnosis of MCT (Clarke et al. 1988; Ohta et al. 1984).

Time of Examination

Renal scintigraphy:	1–3 h after intravenous injection 6 h postinjection in the case of severely impaired kidney function
Scintigraphy of MCT:	2–4 h after intravenous injection

Recommended Activities for Indications

Renal scintigraphy:	37–120 MBq (1–3.2 mCi) of ^{99m}Tc(III)-DMSA complex (0.3–1 mg DMSA)
Scintigraphy of MCT:	370 MBq (10 mCi) of ^{99m}Tc(V)-DMSA complex

The manufacturer's instructions should be followed.

Pediatric Dose. The amount of radioactivity for infants and children is based on body weight (static imaging). Recommendations by the Pediatric Task Group of the European Association of Nuclear Medicine (EANM), based on body weight should be followed (see Appendix 1, Table A1.2).

Additional Information

Because of the slow transfer of activity from blood to kidney, imaging should be delayed for 3 h after injection.

Reduced cortical accumulation is observed in the presence of ammonium chloride, sodium bicarbonate, and mannitol based on experimental data (Yee et al. 1981).

Medication with angiotensin-converting enzyme (ACE)-inhibitors (captopril) may cause reduced uptake in the affected kidney (Hovinga et al. 1989; Kopecky et al. 1990).

A marked increase in hepatic activity may result from poor labeling conditions (Ikeda et al. 1976; Taylor et al. 1980).

The patient should be adequately hydrated before ^{99m}Tc-DMSA scintigraphy.

Quality Control

Radiochemical Purity. ^{99m}Tc-DMSA complex is described in the *Ph. Eur.* (Council of Europe 2005). Thin-layer chromatography (TLC) on silica gel (SG) fiberglass sheets in two different solvents is recommended for the detection of free ^{99m}Tc-pertechnetate and reduced, hydrolized ^{99m}Tc activity. Free ^{99m}Tc-pertechnetate is measured in methyl ethyl ketone (MEK) at the solvent front, reduced, hydrolized ^{99m}Tc activity is identified in saline at the start. The radiochemical purity of ^{99m}Tc-DMSA should not be less than 95%. The amount of free ^{99m}Tc-pertechnetate should not exceed 2% of the measured radioactivity (*Ph. Eur.*).

Thin-layer chromatography (*Ph. Eur.*)		
Stationary phase:	Gelman ITLC-SG (fiberglass), 2×9.5 cm	
Solvent:	Methyl ethyl ketone (MEK)	
Developing time:	10 min	
R_f values:	^{99m}Tc-DMSA complex:	0.0–0.1 (>95%)
	^{99m}Tc reduced, hydrolized:	0.0–0.1
	^{99m}Tc-pertechnetate:	0.9–1.0 (<2%)

Reduced, hydrolized ^{99m}Tc activity is identified in saline at the origin. ^{99m}Tc-pertechnetate should not exceed 2% of the measured radioactivity. The sum of measured impurities (F+H) in both solvents must not exceed 5% of the total radioactivity.

Quantification of labeled components. Each chromatogram is measured (TLC linear analyzer or gamma counter), and the regional radioactivities are expressed as a percentage of the total recovered counts:

$$^{99m}\text{Tc-DMSA}\ (\%) = 100 - (F + H)$$

where F (%) = free ^{99m}Tc-pertechnetate, and H (%) = hydrolized ^{99m}Tc activity.

Results of analysis *(12 samples)*
Results were obtained by thin-layer chromatography in two solvent systems, at different times after labeling.

Labeling and stability	15 min (%)	6 h (%)
^{99m}Tc-DMSA complex	98.3 ± 0.13	97.8 ± 0.28
^{99m}Tc-Na-pertechnetate	0.8±0.22	0.9±0.21
^{99m}Tc-reduced, hydrolized	1.0±0.04	1.0±0.12

Bioassay of ^{99m}Tc-DMSA Complex. To assure high renal uptake, the *Ph. Eur.* recommends a physiological test in rats. The required organ uptake values at 1 h after intravenous injection (at least two of three rats) are more than 40% of the applied radioactivity is measured in the kidneys, less than 10% in the liver, < 2% in the stomach, and < 5% in the lung.

Pharmacokinetic Data

After intravenous injection, the ^{99m}Tc(III)-DMSA complex is taken up in the renal parenchyma (24% at 1 h), showing high cortical affinity (Lin et al. 1974). Uptake is related to renal cortical perfusion; the plasma clearance half-time in patients with normal kidney function is 56 min (Enlander et al. 1974).

In patients, both kidneys are visualized 1 h after intravenous injection, with significant radioactivity in the bladder and negligible amounts in the liver, stomach, thyroid gland, or other organs. Most of the circulating ^{99m}Tc-DMSA is loosely bound to plasma proteins (Arnold et al. 1975). At 1 h after injection, approximately 25% of the injected radioactivity is measured in the proximal tubules, 30% in plasma, and 10% in the urine (Arnold et al. 1975; Bingham and Maisey 1978; Handmaker et al. 1975; Lin et al. 1974). Renal extraction is 4–5% per renal passage; the maximum accumulation in the renal cortex (48.3 ± 3.0%) is reached 3 h after injection, 20% remain in the plasma, less than 10% in liver and muscle, and a urinary excretion of 11.4%. ^{99m}Tc-DMSA is exclusively excreted in the urine as the unchanged molecule; in the presence of renal tubular acidosis, urinary excretion is increased (de Lange et al. 1989). Urinary excretion in 24 h amounts to approximately 30% (Arnold et al. 1975).

Renal accumulation is unaffected by probenecid and *para*-aminohippuric acid (PAH) (Lee and Blaufox 1985). Uptake is decreased by ACE inhibitors in the presence of renal artery stenosis (Hovinga et al. 1989; Kopecky et al. 1990).

In renal failure, ^{99m}Tc-DMSA activity accumulates in the liver, the gallbladder, and the gut.

Pentavalent ^{99m}Tc(V)-DMSA differs structurally and has been evaluated for imaging medullary carcinoma of the thyroid (Ohta et al. 1984; Ramamoorthy et al. 1987). The sensitivity of lesion detection is 95%; no false positive uptake of ^{99m}Tc(V)-DMSA was seen in nine patients with a histologic diagnosis of medullary carcinoma. Accumulation of ^{99m}Tc(V)-DMSA was seen in both bone and soft tissue metastases. In comparison, ^{99m}Tc-MDP detected all known metastases in bone, but none in soft tissue (Clarke et al. 1988).

Radiation Dose

The most exposed organs are the kidneys, bladder wall, adrenals, liver, and spleen. The effective (whole body) dose equivalent is 0.016 mSv/MBq. The effective dose in adults (70 kg) resulting from an intravenous injection of 70 MBq (1.9 mCi) of ^{99m}Tc(III)-DMSA complex for renal scintigraphy is 1.12 mSv. The dose to the kidneys (renal cortex) after intravenous injection of 70 MBq (1.9 mCi) of ^{99m}Tc(III)-DMSA complex is 11.9 mGy.

The effective dose in adults (70 kg) resulting from an intravenous injection of 370 MBq (10 mCi) of ^{99m}Tc(V)-DMSA complex for scintigraphy of medullary carcinoma is 3.0 mSv (Clarke et al. 1988).

Storage and Stability

Storage. The lyophilized kit is stored at 2–8 °C in the refrigerator, protected from light.

Stability. The ^{99m}Tc(III)-DMSA injection solution is stable for 4 h at room temperature.

References

Arnold RW, Subramanian G., McAfee JG, Blair RJ, Thomas FD (1975) Comparison of ^{99m}Tc complexes for renal imaging. J Nucl Med 16:357–367

Bingham JB, and Maisey MN (1978) An evaluation of the use of ^{99m}Tc-dimercaptosuccinic acid (DMSA) as a static renal imaging agent. Br J Radiol 51:599–607

Clarke SEM, Lazarus CR, Wraight P, Sampson C, Maisey MN (1988) Pentavalent ^{99m}Tc-DMSA, ^{131}I-MIBG, and ^{99m}Tc-MDP. An evaluation of three imaging techniques in patients with medullary carcinoma of the thyroid. J Nucl Med 29:33–38

Council of Europe (2005) Technetium [^{99m}Tc] succimer injection. European pharmacopeia monograph no. 643. Council of Europe, Maisonneuve, Sainte-Ruffine, p 1229

Enlander D, Weber PM, dos Remedios LV (1974) Renal cortical imaging in 35 patients: superior quality with ^{99m}Tc-DMSA. J Nucl Med 15:743–749

Handmaker H, Young BW, Lowenstein JM (1975) Clinical experience with ^{99m}Tc-DMSA (dimercaptosuccinic acid), a new renal imaging agent. J Nucl Med 16:28–32

Hovinga TKK, deJong PE, Piers DA, Beekhuis H, van-der Hem GK, de Zeeuw D (1989) Diagnostic use of angiotensin-converting enzyme inhibitors in radioisotope evaluation of unilateral renal artery stenosis. J Nucl Med 30:605–614

Ikeda I, Inoue O, Kurata, K (1976) Chemical and biological studies on ^{99m}Tc-DMSA-II: effect of Sn(II) on the formation of various Tc-DMSA complexes. Int J Appl Radiat Isot 27:681–688

Ikeda I, Inoue O, Kurata K (1977 a) Chemical and biological studies on ^{99m}Tc-DMSA-I: formation of complex by four different methods. Int. J Nucl Med Biol. 4:56–65

Ikeda I, Inoue O, Kurata K (1977b) Preparation of various ^{99m}Tc-dimercaptosuccinate complexes and their evaluation as radiotracers. J Nucl Med 18:1222–1229
International Commission on Radiological Protection (1987) Technetium-DMSA. In: Annals of the ICRP, radiation dose to patients from radiopharmaceuticals, biokinetic models and data. ICRP publication 53, vol. 18, no.1–4. Pergamon, Oxford, pp 185–186
Johannsen B, Spies H, Syhre R (1979) Studies on complexation of ^{99m}Tc with dimercaptosuccinic acids with regard to organ specificity of ^{99m}Tc-radiopharmaceuticals. Eur J Nucl Med 4:148
Kopecky RT, McAfee JG, Thomas FD, Anderson HR Jr, Hellwig B, Roskopf M, Patchin D (1990) Enalaprilat-enhanced renography in a rat model of renovascular hypertension. J Nucl Med 31:501–507
Lange MJ de, Piers DA, Kosterink JGW, van Luijk WHJ, Meijer S, de Zeeuw D, van-der Hem GJ (1989) Renal handling of technetium-99m-DMSA: evidence for glomerular filtration and peritubular uptake. J Nucl Med 30:1219–1223
Lee HB, Blaufox MD (1985) Mechanism of renal concentration of technetium-99m glucoheptonate. J Nucl Med 26:1308–1313
Lin TH, Khentigam A, Winchell HS (1974) A ^{99m}Tc-chelate substitute for organoradiomercurial renal agents. J Nucl Med 15:34–35
Ohta H, Yamamoto K, Endo K, Mori T, Hamanaka D, Shimazu A, Ikekubo K, Makimoto K, Iida Y, Konishi J, Morita R, Hata N, Horiuchi K, Yokoyama A, Torizuka K, Kuma K (1984) A new agent for medullary carcinoma of the thyroid. J Nucl Med 25:323–325
Ramamoorthy N, Shetye SV, Pandey PM, Mani RS, Patel MC, Patel RB, Ramanathan P, Krishna BA, Sharma SM (1987) Preparation and evaluation of ^{99m}Tc(V)-DMSA complex: studies in medullary carcinoma of thyroid. Eur J Nucl Med 12:623–628
Saha GB (1997) Characteristics of specific radiopharmaceuticals. In: Fundamentals of nuclear pharmacy, 4th edn. Springer, Berlin Heidelberg New York, p 125
Taylor A Jr, Lallone RL, Hagan PL (1980) Optimal handling of dimercaptosuccinic acid for quantitative renal scanning. J Nucl Med 21:1190–1193
United States Pharmacopeial Convention (2005) Official Monographs: USP 28, technetium Tc-99m succimer injection. United States Pharmacopeia (USP) 28–national formulary (NF) 23, p 1864
Van Duzee BF, Bugaj JE (1981) The effect of total technetium concentration on the performance of a skeletal imaging agent. Clin Nucl Med 6(Suppl):148
Westera G, Gadze A, Horst W (1985) A convennient method for the preparation of ^{99m}Tc(V)-DMSA. Int J Appl Radiat Isot 36:311–312
Yee CA, Lee HB, Blaufox MD (1981) ^{99m}Tc-DMSA renal uptake: influence of biochemical and physiologic factors. J Nucl Med 22:1054–1058

12.8.2 ^{99m}Tc-DPTA (Diethylenetriaminepentaacetate)

J. Környei and I. Zolle

Chemical name	**Chemical structure**
Diethylenetriaminepentaacetate (DTPA) as Calcium trisodium salt Pentetate (*Ph. Eur.; USP*) Technetium Tc 99m pentetate injection (*Ph. Eur.; USP*) ^{99m}Tc-DTPA complex	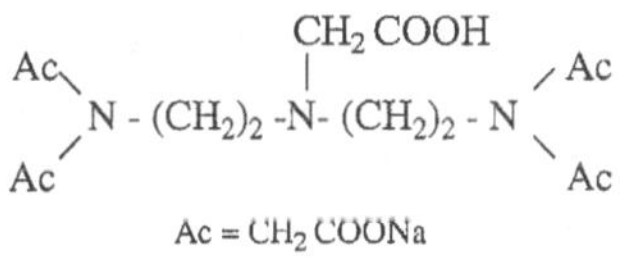 Pentetate
Kit components	**Commercial products**
Amerscan Pentetate II: Calcium trisodium diethylene-triaminepentaacetate 20.6 mg Tin(II)-chloride dihydrate 0.25 mg Sodium *para*-amino-benzoate 2 mg	Amerscan Pentetate II — GE Healthcare TechneScan DTPA — Mallinckrodt/Tyco TechneScan DTPA/Aerosol PentaCis (TCK-6) — CIS Bio DTPAScint — Bristol-Myers Squibb DTPA — Rotop

Preparation

The kit contains the lyophilized ingredients in a multidose vial. Labeling is performed by adding 2–10 ml of ^{99m}Tc eluate containing up to 11.1 GBq (300 mCi) under aseptic conditions. After sterile sodium ^{99m}Tc-pertechnetate has been added to the vial, the powder is dissolved by inverting the vial. The reaction is allowed to proceed at room temperature for 10–30 min. Kit composition varies between manufacturers; therefore, instructions should be followed.

^{99m}Tc-diethylenetriaminepentaacetate (DTPA) complex is a sterile, pyrogen-free, clear, colorless solution suitable for intravenous injection. The pH should be between 4.0 and 7.5 (*European Pharmacopeia* [*Ph. Eur.*]).

Description of the Kit

The ^{99m}Tc-DTPA complex for intravenous injection is used for renal studies (Atkins et al. 1971; Klopper et al. 1972). DTPA forms a negatively charged complex with reduced ^{99m}Tc-technetium in neutral or weakly acidic solutions (Russel et al. 1980). The exact oxidation state is not known, although several valency states have been suggested (III–V). Quality control is essential when ^{99m}Tc-DTPA is used for the measurement of glomerular filtration rate (GFR), since commercial kits contain reduced, hydrolized ^{99m}Tc activity as an impurity (Carlsen et al. 1988). Some formulations use gentisic acid as a stabilizer (Tofe et al. 1980).

The ^{99m}Tc-DTPA complex may also be used as an aerosol; however, the kit composition suitable for nebulization differs from the renal agent. TechneScan DTPA/Aerosol contains 1.25 mg of DTPA and 0.12 mg tin(II)-chloride as dihydrate. Gentisic acid (0.25 mg) is used for stabilization. Labeling is performed by injecting less than 0.1 ml of ^{99m}Tc eluate, corresponding to at least 550 MBq (15 mCi) of radioactivity into the TechneScan DTPA/Aerosol vial, and then water for injection is added to obtain 1 ml of solution. After 15 min at room temperature, 0.5 ml of ethanol (98%) is added to the labeled product. The labeling yield is >95%. The kit TechneScan DTPA/Aerosol must not be used for intravenous injection.

^{99m}Tc-DTPA/Aerosol is a clear or slightly opalescent, aqueous solution, and the pH is 4.0–5.0.

Fractionated elution of the generator is required to obtain a high activity concentration of ^{99m}Tc eluate for labeling. ^{99m}Tc eluate used for labeling should be obtained from a generator that is eluted daily (Van Duzee and Bugaj 1981). Not more than 0.5 ml (1,110 MBq; 30 mCi) of the labeled product should be used for nebulization (to deliver 150 MBq (4 mCi) to the lung).

For inhalation of the ^{99m}Tc-DTPA complex, the vial is transferred to an aerosol generator. Ultrasonic nebulization of aqueous ^{99m}Tc-DTPA complex produces particles with an aerodynamic diameter of 0.5 μm (Wagner 1995).

Clinical Applications

- Renal studies providing both anatomical and functional information
- Determination of the GFR
- Cerebral scintigraphy based on leaks in the blood–brain barrier (BBB)
- Localization of inflammatory bowel disease
- Inhalation scintigraphy to measure regional lung ventilation

Measurement of separate kidney function is particularly indicated for the management of unilateral and bilateral uropathy, staging of disease, assessment of unilateral compensatory hypertrophy, and in cases where x-ray examinations (intravenous pyelography, etc.) are contraindicated (Hilson et al. 1976; Nielsen et al. 1977; Russel 1985; Wassner 1981).

^{99m}Tc-DTPA has been used for conventional cerebral scintigraphic studies for the detection of vascular and neoplastic brain lesions (Hauser et al. 1970).

^{99m}Tc-DTPA accumulates in inflammatory lesions of the gastrointestinal tract and has been used for the detection of bleeding sites. An increase in tracer activity is seen almost immediately after bolus injection in the affected segments of bowel and persists for hours (Kadir and Strauss 1979).

Inhalation scintigraphy using ^{99m}Tc-DTPA aerosol is performed for the assessment of pulmonary ventilation in patients with chronic obstructive lung disease (Coates and O'Brodovich 1986; Santolicandro and Giuntini 1979), and for the differential diagnosis of acute pulmonary embolism in combination with lung perfusion scintigraphy (Taplin and Chopra 1978; Wagner 1995). When inhaled, ^{99m}Tc-DTPA aerosol particles serve as an indicator of regional ventilation. Moist aerosols show a lung penetration between 1 and 10% of the nebulized activity. Labeled aerosols are also used to measure mucociliary clearance (Agnew 1991; Wanner 1977).

Time of Examination

Renal dynamic studies:	Collection of frames should start immediately after the intravenous injection. The total time for one examination is 30–60 min.
Brain scintigraphy:	1–3 h after intravenous injection
Abdominal images:	Images are obtained at 5, 10, 15, and 25 min; in some patients, delayed images are obtained at intervals from 45 min–4 h after the ^{99m}Tc-DTPA bolus
Inhalation scintigraphy:	This should start immediately after inhalation of the aerosol

Recommended Activities for Indications

Renal studies:	111–185 MBq (3–5 mCi) (7–20 mg DTPA)
Brain scintigraphy:	300–500 MBq (8.1–13.5 mCi) (7–20 mg DTPA)
Abdominal scintigraphy:	185–370 MBq (5–10 mCi)
Inhalation scintigraphy:	150 MBq (4 mCi) inhaled radioactivity (nebulized activity 1,110 MBq resp., 30 mCi). Breathing time 3–5 min, with an oxygen supply of 7–10 l/min

The manufacturer's instructions should be followed.

Pediatric Dose. The amount of radioactivity for infants and children is based on either body weight (static imaging) or body surface area (dynamic examinations). Recommendations by the Pediatric Task Group of the European Association of Nuclear Medicine (EANM) based on body weight should be followed (see Appendix 1, Table A2.2).

Additional Information

Free ^{99m}Tc-pertechnetate adds to increased background activity. Reduced, hydrolized ^{99m}Tc activity is bound to plasma proteins in circulating blood, interfering with the measurement of plasma clearance (Rehling 1988). Therefore, quality control of ^{99m}Tc-DTPA chelate should be performed before administration to patients. When ^{99m}Tc-DTPA is used to measure GFR, it is necessary to remove the protein-bound impurity by passing the plasma through an ultrafilter before the measurement of radioactivity (Rowell et al. 1986).

^{99m}Tc-DTPA is specific for the diagnosis of renal artery stenosis (Gruenewald and Collins 1983). Angiotensin-converting enzymes (ACE) inhibitors such as captopril reduce filtration pressure and thus cause a fall in GFR, as manifested by decreased renal uptake of ^{99m}Tc-DTPA. In patients who do not have renovascular hypertension, the GFR of each kidney is unchanged.

Furosemide (Lasix) will enhance the detection of a dilated collecting system by rapid washout of the radiotracer from the pelvis and ureter, while in the case of urinary tract obstruction, no change of pelvic retention and in the shape of the time–activity curve is seen (O'Reilly 1992).

The patient should be adequately hydrated before ^{99m}Tc-DTPA scintigraphy.

^{99m}Tc-DTPA has been used for studies of gastroesophageal reflux and gastric emptying. In this case, scintigraphy is performed after oral administration of ^{99m}Tc-DTPA (10–20 MBq, 0.3–0.6 mCi) in a suitable liquid (300 ml), according to local practice. Sequential scintigraphy and static imaging may be performed. The patient's stomach is imaged at 5 and 10 min after drinking the liquid, and then every 5–15 min until the stomach activity has reached half the original value. The gastric emptying half-time is determined from a plot of percent activity versus time. Normal values are 10–15 min; delayed emptying is observed with gastric ulcers, pyloric stenosis, vagotomy, and in the case of malignancy (Chadhuri 1974).

Quality Control

Radiochemical Purity. ^{99m}Tc-DTPA complex is described in the *Ph. Eur.* (Council of Europe 1997). Thin-layer chromatography (TLC) on silica gel (SG) fiberglass sheets in two different solvent systems is recommended for the separation of free ^{99m}Tc-pertechnetate and reduced, hydrolized ^{99m}Tc activity.

- System I: Free ^{99m}Tc-pertechnetate is measured in methyl ethyl ketone (MEK) at the solvent front.
- System II: Reduced, hydrolized ^{99m}Tc activity is separated in saline at the start. The radiochemical purity of ^{99m}Tc-DTPA chelate should not be less than 95% (*Ph. Eur.*).

Thin-layer chromatography		
Sytem I (MEK)	Free ^{99m}Tc-pertechnetate at the solvent front	(F)
System II (Saline)	Reduced, hydrolized ^{99m}Tc activity at the origin	(H)
	^{99m}Tc-DTPA complex (%) = 100–%(F+H)	

The *Ph. Eur.* states that the sum of measured impurities in both solvents must not exceed 5% of the total radioactivity. Reduced, hydrolized ^{99m}Tc activity should be analyzed before ^{99m}Tc-DTPA is used for GFR measurements:

$$^{99m}\text{Tc-DTPA complex (\%)} = 100 - (F + H)$$

F (%) = ^{99m}Tc-Na-pertechnetate (free)
H (%) = hydrolized ^{99m}Tc-activity.

Results of analysis *(12 samples)*
Results were obtained by thin-layer chromatography in two solvent systems, at different times after labeling.

Labeling and stability	15 min (%)	6 h (%)
^{99m}Tc-DTPA complex	98.5 ± 0.27	97.2 ± 0.32
^{99m}Tc-reduced, hydrolized	0.6 ± 0.12	0.8 ± 0.25
^{99m}Tc-Na-pertechnetate	0.8 ± 0.06	1.0 ± 0.17

Pharmacokinetic Data

After intravenous injection, ^{99m}Tc-DTPA penetrates capillary walls to enter the extravascular space within 4 min (McAfee et al. 1979). Because of its hydrophilicity and negative charge, ^{99m}Tc-DTPA is excluded from cells and is confined to the extracellular space. ^{99m}Tc-DTPA is removed from the circulation exclusively by the kidneys; renal clearance is unaffected by urine flow and by the administration of probenecid (Klopper et al. 1972). However, only the filtered fraction (20%) of the total renal plasma flow is excreted by glomerular filtration (Barbour et al. 1976).

The plasma disappearance of ^{99m}Tc-DTPA can be described by a double exponential function (McAfee et al. 1979). One component (0.99) is excreted with a biological half-time of 100 min, the rest (0.01) with a half-time of 7 days. The fraction excreted by the kidneys is 1.0, and the renal transit time is 5 min. For abnormal cases, the retention half-time of the major component is increased (1,000 min) and so is the transit time (20 min).

^{99m}Tc-DTPA cannot pass through the intact BBB, but in areas of the brain where structural defects permit diffusion, uptake has been used as an indicator of vascular and neoplastic brain lesions (Hauser et al. 1970).

Inhalation of ^{99m}Tc-DTPA as an aerosol shows free diffusion of the particles (diameter of 0.5 µm) to the lung periphery and alveolar retention, larger droplets (>1 µm) sediment preferentially in the trachea and upper bronchial tree (Coates and O'Brodovich 1986). In the lung, the label is released with a biological half-time of approximately 1 h. ^{99m}Tc-DTPA reaching the blood is excreted by the kidneys.

Radiation Dose

The most exposed organs are the kidneys and bladder wall. The effective (whole body) dose equivalent is 0.0063 mSv/MBq (International Commission on Radiological Protection 1987 a). The effective dose in adults (70 kg) resulting from intravenous injection of 185 MBq (5 mCi) of ^{99m}Tc-DTPA complex is 1.2 mSv.

Calculations of the absorbed radiation dose resulting from inhalation of ^{99m}Tc-DTPA are based on technetium-labeled aerosols; it is assumed that the label is released fast, and that ^{99m}Tc-DTPA reaching the blood is excreted by the kidneys. The effective (whole body) dose equivalent is 0.007 mSv/MBq (International Commission on Radiological Protection 1987b). The effective dose in adults (70 kg) resulting from inhalation of 150 MBq (4 mCi) of ^{99m}Tc-DTPA is 1.05 mSv. The dose to the bladder wall after inhalation of 150 MBq (4 mCi) of ^{99m}Tc-DTPA is 7.0 mGy.

Storage and Stability

Storage. The lyophilized kit is stored at 2–8 °C in the refrigerator. ^{99m}Tc-DTPA injection solution is kept at room temperature with adequate shielding.

Stability. The ^{99m}Tc-DTPA injection is stable within 6–8 h after preparation.

References

Atkins HL, Eckelman WC, Hauser W, Klopper JF, Richards P (1971) Evaluation of glomerular filtration rate with ^{99m}Tc-DTPA. J Nucl Med 12:338

Agnew JE (1991) Characterizing lung aerosol penetration. J Aerosol Med 4:237–250

Barbour GL, Crumb CK, Boyd CM, Reeves RD, Rastogi SP, Patterson RM (1976) Comparison of inulin, iothalamate and Tc-99m-DTPA for measurement of glomerular filtration rate. J Nucl Med 17:317–320

Carlsen JE, Moller MH, Lund JO, Trap-Jensen J (1988) Comparison of four commercial Tc-99m-(Sn)-DTPA preparations used for the measurement of glomerular filtration rate. J Nucl Med 21:126–129

Chadhuri TK (1974) Use of 99mTc-DTPA for measuring gastric emptying time. J Nucl Med 15:391–395

Coates G, O'Brodovich H (1986) Measurement of pulmonary epithelial permeability with ^{99m}Tc-DTPA aerosol. Semin Nucl Med 16:275–284

Council of Europe (2005) Technetium ^{99m}Tc pentetate injection. European Pharmacopeia 5.0, monograph no 642. Council of Europe, Maisonneuve, Sainte-Ruffine, pp 1223

Gruenewald SM, Collins LT (1983) Renovascular hypertension: quantitative renography as a screening test. Radiology 149:287–291

Hauser W, Atkins HL, Nelson KG, Richards P (1970) Technetium-99m-DTPA: a new radiopharmaceutical for brain and kidney scanning. Radiology. 94:679–684

Hilson AJW, Mistry RD, Maisey MN (1976) Tc-99m-DTPA for the measurement of glomerular filtration rate. Br J Radiol 49:794–796

International Commission on Radiological Protection (1987a) Technetium-DTPA. In: Annals of the ICRP, radiation dose to patients from radiopharmaceuticals, biokinetic models and data. ICRP publication 53, vol 18, no 1–4. Pergamon, Oxford, pp 187–190

International Commission on Radiological Protection (1987b) Technetium-labelled aerosols. In: Annals of the ICRP, radiation dose to patients from radiopharmaceuticals, biokinetic models and data. ICRP publication 53, vol 18, no 1–4. Pergamon, Oxford, pp 217–219

Kadir S, Strauss WH (1979) Evaluation of inflammatory bowel disease with Tc-99m-DTPA. Radiology 130:443–446

Klopper JF, Hauser W, Atkins HL, Eckelman WC, Richards P (1972) Evaluation of Tc-99m-DTPA for the measurement of glomerular filtration rate, J Nucl Med 13:107–110

McAfee JG, Gagne G, Atkins HL, Kirchner PT, Reba RC, Blaufox MD, Smith EM (1979) Biological distribution and excretion of DTPA labeled with Tc-99m and In-111. J Nucl Med 20:1273–1278

Nielsen SP, Moller ML, Trap-Jensen J (1977) Tc-99m-DTPA scintillation-camera renography: a new method for estimation of single-kidney function, J Nucl Med 18:112–117

O'Reilly PH (1992) Diuresis renography. Recent advances and recommended protocols. Br J Urol 69:113–120

Rehling M (1988) Stability, protein binding and clearance studies of ^{99m}Tc-DTPA. Evaluation of a commercially available dry-kit. Scand J Clin Lab Invest 48:603–609

Rowell KL, Kontzen FK, Stutzman ME, Caranto R, Barber JM, Russel CD, Dubovsky EV, Scott JW (1986) Technical aspects of a new technique for estimating glomerular filtration rate using Technetium-99m-DTPA. J Nucl Med Tech 14:196–198

Russell CD (1985) Radiopharmaceuticals used to assess kidney function and structure. Tauxe WN, Dubovsky EV (eds) Nuclear medicine in clinical urology and nephrology. Appleton-Century-Crofts, Norwalk, Conn., pp 7–31

Russel CD, Crittenden RC, Cash AG (1980) Determination of net ionic charge on Tc-99m-DTPA and Tc-99m-EDTA by a column ion-exchange method. J Nucl Med 21:354–360

Santolicandro A, Giuntini C (1979) Patterns of deposition of labeled monodispersed aerosols in obstructive lung disease. J Nucl Med All Sci 23:115

Taplin GV, Chopra SK (1978) Lung perfusion-inhalation scintigraphy in obstructive airway disease and pulmonary embolism. Radiol Clin N Am 16:491–513

Tofe AJ, Bevan JA, Fawzi MB, Francis MD, Silberstein EB, Alexander GA, Gunderson DE, Blair K (1980) Gentisic acid: a new stabilizer for low tin skeletal imaging agents. Concise communication. J Nucl Med 21:366–370

United States Pharmacopeial Convention (2005) Official Monographs: USB 28, technetium Tc-99m pentetate injection. United States Pharmacopeia (USP) 28–national formulary (NF) 23, p 1860

Van Duzee BF, Bugaj JE (1981) The effect of total technetium concentration on the performance of a skeletal imaging agent. Clin Nucl Med 6(Suppl):148

Wagner HN Jr (1995) Regional ventilation and perfusion. In: Wagner HN Jr, Szabo S, Buchanan JW (eds) Principles of nuclear medicine, 2nd edn. Saunders, Philadelphia, pp 887–895
Wanner A (1977) Clinical aspects of mucociliary transport. Amer Rev Respir Dis 116:73–125
Wassner SJ (1981) Assessment of glomerular filtration rate using single injection of technetium Tc-99m pentetate. Am J Dis Child 135:374–375

12.8.3 ^{99m}Tc-EC (Ethylene Dicysteine)

J. Környei

Chemical name	**Chemical structure**
N,N-Ethylene-L,L-dicysteine (L,L-EC) Tc(V)oxo-*N,N*-ethylene dicysteine Technetium-99m ethylenedicysteine ^{99m}Tc(V)-L,L-ethylenedicysteine (EC) ^{99m}Tc-L,L-EC complex	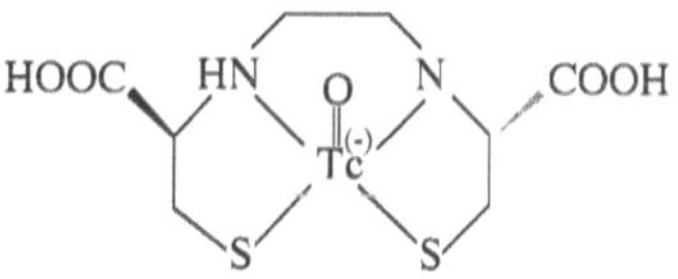 Tc(V)oxo-L,L-EC

Kit components		**Kit components**	
Vial A:		Vial B:	
N,N,-Ethylene-L,L-dicysteine	2.0 mg	Stannous chloride dihydrate	0.2 mg
Ascorbic acid	0.5 mg	Tartaric acid	48 mg
Sodium-ethylenediamine tetraacetic acid (EDTA)	0.35 mg	Ascorbic acid	12 mg
Disodium hydrogen phosphate	13.3 mg	**Commercial products**	
Mannitol	30 mg	EC in vivo kit	Institute of Isotopes (Hu)
Vial C:		(Kit is registered in Hungary)	
Stannous chloride dihydrate	0.2 mg		
Potassium dihydrogen phosphate	18.2 mg		
Ascorbic acid	12 mg		

Preparation

One kit consists of three vials, A, B, and C, containing the lyophilized ingredients. Labeling is performed under aseptic conditions as follows.

Procedure:

- Inject into vial A sufficient ^{99m}Tc activity (0.8–1.6 GBq; resp., 22-43 mCi) in a volume of 2.0 ml.
- Dissolve the content of vial B in 2.0 ml saline. After dissolution, withdraw 0.5 ml and inject into vial A. The resulting pH should not be below 12.0.

- Let react for 15 min at room temperature.
- Dissolve the content of vial C in 1.0 ml saline and transfer this solution to vial A.

^{99m}Tc-EC complex is a sterile, pyrogen-free, clear, colorless solution suitable for intravenous injection. The pH value is 5.5–8.0. ^{99m}Tc-EC injection solution should be used within 3 h after labeling.

Description of the Kit

Vial A contains 2 mg of lyophilized ethylene-L,L-dicysteine (L,L-EC) and several reagents to facilitate labeling with ^{99m}Tc-pertechnetate at alkaline pH. The eluate should be fresh, preferably obtained from a ^{99}Mo/^{99m}Tc generator eluted regularly, every 24 h. If necessary, the appropriate radioactivity concentration is adjusted with 0.9% NaCl solution for injection. Labeling is initiated by the addition of 0.5 ml of reducing agent containing 50 µg of stannous chloride from vial B. After 15 min incubation at room temperature, 1 ml of the buffer solution containing ascorbic acid as a stabilizer is added from vial C. Vial A contains 3.5 ml of ^{99m}Tc-EC injection solution suitable for not less than 3 patient doses.

The original kit composition reported by Verbruggen et al. (1992) consisted of the lyophilized residue of a solution of 0.5 mg of L,L-EC and 0.1 mg of stannous chloride dihydrate in 1 ml of phosphate buffer (0.05 M, pH 12.0). Labeling was performed by adding 1.850 MBq ^{99m}Tc-pertechnetate in a volume of 5 ml to the reaction vial. After incubation for at least 1 min at room temperature, the pH was adjusted with 0.2 ml of 0.5 *M* phosphate buffer (pH 5). The final pH of ^{99m}Tc-EC injection solution was 7.0–8.0; the complex was stable for 8 h. The advantage of this formulation is high specific activity labeling, keeping the amount of injected ligand below 15 µg (Stoffel et al. 1994).

At alkaline pH, an N_2S_2 complex is formed with the EC-ligand and pentavalent technetium, having a TcO core and two carboxylate functions. After neutralization, ^{99m}Tc-EC is a negatively charged complex with the chemical formula $[TcO\ EC]^{-1}$.

Clinical Applications

- Dynamic studies providing information on renal function (camera renography)
- Determination of the tubular extraction rate (TER)
- Examination of renal function in patients with transplanted kidneys

Since the introduction of ^{99m}Tc-L,L-EC as a tubular radiotracer (Verbruggen et al. 1992), clinical verification of the pharmacokinetic parameters has been obtained in normal controls (Van Nerom et al. 1993) and in patients (Gupta et al. 1995; Kabasakal et al. 1995), using standard procedures and also by comparison with the reference compounds *ortho*-iodohippurate (OIH) labeled with iodine-125 and with ^{99m}Tc-mercaptoacetyltriglycine (MAG_3) (Kabasakal et al. 1995; Özker et al. 1994; Stoffel et al. 1996). ^{99m}Tc-EC complex showed high similarity with OIH with respect to plasma clearance and elimination half-times (Özker et al. 1994; Van Nerom et al. 1993). ^{99m}Tc-EC was also evaluated for the diagnosis of renal artery stenosis in patients with suspected renovascular hypertension (Kibar et al. 1997).

Time of Examination

Dynamic renal studies: Collection of frames should start immediately after the intravenous injection of ^{99m}Tc-EC. The total time for one examination is approximately 20 min.

Recommended Activities for Indications

Dynamic renal studies: 90–120 MBq (2.4–3.2 mCi), injected intravenously
0.3 0.7 mg of ^{99m}Tc-EC complex
The manufacturer's instructions should be followed.

Pediatric Dose. The amount of radioactivity for infants and children is based on either body weight (static imaging) or body surface area (dynamic examinations). Recommendations by the Pediatric Task Group of the European Association of Nuclear Medicine (EANM) based on body weight should be followed (see Appendix 1, Table A1.2).

Additional Information

The patient should be adequately hydrated before ^{99m}Tc-EC renography.

Certain medication may affect renal clearance even in patients with normal kidney function.

Slower washout is observed with angiotensin-converting enzymes (ACE) inhibitors.

Tubular secretion is affected by the administered amount of ^{99m}Tc-EC. The single dose should not exceed 0.7 mg of EC (use kit for three patients), since secretion may be delayed; in cases of impaired renal function, a saturation effect may be observed. Higher doses (entire kit content) should be avoided since the diagnostic value of the renal examination may be biased.

^{99m}Tc-EC complex does not contribute to hepatobiliary activity; thus, ^{99m}Tc-EC is especially suited for the examination of renal function in patients with kidney transplants.

Quality Control

Radiochemical Purity. The radiochemical purity of the ^{99m}Tc-EC complex is measured by thin-layer chromatography (TLC), using one solvent system to separate ^{99m}Tc-pertechnetate (SF) and reduced ^{99m}Tc activity (start) from the ^{99m}Tc-EC complex (a more rapid analysis using two solvent systems has been described by Stoffel et al. 1994). The radiochemical purity of the ^{99m}Tc-EC complex should not be less than 95%.

Method Recommended by the Manufacturer

Thin-layer chromatography	
Stationary phase:	Gelman ITLC SG-60 plates 1.5×20 cm
Solvent:	96% Ethanol
Developing time:	2.5 h (A more rapid analysis using 2 solvent systems has been described by Stoffel et al. [1994].)
R_f values:	^{99m}Tc reduced, hydrolized: 0.0–0.1 ^{99m}Tc-EC complex: 0.45–0.55 (>95%) ^{99m}Tc-Na-pertechnetate: 0.9–1.0

The limit of radiochemical impurities at expiry (3 h) is <10%.

Results of analysis *(12 samples)*

Labeling and stability	15 min (%)	3 h (%)
^{99m}Tc-EC complex	98.7±0.65	99.4±0.2
^{99m}Tc-Na-pertechnetate	<0.3	<0.2
^{99m}Tc-reduced, hydrolized	<1.0	<0.4

Pharmacokinetic Data

The elimination of the ^{99m}Tc-L,L-EC complex from the blood is rapid. At 2–3 min post-injection, no heart and liver activities (due to perfusion) are observed. This is in agreement with the protein binding data of ^{99m}Tc-EC (31%) (Van Nerom et al. 1993), which is considerably less than of ^{123}I-OIH (74.1±3.9%) and approximately 30% of ^{99m}Tc-MAG$_3$ (91.3±1.9%) (Bubeck et al. 1990).

^{99m}Tc-EC is excreted by tubular secretion in the kidneys; the TER is 69–75% of the reference *para*-amino-hippurate (PAH). The maximum of radioactivity in the kidneys is observed at 4.4±0.3 min after intravenous injection. The washout of ^{99m}Tc-EC is rapid but slightly slower than of OIH, indicated by the biological half-times of ^{99m}Tc-EC 6.3–7.6 min and OIH 4.7–6.3 min (Özker et al. 1994).

A comparison of ^{99m}Tc-L,L-EC with ^{125}I-OIH in normal volunteers and in a wide range of renal function values has produced data of the mean plasma clearance with a close correlation between ^{99m}Tc-L,L-EC and ^{125}I-OIH ($r=0.99$, $p<0.001$). The mean clearance of ^{99m}Tc-L,L-EC amounted to 70.6±6.2% of the ^{125}I-OIH clearance with a range of 34.7–103.8% (Stoffel et al. 1994). The total volume of distribution of both tracers was not significantly different, corresponding to approximately 20% of the body weight.

When ^{99m}Tc-EC was compared with ^{99m}Tc-MAG$_3$, it showed a higher plasma clearance and considerably lower activity in the liver; there was no significant activity in bowel and gallbladder even in patients with impaired kidney function (Verbruggen et al. 1992).

Radiation Dose

Kidneys, bladder wall, and adrenals are most exposed organs. Calculations of the effective dose are based on dose equivalents for technetium-MAG_3 (International Commission on Radiological Protection 1991). Depending on the functional state of the kidneys the effective dose (mSv/MBq) is given as:

Normal function: 0.0073
Abnormal function: 0.0063
Acute unilateral: 0.01

Accordingly, the effective (whole body) dose resulting from 111 MBq of intravenously injected ^{99m}Tc-EC complex is 0.81 mSv (normal function), 0.70 mSv (abnormal function), and 1.11 mSv (acute unilateral obstruction).

Storage and Stability

Storage. The lyophilized kit is stored at 2–8 °C in the refrigerator. ^{99m}Tc-EC injection solution is kept at room temperature with adequate shielding.

Stability. The ^{99m}Tc-EC injection solution is stable for 3 or 8 h after preparation, depending on the kit formulation.

References

Bubeck B, Brandau W, Weber E, Kaelble T, Parekh N, Georgi P (1990) Pharmacokinetics of technetium-99m-MAG_3 in humans. J Nucl Med 31:1285–1293

Gupta NK, Bomanji JB, Waddington W, Lui D, Costa DC, Verbruggen AM, Ell PJ (1995) Technetium-99m-L,L-ethylenedicysteine scintigraphy in patients with renal disorders. Eur J Nucl Med 22:617–624

International Commission on Radiological Protection (1991) Technetium-labeled mercaptoacetyltriglycine (MAG_3) In: Annals of the ICRP, radiological protection in biomedical research. ICRP publication 62, vol. 22, no. 3. Pergamon, Oxford, pp 15–19

Kabasakal L, Atay S, Vural VA, Özker K, Sönmezoglu K, Demir M, Uslu I, Isitman AT, Önsel C (1995) Evaluation of technetium-99m-ethylenedicysteine in renal disorders and determination of extraction ratio. J Nucl Med 36:1398–1403

Kabasakal L, Turoglu HT, Önsel C, Özker K, Uslu I, Atay S, Cansiz T, Sönmezoglu K, Altiok E, Isitman AT, Kapicioglu T, Urgancioglu I (1995) Clinical comparsion of technetium-99m-L,L-EC, technetium-99m-MAG3 and iodine-125-OIH in renal disorders. J Nucl Med 36:224–228

Kibar M, Tutus A, Paydas S, Reyhan M (1997) Captopril-enhanced technetium-99m-L,L-ethylenedicysteine renal scintigraphy in patients with suspected renovascular hypertension: comparative study with Tc-99m-MAG_3. J Nucl Med 6:132–137

Özker K, Önsel C, Kabasakal L, Sayman HB, Uslu I, Bozluolcay S, Cansiz T, Kapicioglu T, Urgancioglu I (1994) Technetium-99m L,L-ethylenedicysteine: a comparative study of renal scintigraphy with Tc-99m-MAG-3 and I-131-OIH in patients with obstructive renal disease. J Nucl Med 35:840–845

Stoffel M, Jamar F, Van Nerom C, Verbruggen AM, Mourad M, Leners N, Squifflet JP, Beckers C (1994) Evaluation of technetium-99m-L,L-EC in renal transplant recipients: comparative study with technetium-99m-MAG_3 and iodine-125-OIH. J Nucl Med 35:1951–1958

Stoffel M, Jamar F, Van Nerom C, Verbruggen AM, Besse T, Squifflet JP, Beckers C (1996) Technetium-99m-L,L-ethylenedicysteine clearance and correlation with iodine-125 *ortho*-iodohippurate for determination of effective renal plasma flow. Eur J Nucl Med 23:365–370

Van Nerom CG, Bormans GM, De Roo MJ, Verbruggen AM (1993) First experience in healthy volunteers with technetium-99m L,L-ethylenedicysteine, a new renal imaging agent. Eur. J Nucl Med 20:738–746

Verbruggen AM, Nosco DL, Van Nerom CG, Bormans GM, Adriaens PJ, De Roo MJ (1992) Technetium-99m-L,L-ethylenedicysteine: a renal imaging agent. I. Labeling and evaluation in animals. J Nucl Med 33:551–557

12.8.4 ^{99m}Tc-MAG$_3$ (Mercaptoacetyltriglycine)

F. Rakiás and I. Zolle

Chemical name	**Chemical structure**
Benzoylmercapto-acetyltriglycine: Betiatide (*Ph. Eur.; USP*) Tc(V)oxo-*N,N,N*-mercaptoacetyltriglycine: Mertiatide (*Ph. Eur.; USP*) Technetium Tc 99m mertiatide injection (*Ph. Eur.; USP*) ^{99m}Tc(V)O-mercaptoacetyltriglycine complex ^{99m}Tc-MAG$_3$ complex	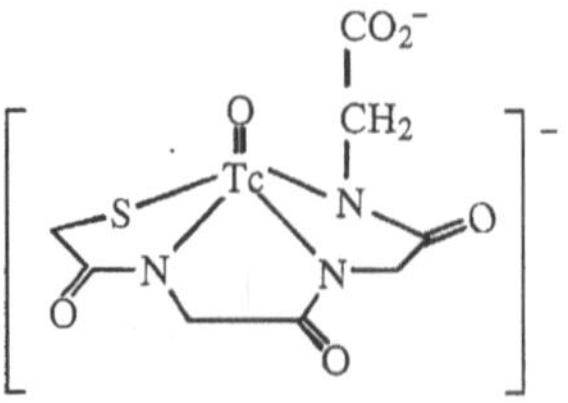 ^{99m}Tc(V)-MAG$_3$ complex
Kit components	**Commercial products**
Betiatide 1.0 mg Stannous chloride dehydrate 0.04 mg Disodium tartrate 16.9 mg	TechneScan MAG$_3$ Mallinckrodt/Tyco

Preparation

The kit contains the lyophilized, sterile, pyrogen-free, inactive ingredients in a nitrogen atmosphere, ready for labeling with ^{99m}Tc-sodium pertecnetate (*European Pharmacopeia* [*Ph. Eur.*]). Labeling is performed according to the instructions given by the manufacturer (Mallinckrodt 1992):

- Method 1: Use 3 ml of fresh eluate; the activity should not exceed 1.11 GBq (30 mCi). The calculated amount of ^{99m}Tc activity is diluted to a volume of 10 ml with saline. Add this volume to a TechneScan mercaptoacetyltriglycine (MAG$_3$) vial, using a thin hypodermic needle (20 gauge or higher). Place the kit in upright position into a boiling water bath for 10 min. After heating, cool the vial in cold water. ^{99m}Tc-MAG$_3$ injection solution (10 ml) is stable for 4 h.

- Method 2: To obtain higher activity concentrations, use 1 ml of fresh eluate for labeling and dilute to 4 ml with saline. The activity should not exceed 925 MBq (25 mCi). Follow the procedure as described above. This preparation (4 ml) is stable for 1 h.

^{99m}Tc-MAG$_3$ is a clear or slightly opalescent, colorless, sterile solution for intravenous injection. The resulting pH should be 5.0–7.5. ^{99m}Tc-MAG$_3$ injection solution may be used for one or multiple administrations.

Description of the Kit

^{99m}Tc-mertiatide is obtained upon heating in a boiling water bath with a labeling yield of approximately 96%. One molecule of *S*-benzoyl-protected betiatide reacts with reduced ^{99m}Tc in the +5 oxidation state to form a negatively charged ^{99m}Tc-MAG$_3$ complex. Initially, at pH 5.5, a labile ^{99m}Tc-tartrate complex is formed, which is readily exchanged by the benzoylmercaptoacetyl triglycine ligand during heating at 95 °C (ligand exchange). Heating in a boiling water bath is required, because at room temperature formation of the ^{99m}Tc-MAG$_3$ complex is slow (52% in 2 h) (Fritzberg et al. 1986). The heating process also increases the rate of hydrolysis of the protecting benzoyl group and the conversion of a bis ligand complex, which may also be formed, to single-ligand ^{99m}Tc-MAG$_3$.

The core ligand system N_3S and an additional carboxylate group are responsible for high tubular specificity and a high excretion rate. In addition, the triamide monomercaptide ligand does not form stereoisomers that required high-performance liquid chromatography (HPLC) purification (Fritzberg et al. 1986). ^{99m}Tc-MAG$_3$ is a tetradentate chelate of oxotechnetium as described for the $TcON_2S_2$ *N,N*-bis(mercaptoacetyl) ethylenediamine (DADS) ligands (Davison et al. 1981).

The effect of stereoisomers and of byproducts on kit preparation has been investigated extensively (Brandau et al. 1988 a, b; Bubeck et al. 1986; Coveney and Robbins 1987; Fritzberg et al. 1982).

Labeling should be performed with eluates with the highest possible radioactive concentration. In order to obtain a high yield and to avoid labeled impurities, the ^{99m}Tc eluate used for labeling must be obtained from a ^{99m}Tc generator that is eluted daily at a 24-h interval (Van Duzee and Bugaj 1981). Moreover, only eluates from a ^{99m}Tc generator that has been in use for no longer than 5 days are suitable. Dilutions should be performed with saline.

Clinical Applications

Intravenous injection: Renal imaging/renography

- To obtain anatomical and functional information
- To demonstrate adequate renal perfusion
- To evaluate renal tubular function
- To determine the tubular extraction rate (TER)
- As a control after surgical intervention (kidney transplants)
- To evaluate renal artery stenosis and obstructive uropathy
- To diagnose urinary obstruction in infants

The similarity of ^{99m}Tc-MAG$_3$ with ^{131}I-OIH (*o*-iodohippurate) has been demonstrated in experimental animals (Fritzberg et al. 1986) and in volunteers (Taylor et al. 1986).

Since glomerular filtration accounts for less than 2% of the total clearance, ^{99m}Tc-MAG$_3$ has been recommended for determination of the TER (Bubeck et al. 1987, 1990).

^{99m}Tc-MAG$_3$ is used for the evaluation of nephrological and urological disorders, in particular for the study of renal perfusion, relative kidney function, and characterization of urinary flow. Renovascular hypertension is diagnosed after pharmacological intervention (captopril test) (Kletter 1988).

Time of Examination

Renal dynamic studies:	Collection of frames should start immediately after the intravenous injection of. The total time for one examination is approximately 20 min.

Recommended Activities for Indications

Renal clearance (plasma clearance):	10–20 MBq (0.27–0.54 mCi)
Renography:	60–80 MBq (1.6–2.2 mCi)
Functional scintigraphy:	150–200 MBq (4–5 mCi), 0.3–1 mg of ^{99m}Tc-MAG$_3$ 100 MBq (2.7 mCi) maximum recommended activity (Administration of Radioactive Substances Advisory Committee [ARSAC])

Depending on the parameters to be studied and the method to be used, 37–185 MBq (1–5 mCi) are used in adults (70 kg). Studies of renal blood flow or transport through the ureters generally require a larger dose than do studies of intrarenal transport; whereas renography requires smaller activities than dynamic sequential scintigraphy.

Pediatric Dose. The amount of radioactivity for infants and children is based on either body weight (static imaging) or body surface area (dynamic examinations). Recommendations by the Pediatric Task Group of the European Association of Nuclear Medicine (EANM) based on body weight should be followed (see Appendix 1, Table A1.2).

Additional Information

The patient should be adequately hydrated before injecting ^{99m}Tc-MAG$_3$. Insufficient hydration of the patient will affect the renal excretion rate. In order to reduce the radiation exposure to the bladder wall, the patient should be asked to drink water during the following hours.

Small amounts of ^{99m}Tc-labeled impurities may be formed during the labeling process, which accumulate in the liver and contribute to hepatobiliary activity. Dynamic renal studies may be affected in the late phase (after 30 min) due to an overlap of activity in the region of interest.

ACE inhibitors such as captopril may facilitate the differential diagnosis of renovascular hypertension; diuretics such as furosemide (Lasix) cause rapid washout of the radiotracer or demonstrate urinary tract obstruction (Kletter 1988).

Quality Control

Radiochemical Purity. The *Ph. Eur.* (Council of Europe 2005) requires paper chromatography using acetonitrile/water (60:40, v/v) to determine reduced, hydrolized technetium at the start, which should not exceed 2% of the measured radioactivity. For identification of ^{99m}Tc-mertiatide and ^{99m}Tc-Na-pertechnetate, high performance liquid chromatography (HPLC) is recommended. The radiochemical purity of the ^{99m}Tc-MAG_3 complex should not be less than 94%.

HPLC method

Gradient elution for the simultaneous determination of free ^{99m}Tc-pertechnetate and ^{99m}Tc-mertiatide.

- The column is a Merck octadecylsilyl-silica (5 µm), 25 cm×4 mm
- Flow rate is 1.0 ml/min
- Mobile phase A is a 19:1 mixture of phosphate solution (1000 parts 0.01 M NaH_2PO_4 and 114 parts 0.01 M Na_2HPO_4, adjusted to pH 6.0) and ethanol.
- Mobile phase B is a 1:9 mixture of water and methanol.
- Inject 20 µl of the test solution into the chromatograph and record the chromatograms by gradient elution. Use a gradient elution program.
- Retention times: ^{99m}Tc-pertechnetate, 1.8–2.2 min; ^{99m}Tc-mertiatide, 10–14 min

*Requirements (*Ph. Eur.*).* The sum of activities eluted before the major peak activity (hydrophilic impurities including ^{99m}Tc-pertechnetate) should not exceed 3% of the sum of all peaks, and the sum of activities eluted after the major peak activity (lipophilic impurities) should not be more than 4% of the sum of all peaks. The radiochemical purity of the ^{99m}Tc-MAG_3 complex should not be less than 94%.

Results of HPLC analysis of the ^{99m}Tc-MAG_3 complex, performed at different times after preparation:

Eluted peak activity	TOC (%)	4 h (%)
Mertiatide	≥96.0	≥95.0
Total front fractions	≤3.0	≤3.0
Methanol fraction	≤4.0	≤4.0

Methods Recommended by the Manufacturer. Thin-layer chromatography (TLC) on silica gel reversed phase plates (migration distance 10–15 cm) and acidified methanol-saline is used as solvent for the analysis of three components, namely the ^{99m}Tc-MAG_3 complex (R_f=0.4–0.5); reduced, hydrolized technetium at the start; and unbound ^{99m}Tc-Na-pertechnetate at the solvent front. The radiochemical purity of the ^{99m}Tc-MAG_3 complex should not be less than 90%.

Thin-layer chromatography		
Stationary phase:	Merck RP-18 F_{254} plates, 1.5×20 cm	
Solvent:	Methanol/saline/acetic acid (45:54:1, v/v/v)	
Developing time:	1.5 h	
R_f values:	^{99m}Tc-reduced, hydrolized:	0.0–0.1
	^{99m}Tc-mertiatide:	0.1 0.5 (> 90%)
	^{99m}Tc-Na-pertechnetate:	0.9–1.0

Quantification of labeled components. Each chromatogram is measured (TLC linear analyzer, or gamma counter) and the regional radioactivities are expressed as a percentage of the total recovered counts.

Results of analysis *(12 samples):*

Labeling and stability	15 min (%)	3 h (%)
^{99m}Tc-MAG_3 complex	98.3 ± 0.30	96.2 ± 0.34
^{99m}Tc-Na-pertechnetate	1.0 ± 0.45	1.0 ± 0.39
^{99m}Tc reduced, hydrolized	0.6 ± 0.27	2.7 ± 0.54

Column chromatography. This method is based on the extraction by Sep-Pak C18 cartridges (Waters, light) and elution with 50% ethanol/saline.

Procedure:

1. Rinse the cartridge first with 5 ml ethanol, and then with 5 ml 0.001 N hydrochloric acid.
2. Mix 5 ml ethanol and 5 ml saline.
3. Apply 0.1–0.3 ml of the ^{99m}Tc-MAG_3 injection solution onto the column.
4. Rinse the column with 5 ml 0.001 N HCl into vial A.
5. Elute cartridge with 5 ml of ethanol/saline into vial B.

Vials A and B and the eluted cartridge are measured for ^{99m}Tc activity in a dose calibrator.

Use the sum of eluted radioactivities as 100%:

$$^{99m}\text{Tc-MAG}_3 \text{ complex } (\%) = \frac{\text{B} \times 100}{\text{Sum of A} + \text{B} + \text{SEP-PAK}}$$

The radiochemical purity of ^{99m}Tc-mertiatide should not be less than 90%.

Pharmacokinetic Data

After intravenous injection, ^{99m}Tc-mertiatide is rapidly distributed in the extracellular fluid and excreted entirely by the renal system. The maximum renal accumulation of radioactivity is observed at 3–4 min (207 ± 80 s) after intravenous injection (Taylor et al. 1986).

The elimination from plasma is described by two half-times, namely, 3.2 and 16.9 min. The mean parenchymal transit time is approximately 4 min (100–270 s) (Jafri et al. 1988).

At 3 h postinjection the activity in blood is less than 1% of the injected dose. Urinary excretion of ^{99m}Tc-MAG_3 is 70% in 30 min (normal renal function), and 99.9 ± 4.3% at three h postinjection. ^{99m}Tc-MAG_3 is not metabolized and is excreted unchanged (Taylor et al. 1986).

The mechanism of excretion is predominantly based on renal tubular secretion. The TER of ^{99m}Tc-MAG_3 is 0.55 of the reference *para*-amino-hippurate ([PAH] 1.0); in comparison, *ortho*-iodohippurate (OIH) shows a value of 0.83 (Bubeck et al. 1987). Using this coefficient (0.55), ^{99m}Tc-MAG_3 may substitute for OIH as an indicator of effective renal plasma flow (ERPF) (Bubeck et al. 1990).

The renal clearance is dependent on the functional state of the kidneys and the urogenital system; however, there is a good correlation between the clearance of ^{99m}Tc-MAG$_3$ and ^{131}I-OIH in patients with seriously impaired renal function or transplant kidneys as well as in patients with normal tubular function (Bubeck et al. 1988c). Simultaneously obtained clearance data showed striking differences between plasma clearance and plasma concentration. The plasma clearance of ^{99m}Tc-MAG$_3$ was about one half that of ^{131}I-OIH, but its plasma concentration was much higher in concordance with the high protein binding; the net effect being a similar urinary excretion and qualitatively similar renogram curves (Russel et al. 1988). Using a two-compartment model, Taylor et al. (1986) reported a ^{99m}Tc-MAG$_3$/OIH plasma-clearance ratio of 0.69 during the first 30 min postinjection, and Bubeck et al. (1990) at steady state a ratio of 0.67.

^{99m}Tc-mertiatide shows high binding to plasma proteins ($91.3 \pm 1.9\%$). Glomerular filtration accounts for less than 2% of the total clearance (Bubeck et al. 1987, 1990).

Radiation Dose

The kidneys, bladder wall, and adrenals are most exposed organs. MAG$_3$ is rapidly distributed in the extracellular fluid and excreted entirely by the renal system. The renal transit time is approximately 4 min, as is for OIH (Bubeck et al. 1987).

Calculations of the effective dose are based on dose equivalents for technetium-MAG$_3$ (International Commission on Radiological Protection 1991). Depending on the functional state of the kidneys the effective dose (mSv/MBq) is given as:

- Normal function: 0.0073
- Abnormal function: 0.0063
- Acute unilateral blockage: 0.01

Accordingly, the effective (whole body) dose resulting from 100 MBq (2.7 mCi) of intravenously injected ^{99m}Tc-MAG$_3$ complex is 0.73 mSv (normal function), 0.63 mSv (abnormal function), and 1.0 mSv (acute unilateral obstruction).

The doses resulting from 100 MBq (2.7 mCi) to exposed organs (mGy) are:

- Bladder 11.0, kidneys 0.34, uterus 1.2 (normal function)
- Bladder 8.3, kidneys 1.4, uterus 1.0 (abnormal function)
- Adrenals 1.1, bladder 5.6, kidneys 20.0 (acute, unilateral obstruction)

When renal function is bilaterally impaired, it is assumed that the clearance rate of MAG$_3$ is one tenth of that for the normal case, the renal transit time is increased to 20 min, and a fraction of 0.04 is taken up in the liver.

In the case of acute unilateral renal blockage, it is assumed that a fraction of 0.5 of administered MAG$_3$ is taken up by one kidney and slowly released to the blood with a half-time of 5 days, and subsequently excreted by the other kidney, which is assumed to function normally (International Commission on Radiological Protection 1991).

Unilateral renal blockage causes a shift in the excretion pattern, with the highest radiation absorbed dose delivered to the kidneys, namely, 0.20 mGy/MBq.

Consequently, the effective (whole body) dose value is 0.01 mSv/MBq, and the effective dose in adults (70 kg) resulting from intravenous injection of 150 MBq (4 mCi) of ^{99m}Tc MAG$_3$ complex is 1.5 mSv.

Storage and Stability

Storage. The TecneScan MAG_3 kit is stored at 2–8 °C, protected from light until use. ^{99m}Tc-MAG_3 injection solution is kept at room temperature with adequate shielding.

Stability. ^{99m}Tc-MAG_3 injection is stable for 1 or 4 h after labeling, depending on the volume of preparation (4 or 10 ml). If the injection solution is used for multiple administrations, the vial should be kept in the refrigerator.

References

Bubeck B, Brandau W, Dreikorn K, Steinbächer M, Eisenhut M, Trojan H, zum Winkel K (1986) Clinical comparison of I-131 o-iodohippurate with Tc-99m-CO_2-DADS-A and Tc-99m-MAG_3 by simultaneous double tracer measurement. Nucl Compact 17:135–138

Bubeck B, Brandau W, Eisenhut M, Weidenhammer K, Georgi P (1987) The tubular extraction rate (TER) of Tc-99m-MAG_3: a new quantitative parameter of renal function. Nuc Compact 18:260–267

Brandau W, Bubeck B, Eisenhut M, Taylor DM (1988a) Technetium-99m labeled renal function and imaging agents: III. Synthesis of Tc-99m-MAG_3 and biodistribution of by-products. Appl Radiat Isot 39:121–129

Bubeck B, Brandau W, Reinbold F, Dreikorn K, Steinbächer M, Eisenhut M, Georgi P (1988b) Technetium-99m labeled renal function and imaging agents: I clinical evaluation of Tc-99m-CO_2-DADS-A (Tc-99m-*N,N*-bis-(mercapto-acetyl)-2,3-diamino-propanoate). Nucl Med Biol 15:99–108

Bubeck B, Brandau W, Steinbächer M, Reinbold F, Dreikorn K, Eisenhut M, Georgi P (1988c) Technetium-99m labeled renal function and imaging agents: II. Clinical evaluation of ^{99m}Tc-MAG_3 (^{99m}Tc-mercaptoacetylglycylglycylglycine). Nucl Med Biol 15:109–118

Bubeck B, Brandau W, Weber E, Kaelble T, Parekh N, Georgi P (1990) Pharmacokinetics of technetium-99m-MAG_3 in humans. J Nucl Med 31:1285–1293

Council of Europe (2005) Technetium ^{99m}Tc mertiatide injection. European pharmacopeia 5.0, monograph no 1372. Council of Europe, Maisonneuve, Sainte-Ruffine, pp 1220

Coveney JR, Robbins MS (1987) Comparison of technetium-99m MAG_3 kit with HPLC-purified technetium-99m MAG_3 and OIH in rats. J Nucl Med 28:1881–1887

Davison A, Jones A, Orvig C, Sohn M (1981) A new class of oxotechnetium (+5) chelate complexes containing a $TcON_2S_2$ core. Inorg Chem 20:1629–1632

Fritzberg AR, Kuni CC, Klingensmith WC III, Stevens J, Whitney WP (1982) Synthesis and biological evaluation of Tc-99m *N,N*-bis (mercaptoacetyl)-2,3-diaminopropanoate: a potential replacement for [^{131}I]-*o*-iodohippurate, J Nucl Med 23:592–598

Fritzberg AR, Kasina S, Eshima D, Johnson DL (1986) Synthesis and biological evaluation of technetium-99m-MAG_3 as a hippuran replacement. J Nucl Med 27:111–116

International Commission on Radiological Protection (1991) Technetium-labelled mercaptoacetyltriglycine (MAG_3) In: Annals of the ICRP, radiological protection in biomedical research. ICRP publication 62, vol 22, no 3. Pergamon, Oxford, pp 15–19

Jafri RA, Britton KE, Nimmon CC, Solanki K, Al-Nahas A, Bomanji J, Fettich J, Hawkins LA (1988) Technetium-99m MAG_3, a comparison with iodine-123 and iodine-131 *ortho*-iodohippurate, in patients with renal disorders. J Nucl Med 29:147–158

Kletter K (1988) Neue Aspekte der nuklearmedizinischen Nierenfunktionsdiagnostik: Verbesserte Aussagekraft durch pharmakologische Intervention und quantitative Analyseverfahren [in German]. Wien Klin Wochenschr 100 (Suppl 177)

Mallinckrodt (1992) TechneScan MAG_3 product information. Malinckrodt, Phillipsburg, NJ

Russel CD, Thorstad B, Yester MV, Stutzman M, Baker T, Dubovsky EV (1988) Comparison of technetium-99m MAG_3 with iodine-131 hippuran by a simultaneous dual channel technique. J Nucl Med 29:1189–1193

Taylor A, Eshima D, Fritzberg AR, Christian PE, Kasina S (1986) Comparison of iodine-131 OIH and technetium-99m MAG_3 renal imaging in volunteers. J Nucl Med 27:795–803

United States Pharmacopeial Convention (2005) Official Monographs: USP 28, technetium Tc-99m mertiatide injection. United States Pharmacopeia (USP) 28–national formulary (NF) 23, p 1858

Van Duzee BF, Bugaj JE (1981) The effect of total technetium concentration on the performance of a skeletal imaging agent. Clin Nucl Med 6(Suppl):P148

12.9 ^{99m}Tc-Labeled Hepatobiliary Agents

12.9.1 ^{99m}Tc-IDA (Iminodiacetic Acid) Derivatives

I. Zolle and A.G. Bratouss

Accepted Chemical Names

N-(2,6-diethylphenylcarbamoylmethyl)-iminodiacetic acid (Etifenin, Ph. Eur.)

N-(2,6-dimethylphenylcarbamoylmethyl)-iminodiacetic acid (Lidofenin, USP)

N-(2,4,6-trimehtyl-3-bromophenylcarbamoylmethyl)-iminodiacetic acid (Mebrofenin)

N-(2,6-diisopropylphenylcarbamoylmethyl)-iminiodiacetic acid (Disofenin)

Kit Components*

Etifenin	20.0 mg
Mebrofenin	40.0 mg
Disofenin	20.0 mg
Tin(II)-chloride-dihydrate	0.2–0.6 mg

Listed Trade Names

Etifenin – EHIDA (Rotop)
as disodium salt – Hepatobida

Mebrofenin – TCK-22 (CIS Bio)
CholeCIS
Choletec (Bracco)
Bridatec (Sorin)

Disofenin – DISIDA
Hepatolite

Chemical Structures

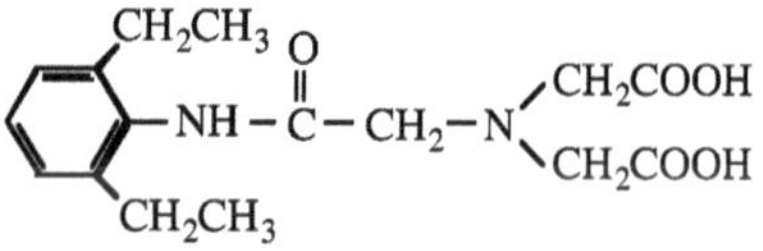

Etifenin

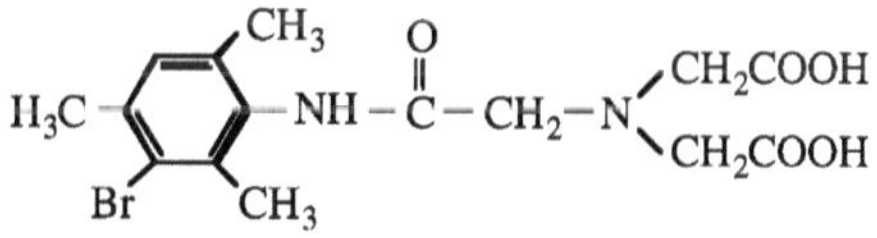

Mebrofenin

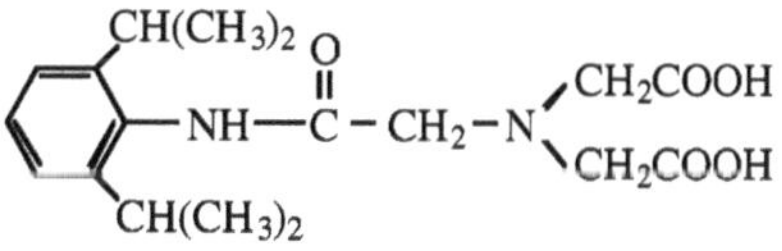

Disofenin

* *The composition of the kit components may vary with new manufacturers*

Preparation

The kit contains the lyophilized, sterile ingredients in a multidose vial. Labeling with ^{99m}Tc eluate is carried out under aseptic conditions by adding 1–5 ml of sterile ^{99m}Tc-sodium pertechnetate to the vial (0.3–1.5 GBq resp., 8.1–40.5 mCi). The reaction is allowed to proceed at room temperature for 15 min. ^{99m}Tc-iminodiacetic acid (IDA) preparations are sterile, pyrogen-free, clear, colorless solutions suitable for intravenous injection. The resulting pH should be 4.0–6.0 (TCK-22), 4.2–5.7 (Choletec), and 6.5–7.5 (Bridatec).

Description of the Kit. Etifenin is officinal in the *European Pharmacopeia* (*Ph. Eur.*) Diethyl-IDA or EHIDA is a registered radiopharmaceutical (kit) in Europe. Three IDA derivatives (lidofenin, disofenin, and mebrofenin) have been included in the *USP* since the first hepatobiliary agent dimethyl-IDA, with the generic name lidofenin, was introduced (Loberg et al. 1976).

^{99m}Tc-IDA complexes are formed easily with reduced technetium at room temperature. The size of the substituents attached to the phenyl ring does affect the labeling yield, radiochemical purity, and stability (Loberg et al. 1976; Ryan et al. 1977). Labeling is also affected by pH and the ligand concentration, showing higher labeling at low pH (Nunn and Schramm 1981). The amount of tin has no affect on the labeling yield; however, high labeling is depending on maintaining tin in the reduced state. Isotopic dilution is observed with higher concentrations of ^{99}Tc in the first eluate of new generators, reducing the labeling efficiency (Ponto et al. 1987). Increasing the reaction time to 30 min will ensure maximum labeling.

Clinical Applications

Intravenous injection: Hepatobiliary scintigraphy

- To evaluate hepatocyte function
- To demonstrate patency or obstruction of cystic duct
- To rule out acute cholecystitis
- To demonstrate common bile duct obstruction
- As a control after surgical intervention
- To verify hepatic bile duct atresia in infants

Time of Examination. Hepatobiliary scintigraphy is started 5 min after the intravenous injection. Thereafter, scintigrams are taken at 10-min intervals up to 60 min. If the gallbladder is not visualized within 60 min, delayed images up to 4 h after administration of the radiotracer are obtained.

Recommended Activities for Indications

Hepatobiliary scintigraphy: 75–185 MBq, injected intravenously
150 MBq maximum recommended activity
1–2.6 MBq/kg body weight (normal bilirubin level)
$<$ 0.5 mg etifenin/kg body weight

Pediatric Dose. The amount of radioactivity for infants and children administered for hepatobiliary scintigraphy is based on body weight, using the scaling factors given in Appendix 1, Table A1.2. The minimum value of the scaling factor is 0.1, since activities below 10% of the value of the adult radioactivity have resulted in poor image quality.

Additional Information

The patient should not eat 2–6 h prior to the hepatobiliary scintigraphy, because hepatocyte clearance of the radiotracer and parenchymal transit time is affected by the ingestion of food. The gallbladder cannot be visualized in 65% of cases within the first 60 min of injection of the ^{99m}Tc-IDA complex, even if the cystic duct is patent (Fink-Bennett 1995). Gallbladder contractility can be provoked with a fatty meal or intravenous cholecystokinin.

Persistent nonvisualization of the gallbladder during a 4-h period is suggestive of acute cholecystitis; whereas delayed visualization of the gallbladder after 1.5 h up to 4 h is more typical of chronic cholecystitis (Weissmann et al. 1979). To enhance gallbladder visualization, cholecystokinin or sincalide have been used to induce contraction and subsequent filling of the gallbladder. The usual intravenous dose of sincalide is 0.02 μg/kg, administered either 30 min before imaging is started or after the nonvisualization of the gallbladder within 1 h. The limitations of this medication in cholescintigraphic studies have been described (Freeman et al. 1981).

Narcotic (opioid) analgesics (morphine, meperidine) and phenobarbital cause a marked increase in biliary tract pressure. The increase of intraluminal common bile duct pressure by morphine (50–60%) has been used to differentiate between acute and chronic cholecystitis. Morphine-augmented cholescintigraphy has been considered superior to conventional hepatobiliary scintigraphy in assessing cystic duct patency because diagnostic results are obtained faster. The increase in common bile duct pressure is sufficient to overcome the resistance to bile flow caused by a sludge-filled gallbladder, so that the radiotracer may penetrate the cystic duct into the gallbladder. A dose of 0.04 mg/kg morphine sulfate diluted to 10 ml with saline is administered intravenously over 3 min, when the gallbladder has not been visualized within 60 min after the administration of the ^{99m}Tc-IDA complex. Serial images are then obtained at 5-min intervals for 30 min. In the case of persistent nonvisualization of the gallbladder after morphine administration, acute cholecystitis is diagnosed (Fink-Bennett 1995). If, however, the gallbladder is visualized 5–30 min after morphine administration, chronic cholecystitis is present. Morphine should not be given to patients with a history of drug abuse, an allergy to morphine, or with pancreatitis.

Phenobarbital enhances the biliary conjugation and excretion of bilirubin. It promotes the excretion of organic anions such as ^{99m}Tc-IDA complexes that are not conjugated by the liver. The hepatic extraction of ^{99m}Tc-IDA complexes and canalicular bile flow are increased; thus, phenobarbital is used in neonates for hepatobiliary imaging primarily to increase the diagnostic accuracy of differentiating between neonatal hepatitis and biliary atresia. The administered dose to increase the rate of excretion of a ^{99m}Tc-IDA complex is approximately 5 mg/kg per day for at least 5 days prior to the imaging procedure (Majd et al. 1981).

Nicotinic acid in high doses may affect the cholescintigraphy because of poor extraction and elimination of the radiotracer (toxic effect on hepatocytes). Total parenteral nutrition may cause delayed or no visualization of the gallbladder, even in patients with no gallbladder disease caused by bile stasis and the formation of thick viscous jelly-like bile. Nonvisualization of the gallbladder may also be caused by hepatic artery infusion during chemotherapy.

Quality Control

Radiochemical Purity. The *Ph. Eur.* (Council of Europe 2005) requires thin-layer chromatography (TLC) on silica gel (SG) fiberglass sheets (migration distance of 10–15 cm) for the analysis of the ^{99m}Tc-EHIDA complex, using 0.9% sodium chloride solution (saline) as solvent. Reduced, hydrolized technetium remains at the start; the ^{99m}Tc-etifenin complex is identified at an R_f of 0.4–0.5; unbound ^{99m}Tc-Na-pertechnetate is measured at the solvent front. The radiochemical purity of the ^{99m}Tc-EHIDA complex should not be less than 95% (*Ph. Eur.*).

Recommended Methods

Thin-layer chromatography. TLC with two different solvent systems is based on the separate identification of labeled impurities (Nunn et al. 1983).

- System I: In saline (20%) unbound ^{99m}Tc-Na-pertechnetate is identified at the solvent front (R_f=1). The sum of activities of reduced, hydrolized technetium and the ^{99m}Tc-IDA complex is measured at the start (R_f=0).
- System II: Using Gelman ITLC-SG and acetonitrile/water as solvent, reduced, hydrolized activity is identified separately at the start (R_f=0).

System I		
Stationary phase:	Gelman ITLC-SA, 1×9.5 cm	
Solvent:	Saturated NaCl solution	
Developing time:	5 min	
R_f values:	^{99m}Tc-IDA complex:	0.0–01
	^{99m}Tc reduced, hydrolized:	0.0–0.1
	^{99m}Tc-Na-pertechnetate:	0.9–1.0

System II		
Stationary phase:	Gelman ITLC-SG, 1×9.5 cm	
Solvent:	Acetonitrile-water (3:1)	
Developing time:	5 min	
R_f values:	**^{99m}Tc reduced, hydrolized:**	0.0–0.1
	^{99m}Tc-IDA complex:	0.9–1.0
	^{99m}Tc-Na-pertechnetate:	0.9–1.0

Quantification of labeled components. Each chromatogram is measured (TLC linear analyzer or gamma counter), and the regional radioactivities are expressed as a percentage of the total recovered counts. The percentage of free ^{99m}Tc Na-pertechnetate (F) and reduced, hydrolized activity (H) are determined and the pure ^{99m}Tc-IDA complex quantified according to the equation:

$$^{99m}\text{Tc-IDA complex (\%)} = 100 - \%(F + H)$$

Results of analysis *(12 samples)*

Results were obtained using the analytical method described by Nunn et al. (1983).

Labeling and stability	15 min (%)	3 h (%)
^{99m}Tc-IDA complex	96.5 ± 1.68	95.2 ± 0.71
^{99m}Tc-Na-pertechnetate	1.6 ± 1.51	2.9 ± 1.02
^{99m}Tc-reduced, hydrolized	1.9 ± 0.61	1.9 ± 0.59

Pharmacokinetic Data

Hepatobiliary excretion of ^{99m}Tc-IDA complexes is governed by molecular size (optimal molecular weight: 700–900 Da) and structural configuration. Two molecules of ligand are coordinated to one technetium atom in the +3 oxidation state. ^{99m}Tc-IDA complexes have a negative charge, high polarity, and show high binding to plasma protein (Loberg and Fields 1978).

Substituents attached to the phenyl ring affect the degree of hepatic uptake and the rate of excretion, as well as urinary elimination of the ^{99m}Tc-IDA complex (Nunn et al. 1983). Approximately 82% of the injected dose of ^{99m}Tc-EHIDA, 88% of ^{99m}Tc-disofenin, and 98% of ^{99m}Tc-mebrofenin are extracted by the hepatocytes and secreted into bile with a hepatic excretion half-time of 37.3, 19.0, and 17.0 min, respectively (Krishnamurthy and Krishnamurthy 1989; Krishnamurthy and Turner 1990). Clinical comparison of diethyl-IDA (etifenin) and diisopropyl-IDA (disofenin) was also reported (Klingensmith et al. 1981). ^{99m}Tc-DISIDA and ^{99m}Tc-mebrofenin show least hepatic retention and are best suited to delineate hepatic biliary anatomy (Krishnamurthy and Krishnamurthy 1989).

^{99m}Tc-IDA complexes are excreted into bile as the original bis-coordinated complex; they are not reabsorbed by epithelia of the biliary ducts nor metabolized, and released into the duodenum unchanged (Krishnamurthy and Krishnamurthy 1989; Loberg et al. 1976). Excretion across the bile canalicular membrane is a carrier-mediated process (Nielson and Rasmussen 1975).

Following intravenous injection, the ^{99m}Tc-IDA complex is bound to plasma protein (mainly albumin) and carried to the liver (Nicholson et al. 1980). Accumulation in the liver involves the same carrier-mediated, non-sodium-dependent organic anion transport processes as for the uptake of bilirubin. In the space of Disse, the albumin-^{99m}Tc-IDA conjugate is dissociated to facilitate active transport of the ^{99m}Tc-IDA complex into hepatocytes (Krishnamurthy and Krishnamurthy 1989). In patients with normal hepatobiliary function, maximal liver uptake is measured at 12 min (^{99m}Tc-mebrofenin, 10.9 ± 1.9 min; ^{99m}Tc-disofenin, 11.5 ± 3.1 min) (Fritzberg 1986). The radioactivity is half this value within approximately 20 min. The gallbladder is well visualized 20 min postinjection. Intestinal activity appears on the average at 15–30 min. The common bile duct may be visualized after 14 min. The upper limit of "normal" for visualization of these structures is 1 h (Weissmann et al. 1979).

The blood clearance curve shows two major half-times of elimination, a fast component with $T_{1/2} = 2.5$ min (23–40%) and a slower component with $T_{1/2} = 17$ min (7–16%); a small percentage is excreted with $T_{1/2} \geq 16$ h (1.4–2.6%). The total radioactivity in blood at 1 and 24 h after the intravenous injection is 3%, and less than 1% of the administered dose, respectively (Brown et al. 1982).

Hepatic extraction efficiency decreases with increasing serum bilirubin levels. With impaired hepatocyte function, high plasma concentrations (>8 mg/100 ml) may inhibit uptake of ^{99m}Tc-IDA complexes. As hepatobiliary excretion decreases, renal accumulation of certain ^{99m}Tc-IDA complexes is observed (Fink-Bennett 1995).

Normally, the renal elimination of IDA complexes is low: cumulative urinary excretion of ^{99m}Tc-mebrofenin is 1.2% in 3 h, and 7.1% in the case of ^{99m}Tc-DISIDA. With a serum bilirubin level of 15 mg/100 ml, urinary excretion increases to approximately 5 and 23%, respectively (Fritzberg 1986). ^{99m}Tc-etifenin, on the other hand, shows a cumulative urinary excretion of 17% in 5 h, with normal bilirubin levels (Nielson and Rasmussen 1975).

Among the many ligands that have been evaluated, ^{99m}Tc-mebrofenin has shown the best in vivo characteristics, namely high hepatocyte extraction (98%) and a fast clearance ($T_{1/2}$ = 17 min), a rapid hepatobiliary-to-bowel transit time, and low urinary excretion. ^{99m}Tc-mebrofenin provides excellent visualization of the common bile duct and the gallbladder at serum bilirubin levels as high as 30 mg/100 ml (Fink-Bennett 1995; Krishnamurthy and Turner 1990).

Healthy persons fasting overnight prior to the examination showed 56% of the radiotracer in the gallbladder; 44% were excreted directly into the small intestine (Brown et al. 1982). The fraction of gallbladder emptying following a whole meal has been measured as 87%.

Nonvisualization of the gallbladder with visualization of the common bile duct and the duodenum in fasting patients within 2 h is diagnostic of acute cholecystitis (Weissmann et al. 1979). It has been suggested to repeat scintigraphy with the ^{99m}Tc-IDA complex 30 min after stimulation of gallbladder contractility with intravenous cholecystokinin. Persistent nonvisualization of the gallbladder confirms cystic duct obstruction, caused by acute cholecystitis. If the gallbladder visualizes after administration of cholecystokinin, chronic cholecystitis is diagnosed (Weissmann et al. 1979). Acute cholecystitis can be virtually excluded if the gallbladder, the common bile duct, and the duodenum are visualized in fasting patients within 1 h after injection of the ^{99m}Tc-IDA complex (Weissmann et al. 1979).

An obstruction or occlusion of the common bile duct affects the clearance of the ^{99m}Tc-IDA complex into the duodenum, resulting in an increased transit time and an abnormal scintigram. The absence of biliary tract visualization in the presence of normal hepatic extraction indicates an acute common bile duct obstruction from a stone in the common bile duct (Fink-Bennett 1995).

Few reports on adverse reactions after intravenous injection of IDA-derivatives are available. The LD_{50} value of etifenin, determined in mice and in rats, is 280 and 270 mg/kg body weight, respectively. The LD_{50} value of mebrofenin is given with 285 mg/kg body weight (in mice) and 250 mg/kg body weight (in rats).

Radiation Dose

The most exposed organs are the gallbladder wall, the upper large intestinal wall, the lower large intestinal wall, the small intestine, and the liver (Brown et al. 1982). The gallbladder receives the highest absorbed dose, 0.18–0.21 mGy/MBq. The relative amount of the delivered dose is strongly dependent on the type of stimulation used to induce gallbladder emptying. In fasting subjects (no gallbladder stimulation), an increase in the radiation absorbed dose to the gallbladder by approximately 170% versus whole-meal gallbladder stimulation is observed (Brown et al. 1982).

Severe hepatocellular disease (high bilirubin levels) will cause a shift in the excretion pattern, with the highest radiation absorbed dose delivered to the urinary bladder (Brown et al. 1982). The effective (whole body) dose equivalent is 0.024 mSv/MBq (International Commission on Radiological Protection 1987). The effective dose in adults (70 kg) resulting from 185 MBq (5 mCi) of intravenously injected ^{99m}Tc-IDA complex is 4.4 mSv.

Storage and Stability

Storage. The lyophilized kit is stored at 2–8 °C.

Stability. The ^{99m}Tc-IDA complex is stable for 6 h after preparation.

References

Brown PH, Krishnamurthy GT, Bobba VR, Kingston E, Turner FE (1982) Radiation-dose calculation for five Tc-99m IDA hepatobiliary agents. J Nucl Med 23:1025–1030

Council of Europe (2005) Technetium ^{99m}Tc etifenin injection. European pharmacopeia 5.0, monograph no 585. Council of Europe, Maisonneuve, Sainte-Ruffine, p 1212

Fink-Bennett D (1995) Gallbladder and biliary ducts. Wagner HN Jr, Szabo Z, Buchanan JW (eds) In: Principles of nuclear medicine, 2nd edn. Saunders, Philadelphia, pp 946–957

Freeman LM, Sugarman LA, Weissmann HS (1981) Role of cholecystokinetic agents in ^{99m}Tc-IDA cholescintigraphy. Sem Nucl Med 11:186–193

Fritzberg AR (1986) Advances in the development of hepatobiliary radiopharmaceuticals. In: Fritzberg AR (ed) Radiopharmaceuticals: progress and clinical perspectives, vol. I. CRC Press, Boca Raton, pp 89–116

International Commission on Radiological Protection (1987) Technetium-labelled iminodiacetic acid (IDA) derivatives. In: Annals of the ICRP, radiation dose to patients from radiopharmaceuticals, biokinetic models and data. ICRP publication 53, vol 18, no 1–4. Pergamon, Oxford, pp 201–205

Klingensmith WC, Fritzberg AR, Spitzer VM, Kuni CC, Shanahan WSM (1981) Clinical comparison of Tc-99m diisopropyl-IDA and diethyl-IDA for evaluation of the hepatobiliary system. Radiology 140:791–795

Krishnamurthy S, Krishnamurthy GT (1989) Technetium-99m-iminodiacetic acid organic anions: review of biokinetics and clinical application in hepatology. Hepatology 9:139–159

Krishnamurthy GT, Turner FE (1990) Pharmacokinetics and clinical application of technetium-99m-labeled hepatobiliary agents. Sem Nucl Med 20:130–149

Loberg MD, Fields AT (1978) Chemical structures of Tc-99m-labeled *N*-(2,6-dimethylphenyl-carbamoylmethyl)iminodiacetic acid (Tc-HIDA). Int J Appl Radiat Isot 29:167–173

Loberg MD, Cooper MD, Harvey EB, Callery PS, Faith WC (1976) Development of new radio-pharmaceuticals based on N-substitution of iminodiacetic acid. J Nucl Med 17:633–638

Majd M, Reba R, Altman RP (1981) Effect of phenobarbital on Tc-99m-IDA scintigraphy in the evaluation of neonatal jaundice. Sem Nucl Med 11:194–204

Nicholson RW, Herman KJ, Shields RA, Testa HJ (1980) The plasma protein binding of HIDA. Eur J Nucl Med 5:311–312

Nielson P, Rasmussen F (1975) Relationship between molecular structure and excretion of drugs. Life Sci 17:1495

Nunn AD, Schramm E (1981) Analysis of Tc-HIDAs and factors affecting their labeling rate, purity, and stability. J Nucl Med 22:P52

Nunn AD, Loberg MD, Conley RA (1983) A structure-distribution-relationship approach leading to the development of Tc-99m-mebrofenin: an improved cholescintigraphic agent. J Nucl Med 24:423–430

Ponto JA, Swanson DP, Freitas JE (1987) Clinical manifestations of radiopharmaceutical formulation problems. In: Hladik WB III, Saha GB, Study KT (eds) Essentials of nuclear medicine science. Williams & Wilkins, Baltimore, pp 268–279
Ryan J, Cooper MD, Loberg MD (1977) Technetium-99m-labeled *N*-(2,6-dimethylphenylcarbamoylmethyl)iminodiacetic acid (Tc-99m-HIDA): a new radiopharmaceutical for hepatobiliary imaging studies. J Nucl Med 18:997–1004
United States Pharmacopeial Convention (2005) Official Monographs: USP 28, technetium Tc 99m disofenin injection, technetium Tc 99m lidofenin injection, technetium Tc 99m mebrofenin injection. United States Pharmacopeia (USP) 28–national formulary (NF) 23, pp 1854–1857
Weissmann HS, Frank MS, Bernstein LH, Freeman LM (1979) Rapid and accurate diagnosis of acute cholecystitis with Tc-99m-HIDA cholescintigraphy. Am J Roentg 132:523–528

12.10 ^{99m}Tc-Labeled Peptides

12.10.1 ^{99m}Tc-Depreotide

I. Zolle

Chemical name	**Chemical structure**
Depreotide-trifluoroacetate ^{99m}Tc(V)O-depreotide Technetium Tc 99m depreotide injection (*USP*)	Depreotide = synthetic cyclic decapeptide: cyclo-(*N*-Me)Phe-Tyr-(D-Trp)-Lys-Val-Hcy(CH_2CO-(*β*-Dap)-Lys-Cys-Lys–NH_2) Hcy = L-homocysteine *β*-Dap = L-1,2-diaminopropionic acid
Kit components	**Commercial products**
Depreotide 47.0 µg resp. 50.0 µg Tin(II)-chloride dihydrate 50.0 µg Sodium D-glucoheptonate (gluceptate) dihydrate 5 mg Edetate disodium dihydrate 0.1 mg HCl and/or NaOH	NeoSpect GE Healthcare NeoTect Berlex Laboratories

Preparation

The commercial kits contain the sterile, lyophilized components including preformed synthetic peptide in a nitrogen atmosphere, as a single dose. Labeling with ^{99m}Tc-pertechnetate injection is carried out under aseptic conditions by adding a volume of exactly 1 ml of eluate to the vial at ambient temperature. The ^{99m}Tc activity should not exceed 1.8 GBq (50 mCi). Before removing the syringe, an equal volume of headspace should be withdrawn to normalize the pressure in the vial. The shielded vial is agitated carefully for 10 s to dissolve the lyophilized material, and then the vial is placed into a lead-shielded boiling water bath for 10 min. After the labeling procedure, the vial is placed into the lead shield and cooled at room temperature for approximately 15 min.

The vial must not be cooled under running water (product monographs, Amersham Healthcare 2000; Berlex Laboratories 2001).

^{99m}Tc-depreotide is a clear, aqueous solution suitable for intravenous injection, used as a single dose. The pH of the injection solution is 6.0–8.0 (United States Pharmacopeial Convention 2005).

Description of the Kit

Kits NeoSpect and NeoTect have almost identical composition, labeling of the cyclic decapeptide with ^{99m}Tc is performed by ligand exchange of intermediary ^{99m}Tc-glucoheptonate, using stannous ion for reduction and heating for 10 min in a boiling water bath. The injection solution should be inspected visually for particulate matter and discoloration before administration. Oxidative processes interfere with the labeling reaction.

Only eluates from generators eluted regularly within 24 h after the previous elution may be used for labeling.

The amount of ^{99m}Tc activity (up to 1.8 GBq resp., 50 mCi) used for labeling is based on calculations recommending activities of 555–740 MBq (15–20 mCi) for injection at a certain time after labeling, and on using a total volume of 1 ml of ^{99m}Tc-depreotide (47 resp., 50 µg) for one patient.

Kits NeoSpect and NeoTect contain no antimicrobial agents. Saline used for dilutions must be prepared without addition of any bacteriostatic agent (product monographs, Amersham Healthcare 2000; Berlex Laboratories 2001).

Depreotide (P829 Diatide Inc.) is a synthetic peptide comprising two separate domains. The pharmacophore is part of a cyclic hexapeptide, to which a linear tetrapeptide is appended via the thiol group of the homocysteine residue, Hcy(CH_2CO). The linear sequence Ala-Lys-Cys-Lys-NH_2 constitutes the structural requirements for complex formation with technetium. Cyclic configuration avoids reductive cleavage during labeling with Tc-99m, which has been observed with small peptides containing an accessible disulfide bridge (Vallabhajosula et al. 1996).

^{99m}Tc-depreotide contains a triamide-thiol chelate (N_3S), which coordinates oxotechnetium, having a Tc(V)O core.

Clinical Applications

^{99m}Tc-depreotide is valuable for scintigraphic imaging of solitary pulmonary nodules in combination with computer tomography (CT) or chest x-ray, in patients suspected of malignancy.

Lung scintigraphy: Somatostatin receptor imaging of non-small cell lung cancer.

Somatostatin is a neuroregulatory peptide secreted by the hypothalamus. Human somatostatin receptors (ssts) are expressed in the brain, anterior pituitary gland, pancreas, thyroid gland, and in the mucosa of the gastrointestinal tract (Hofland and Lamberts 1997; Patel 1999). Tumors arising from sst-positive cells in these organs contain a high density of ssts (Hofland et al. 2003; Lamberts et al. 1991; Reubi 1996).

The somatostatin analog depreotide consists of ten amino acids containing the key ssts-binding amino acid residues – Tyr-(D-Trp)-Lys-Val – protected in a cyclic configuration, increasing in vivo stability. Depreotide binds with high affinity to sst-positive tumors and their metastases (Vallabhajosula et al. 1996).

The labeled analogue, ^{99m}T-depreotide (P829), has been first used in patients (Virgolini et al. 1998) and evaluated for the detection of solitary pulmonary nodules in a multicenter trial (Blum et al. 1999, 2000). Experience gained with this ssts-binding radioligand in the diagnosis of non–small cell lung cancer has been reported (Menda and Kahn 2002). Comparative studies using ^{99m}T-depreotide and ^{18}F-FDG positron emission tomography (PET) in patients suspected of non–small cell lung cancer demonstrated high sensitivity of both functional imaging methods for detecting lung cancer in primary lesions and in hilar and mediastinal lymph nodes (Kahn et al. 2004). Another study investigated the involvement of regional lymph nodes in 56 patients with lung cancer and in 30 patients with benign lung lesions, using ^{99m}T-depreotide single-photon emission computer tomography (SPECT) and CT to enhance staging accuracy. Scintigraphic results displayed high sensitivity (93.7–99%), demonstrating that a negative result with ^{99m}T-depreotide can exclude regional lymph node metastases with a high degree of probability (Danielsson et al. 2005). Response to endocrine therapy in advanced breast cancer patients could be predicted using sequential ^{99m}Tc-depreotide scintigraphy, thus selecting nonresponders as early as 3 weeks after initiation of treatment (Van den Bossche et al. 2006).

Time of Examination. The optimal time for SPECT imaging is 2–4 h after the intravenous injection.

Recommended Activities for Indications

Lung scintigraphy: 555–740 MBq (15–20 mCi), injected intravenously
$\leq$0.7 µg/kg body weight

Pediatric Dose. The application of ^{99m}Tc-depreotide to patients less than 18 years is not recommended. No data are available for this age group.

Additional Information

The diagnostic application of ^{99m}Tc-depreotide is restricted to a single intravenous injection.

Contraindications are a known hypersensitivity against depreotide, or against another kit component.

^{99m}Tc-depreotide should not be mixed with other drugs or components and should be injected separately.

Attention should be given to patients with reduced kidney function because of decreased renal excretion causing an increased radiation exposure. Attention should also be given to patients with reduced liver function.

Patients should drink sufficient water and should be encouraged to frequent bladder emptying during the first hours after injection.

Since depreotide binds to somatostatin receptors, caution should be paid to patients with insulinoma or diabetes mellitus.

Quality Control

Radiochemical Purity. ^{99m}Tc-depreotide is not described in the *European Pharmacopeia.* Thin-layer chromatography (TLC) is recommended by the manufacturer, using instant (I)TLC-silica gel (SG) strips as a stationary phase and analysis in two solvent systems. Using methanol/1 M ammonium acetate 1:1 (MAM) as mobile phase, the insoluble ^{99m}Tc components are measured at the start (R_f=0–0.4). With saturated NaCl solution as mobile phase, free ^{99m}Tc-sodium pertechnetate, ^{99m}Tc-glucoheptonate, and ^{99m}Tc-edetate move with the solvent front and are measured at R_f=0.75–1.0.

The radiochemical purity of ^{99m}Tc-depreotide should not be less than 90% (United States Pharmacopeial Convention 2005).

The radiochemical purity should be analyzed prior to administration of ^{99m}Tc-depreotide.

Methods Recommended by the Manufacturer

Thin-layer chromatography

Stationary phase:	Gelman ITLC-SG strips (precut to 2.0×10 cm)
Solvent I:	MAM
Solvent II:	Saturated sodium chloride solution (SSCS)

Procedure

Aliquots are spotted and analyzed without drying.

- System I: Gelman ITLC-SG fiberglass plates and MAM as solvent: Reduced, hydrolized ^{99m}Tc-technetium is measured at the start (A); ^{99m}Tc-depreotide and free ^{99m}Tc-Na-pertechnetate as well as the ^{99m}Tc complexes (gluceptate and edetate) move with the solvent front.
- System II: Gelman ITLC-SG fiberglass plates and saturated sodium chloride as solvent: ^{99m}Tc-depreotide and colloidal ^{99m}Tc activity remain at the start; ^{99m}Tc-Na-pertechnetate and the ^{99m}Tc-complexes (gluceptate and edetate) are measured at the solvent front (B).

Table 1. Thin-layer chromatography on silica gel plates using two solvent systems

System I (MAM)	Reduced, hydrolized ^{99m}Tc activity at the start:	(A) <3%
System II (SSCS)	^{99m}Tc-depreotide and colloidal ^{99m}Tc activity at origin Free ^{99m}Tc-pertechnetate and the ^{99m}Tc-complexes (gluceptate and edetate) at the solvent front:	(B)
	A and B represent labeled impurities: <10%	
	^{99m}Tc-depreotide (%)=100–%(A+B)	

Preparation of saturated sodium chloride solution (SSCS)

Five grams of sodium chloride are dissolved in 10 ml distilled water in the TLC tank. Over a period of 10–15 min, the tank is swayed repeatedly. Undissolved sodium chloride should settle at the bottom. If all sodium chloride dissolves, more sodium chloride should be added.

Preparation of MAM

- 1 M ammonium acetate: 3.9 g ammonium acetate is dissolved in 50 ml distilled water, the solution is stable for one month.
- Methanol/1 *M* ammonium acetate (1:1): one part methanol is mixed with one part of the 1 M ammonium acetate solution. This solution should be prepared every day.

Method of analysis (TLC)

1. Fill the TLC tanks (beakers) each with 0.5 ml of MAM or SSCS.
2. Close the tanks (beakers) and allow them to equilibrate.
3. Apply 1 drop of ^{99m}Tc-depreotide (5–10 μl), using a 1-ml syringe with a 21-gauge needle onto each of the Gelman ITLC-SG strips, 1.0 cm from the bottom. Do not allow the spot to dry.
4. Develop each plate in the covered TLC tank, using MAM and SSCS, respectively, until the upper edge of the plates is reached.
5. Allow the plates to dry.
6. Cut the ITLC-SG MAM plate at $R_f = 0.4$ (A).
7. Cut the ITLC-SG SSCS plate at $R_f = 0.75$ (B).
8. Measure the radioactivity of each portion in a counter and record separately.
9. Calculate the percentage of A and B as a fraction of the sum of recovered counts.
10. Calculate the radiochemical purity of ^{99m}Tc-depreotide (Table 1).

Pharmacokinetic Data

After intravenous injection of ^{99m}Tc-depreotide, the elimination from blood is described by three effective half-times, namely $T_{1/2} \leq 5$ min, 45 min, and 22 h. Approximately 12% of injected ^{99m}Tc-depreotide is bound to plasma proteins (5-min plasma sample). Four hours after injection, 71–84% of activity in blood was bound to depreotide, in the urine less, namely, 61–64% (NeoTect product monograph, Berlex Laboratories 2001).

High-affinity binding to somatostatin receptors was demonstrated in vivo using Lewis rats bearing CA20948 rat pancreatic tumor implants, and in vitro using human tumor cell membranes. Tumor uptake in rats was 4.9% injected dose (ID)/g at 90 min after injection, compared with 2.9% in the case of ^{111}In-(diethylene triamine pentaacetate [DTPA])octreotide (Vallabhajosula et al. 1996).

In patients, specific uptake in single pulmonary nodules is seen 1.5–2 h postinjection, also in regional lymph nodes overexpressing somatostatin receptors. Lymphoreticular response affects initially hilar, mediastinal, supraclavicular and axillary nodes, extending to distant nodes and the spleen (Danielsson et al. 2005).

Increased uptake of ^{99m}Tc-depreotide is seen in the spine, sternum and rib ends, and somewhat lower in the hilar and mediastinal regions. Nonspecific mediastinal uptake of ^{99m}Tc-depreotide has been reported (Menda et al. 2001).

^{99m}Tc-depreotide is not metabolized and is excreted unchanged (>90%). The major route of excretion is the renal system. Four hours after injection, 12% of injected radioactivity is measured in the urine (NeoTect product monograph, Berlex Laboratories 2001). The average renal clearance amounts to approximately 0.3 ml/min/kg. The average overall clearance is 2–4 ml/min/kg. External whole-body scintigraphy localized the highest activity in the abdomen (Kahn et al. 2004). Gastrointestinal excretion is ≤5%.

A fraction of the injected activity is retained in the kidneys, indicating proximal tubular reabsorption, as has been observed with ^{111}In-(DTPA)octreotide (Vallabhajosula et al. 1996).

Radiation Dose

^{99m}Tc-depreotide is excreted by the kidneys. The most exposed organs are the kidneys and urinary bladder, as well as the spleen, the liver, and the thyroid gland.

The effective (whole body) dose equivalent for ^{99m}Tc-depreotide was calculated as 0.016 mSv/MBq (International Commission on Radiological Protection 1990). The effective dose in adults (70 kg) resulting from 740 MBq (20 mCi) of intravenously injected ^{99m}Tc-depreotide is 11.84 mSv. The values are calculated assuming a 4.8-h bladder voiding period.

The absorbed radiation dose to the lung resulting from an intravenous injection of 555 MBq (15 mCi) of ^{99m}Tc-depreotide for lung scintigraphy is 7.8 mGy.

Storage and Stability

Storage. Kits should be stored at −10 °C or below. Keep the ^{99m}Tc-depreotide injection solution at 15–25 °C.

Stability. ^{99m}Tc-depreotide injection solution is stable for 5 h.

References

Amersham Healthcare (2000) Product monograph for the NeoSpect kit for the preparation of Tc-99m depreotide. Amersham Healthcare, UK

Berlex Laboratories (2001) Product monograph for the NeoTect kit for the preparation of Tc-99m depreotide. Berlex Laboratories, Wayne, NJ (Diatide, NDA No. 21-012)

Blum JE, Handmaker H, Rinne NA (1999) The utility of a somatostatin-type receptor binding peptide radiopharmaceutical (P829) in the evaluation of solitary pulmonary nodules. Chest 115:224–232

Blum JE, Handmaker H, Lister-James J, Rinne NA (2000) NeoTect solitary pulmonary nodule study group. A multicenter trial with a somatostatin analog ^{99m}Tc-depreotide in the evaluation of solitary pulmonary nodules. Chest 117:1232–1238

Danielsson R, BÅÅth M, Svensson L, Forslöv U, Kölbeck K-G (2005) Imaging of regional lymph node metastases with ^{99m}Tc-depreotide in patients with lung cancer. Eur J Nucl Med Mol Imag 32:925–931

Hofland LJ, Lamberts SWJ (1997) Somatostatin analogs and receptors: diagnostic and therapeutic applications. Cancer Treat Res 89:365–382

Hofland LJ, Lamberts SWJ, Van Hagen PM, Reubi JC, Schaeffer J, Waaijers M, Van Koetsveld PM, Srinivasan A, Krenning EP, Breeman WAP (2003) Crucial role for somatostatin receptor subtype 2 in determining the uptake of [^{111}In-DTPA-D-Phe^{1}]octreotide in somatostatin receptor-positive organs. J Nucl Med 44:1315–1321

International Commission on Radiological Protection (1990) ICRP Publication 60: recommendations of the International Commission on Radiological Protection. In: Annals of the ICRP, vol 21, no 1–3. Pergamon, Oxford

Kahn D, Menda Y, Kernstine K, Bushnell DL, McLaughlin K, Miller S, Berbaum K (2004) The utility of ^{99m}Tc-depreotide compared with F-18 fluorodeoxyglucose positron emission tomography and surgical staging in patients with suspected non-small cell lung cancer. Chest 125:494–501
Lamberts SWJ, Krenning EP, Reubi JC (1991) The role of somatostatin and ist analogs in the diagnosis and treatment of tumors. Endocr Rev 12:450–482
Menda Y, Kahn D (2002) Somatostatin receptor imaging of non-small cell lung cancer with 99m-Tc depreotide. Semin Nucl Med 32:92–96
Menda Y, Kahn D, Bushnell DL, Thomas M, Miller S, McLaughlin K, Kernstine KH (2001) Non-specific mediastinal uptake of ^{99m}Tc-depreotide (NeoTect). J Nucl Med 42(Suppl):304P
Patel YC (1999) Somatostatin and its receptor family. Front Neuroendocrinology 20:157–198
Reubi JC, Schaer JC, Laissue JA, Waser B (1996) Somatostatin receptors and their subtypes in human tumors and in peritumoral vessels. Metabolism 45:39–41
United States Pharmacopeial Convention (2005) Official Monographs: USP 28, technetium Tc 99m depreotide injection. United States Pharmacopeia Convention, Rockville, Md, p 1853
Vallabhajosula S, Moyer BR, Lister-James J, McBride BJ, Lipszyc H, Lee H, Bastidas D, Dean RT (1996) Preclinical evaluation of technetium-99m-labeled somatostatin receptor-binding peptides. JNM 37:1016–1022
Van Den Bossche B, Van Belle S, De Winter F, Signore A, Van de Wiele C (2006) Early prediction of endocrine therapy effect in advanced breast cancer patients using ^{99m}Tc-depreotide scintigraphy. J Nucl Med 47:6–13
Virgolini I, Leimer M, Handmaker H, Lastoria S, Bischof C, Muto P, Pangerl T, Gludovacz D, Peck-Radosavljevic M, Lister-James J, Hamilton G, Kaserer K, Valent P, Dean R (1998) Somastotatin receptor subtype specific and in vivo binding of a novel tumor tracer, ^{99m}Tc-P829. Cancer Res 58:1850–1859

12.11 ^{99m}Tc-Labeled Monoclonal Antibodies

12.11.1 ^{99m}Tc-Arcitumomab

F. Rakiás

Chemical name			
Arcitumomab IMMU-4 Fab′ anti-CEA monoclonal antibody fragments			
Technetium Tc 99m arcitumomab injection (*USP*)			
^{99m}Tc-arcitumomab injection			
^{99m}Tc-CEA-Scan injection			
Kit components		**Commercial products**	
Arcitumomab	1.25 mg	CEA-Scan	Immunomedics Europe
Stannous chloride	0.29 mg		
Potassium sodium tartrate tetrahydrate			
Sodium acetate trihydrate			
Sodium chloride			
Sucrose			

Preparation

The carcinoembryonic antigen (CEA)-Scan kit contains the lyophilized, sterile components in argon atmosphere, ready for aseptic labeling with ^{99m}Tc-sodium pertechnetate. Labeling is carried out by adding 1.0 ml of ^{99m}Tc activity to the vial, not exceeding 1.11 GBq/ml (30 mCi/ml). The vial is allowed to react at room temperature for at least 5 min, and then 1 ml saline is added to the labeled product and mixed well.

^{99m}Tc-CEA-Scan injection is a clear, colorless solution suitable for intravenous injection, used as a single dose (740–1100 MBq; resp., 20–30 mCi). The total volume is 2 ml. The pH of the injection solution is 5.0–7.0. The injection solution should be inspected visually for particulate matter and discoloration. If either is present, the product should be discarded (Immunomedics Europe 2000).

Description of the Kit

Each 3-ml vial contains 1.25 mg arcitumomab (IMMU-4 Fab′ anti-CEA monoclonal antibody fragments, consisting mainly of Fab′, but also containing $F(ab')_2$ at ≤5% of total protein, with heavy- and light-chain fragments), buffered at pH 5.0–7.0. Before labeling, the ^{99m}Tc activity should be adjusted to a final concentration of 1.11 GBq/ml (30 mCi/ml), using a 2-ml vial.

Isotopic dilution is observed with higher concentrations of ^{99}Tc in the first eluate; therefore, only eluates from generators eluted regularly within 24 h after the previous elution may be used for labeling (Ponto et al. 1987). The CEA-Scan kit contains no antimicrobial agents (Immunomedics Europe 2000).

Clinical applications

^{99m}Tc-CEA-Scan injection solution is indicated for the detection of malignant lesions in patients with a history of colorectal carcinoma and with evidence of recurrence and/or metastases. ^{99m}Tc-CEA-Scan is employed as an adjunct to standard noninvasive imaging techniques, such as ultrasonography or computer tomography (CT), in the following situations:

- Patients with evidence of recurrence and/or metastatic carcinoma of the colon or rectum, who are undergoing an evaluation for the extent of disease, such as prior to surgical resection and/or other therapy
- Patients with suspected recurrence and/or metastatic carcinoma of the colon or rectum in association with rising levels of CEA

The specificity of monoclonal antibodies for the detection of heterogenous carcinoembryonic antigen posed a considerable uncertainty (Primus 1983). The metabolism and kinetics of labeled antibodies affected the detection of liver metastases of colorectal cancer (Behr et al. 1995). The utility of external immunoscintigraphy with the IMMU-4 technetium-99m Fab′ antibody fragment was evaluated in patients undergoing surgery for carcinoma of the colon and rectum in a phase III clinical trial (Moffat et al. 1996). Both radioimmunoscintigraphy and computed tomography were used for predicting the resectability of recurrent colorectal cancer (Hughes et al. 1997). The safety and efficacy of repeated administrations of arcitumomab was demonstrated in patients with colorectal cancer (Wegener et al. 2000). Of considerable importance is the diagnosis of recurrence of colorectal carcinoma (Willkomm et al. 2000) and the detection of resectable rectal cancer recurrence by CEA immunoscintigraphy (Lechner et al. 2000).

Time of Examination. A whole-body planar scan at 2–5 h postinjection can be used to localize sites of colorectal cancer.

Immunoscintigraphy, using planar and single-photon emission computer tomography (SPECT) techniques, should be performed preferably 1–5 h after injection.

Recommended Activities for Indications

Planar and SPECT imaging: 750–1100 MBq (20–30 mCi), injected intravenously (30 s)
1 mg of Fab′ fragment

Since the recommended adult dose of Fab fragment injected as a single dose is limited with 1 mg, 1.6 ml of the injection solution may be used for one patient.

Pediatric Dose. Recommendations by the Pediatric Task Group of the European Association of Nuclear Medicine (EANM) based on body weight should be followed (see Appendix 1, Table A1.2).

Additional Information

Patients should be encouraged to drink sufficient water and to empty the bladder before scintigraphy is started. Frequent bladder emptying is recommended to reduce the radiation exposure to the bladder wall.

Patients with known allergies or hypersensitivity to mouse proteins, human anti-mouse antibodies (HAMA) titers should be determined before administration of ^{99m}Tc-CEA-Scan.

Doses of arcitumomab up to 10 mg have not shown any serious adverse reaction.

^{99m}Tc-CEAScan should be used only once in each patient (Immunomedics Europe 2000).

Any remaining portion of the injection solution should be discarded.

Quality Control

Radiochemical Purity

^{99m}Tc-CEA-Scan is not described in the Eur. Pharmacopoeia. Thin-layer chromatography is recommended by the manufacturer using instant (I)TLC-silica gel (SG) fiberglass sheets for the identification of free ^{99m}Tc-Na-pertechnetate, using acetone as solvent. Unbound ^{99m}Tc-pertechnetate moves with the solvent front (R_f=1.0). The insoluble ^{99m}Tc components are measured at the start.

The radiochemical purity of ^{99m}Tc-arcitumomab should not be less than 90% (*USP*).

The analysis should be performed prior to administration of ^{99m}Tc-CEA-Scan in the patient.

Thin-layer chromatography		
Stationary phase:	Gelman ITLC-SG fiberglass, 1×9.5 cm	
Solvent:	Acetone	
Developing time:	5 min	
R_f values:	^{99m}Tc-arcitumomab	0.0–0.1
	^{99m}Tc reduced, hydrolized:	0.0–0.1
	^{99m}Tc-pertechnetate:	0.9–1.0 (< 10%)

Procedure
- A 10 µl sample of the labeled antibody is diluted with 1.5 ml saline, and immediately spotted for thin-layer chromatography.
- When the solvent front is within 1 cm of the top, the strip is removed, dried, and analyzed in a radiochromatogram scanner.
- Otherwise, the strip is cut into half and placed into two test tubes for measurement with a gamma scintillation counter or a dose calibrator.

Calculate the percent impurity as follows:

$$^{99m}\text{Tc-pertechnetate (\%)} = \frac{\text{Activity in upper piece}}{\text{Activity in both pieces}} \times 100$$

Results of analysis *(12 samples)*
Results were obtained by thin-layer chromatography in acetone, at different times after labeling.

Labeling and stability	15 min (%)	3 h (%)
^{99m}Tc-Arcitumomab	95.3 ± 0.32	92.7 ± 0.44
^{99m}Tc-Na-pertechnetate	4.7 ± 0.48	7.3 ± 0.21

Pharmacokinetic Data

Pharmacokinetic studies were performed after the intravenous administration of ^{99m}Tc-CEA-Scan. The elimination from the blood is indicated by 63% of baseline activity at 1 h after infusion, 23% after 5 h, and 7% after 24 h. The distribution half-time is approximately 1 h; the elimination from blood follows a half-time of approximately 13 h. Twenty-eight percent of the administered activity is excreted in the urine during the first 24 h after infusion (Immunomedics Europe 2000).

CEA is expressed in a variety of carcinomas, particularly of the gastrointestinal tract (e.g., Crohn's disease, inflammatory bowel disease, post–radiation therapy to the bowel) and can be detected in the serum. IMMU-4 is specific for the classical 200 000-Da CEA that is found predominantly on the cell membrane. ^{99m}Tc-CEA-Scan complexes the circulating CEA and binds to CEA on the cell surface. Imaging efficacy and safety have been evaluated in four clinical trials to evaluate the presence, location, and extent of colorectal cancer, primarily in the liver and extrahepatic abdominal and pelvic regions.

Radiation Dose

^{99m}Tc-CEA-Scan is excreted by the kidneys. The most exposed organs are the kidneys and urinary bladder, along with the spleen and liver.

The effective (whole body) dose equivalent for ^{99m}Tc-CEA-Scan was calculated as 0.0131 mSv/MBq (International Commission on Radiological Protection 1990). The effective dose in adults (70 kg) resulting from 750 MBq (20.3 mCi) of intravenously injected ^{99m}Tc-CEA-Scan is approximately 9.8 mSv. The values were calculated assuming a 2-h bladder voiding period.

The absorbed radiation dose to the kidneys resulting from an intravenous injection of 750 MBq (20.3 mCi) of ^{99m}Tc-CEA-Scan for colon scintigraphy is calculated as 75 mGy, the absorbed radiation dose to the spleen as 11.9 mGy (based on data obtained by the standard medical internal radiation dose [MIRD] method) (Immunomedics Europe 2000).

Storage and Stability

Storage. Lyophilized kits should be stored at 2–8 °C, and not frozen. The ^{99m}Tc-CEA-Scan injection solution should be kept at 15–25 °C, and not refrigerated or frozen.

Stability. ^{99m}Tc-CEA-Scan injection solution is stable for 4 h.

References

Behr T, Becker W, Hanappel E, Goldenberg DM, Wolf F (1995) Targeting of liver metastases of colorectal cancer with IgG, F(ab')$_2$, and Fab' anti-carcinoembryonic antigen antibodies labelled with ^{99m}Tc: the role of metabolism and kinetics. Cancer Res. 55: 5777s–5785s

Hughes KS, Pinsky CM, Petrelli NJ, Moffat FL Jr, Patt YZ, Hammershaimb L, Goldenberg DM (1997) Use of carcinoembryonic antigen radioimmunodetection and computed tomography for predicting the resectability of recurrent colorectal cancer. Ann Surg 226:621–631

Immunomedics Europe (2000) Product monograph for the CEA-Scan (Arcitumomab) kit for the preparation of Tc-99m CEA-Scan. Immunomedics Europe, Darmstadt, Germany

International Commission on Radiological Protection (1990) ICRP Publication 60: recommendations of the International Commission on Radiological Protection. In: Annals of the ICRP, vol 21, no 1–3. Pergamon, Oxford

Lechner P, Lind P, Goldenberg DM (2000) Can postoperative surveillance with serial CEA immunoscintigraphy detect resectable rectal cancer recurrence and potentially improve tumor-free survival? J Am Coll Surg 191:511–518

Moffat FL Jr, Pinsky CM, Hammershaimb L, Petrelli NJ, Patt YZ, Whaley FS, Goldenberg DM (1996) Immunomedics study group clinical utility of external immunoscintigraphy with the IMMU-4 technetium-99m Fab' antibody fragment in patients undergoing surgery for carcinoma of the colon and rectum: results of a pivotal, phase III trial. J Clin Oncol 14:2295–2305

Ponto JA, Swanson DP, Freitas JE (1987) Clinical manifestations of radiopharmaceutical formulation problems. In: Hladik WB III, Saha GB, Study KT (eds) Essentials of nuclear medicine science. Williams & Wilkins, Baltimore, pp 268–289

Primus FJ, Newell KD, Blue A, Goldenberg DM (1983) Immunological heterogeneity of carcinoembryonic antigen: antigenic determinants on carcinoembryonic antigen distinguished by monoclonal antibodies. Cancer Res 43:686–692

United States Pharmacopeial Convention (2005) Official Monographs: USP 28, technetium Tc-99m arcitumomab injection, United States Pharmacopeia (USP) 28–national formulary (NF) 23, p 1852

Wegener W, Petrelli NJ, Serafini A, Goldenberg DM (2000) Safety and efficacy of arcitumomab imaging in colorectal cancer after repeated administration. J Nucl Med 41:1016–1020
Willkomm P, Bender H, Bangard M, Decker P, Grunwald F, Biersack HJ (2000) FDG PET and immunoscintigraphy with ^{99m}Tc-labeled antibody fragments for detection of the recurrence of colorectal carcinoma. J Nucl Med 41:1657–1663

12.11.2 ^{99m}Tc-Sulesomab

F. Rakiás

Chemical name			
Sulesomab, IMMU-MN3 Fab′-SH antigranulocyte monoclonal antibody fragments			
^{99m}Tc-Sulesomab injection			
Kit components		**Commercial products**	
Sulesomab	0.31 mg	LeukoScan	Immunomedics Europe
Stannous chloride, dihydrate	0.22 mg		
Sodium chloride			
Hydrochloric acid			
Sodium potassium tartrate, tetrahydrate			
Sodium acetate, trihydrate			
Sucrose			

Preparation

The kit contains the lyophilized, sterile, pyrogen-free, inactive components in an argon or nitrogen atmosphere, ready for labeling with ^{99m}Tc-sodium pertechnetate. Labeling is carried out by first adding 0.5 ml of isotonic saline and swirling the content for 30 s; immediately after dissolution, 1 ml ^{99m}Tc-pertechnetate is added to the shielded vial, corresponding to an activity of at least 1,100 MBq (30 mCi).

^{99m}Tc-LeukoScan is a clear, colorless solution suitable for intravenous injection, used as a single dose. The total volume is 1.5 ml. The pH of the injection solution is 5.0–7.0. The injection solution should be inspected visually for particulate matter and discoloration. If either is present, the product should be discarded (Immunomedics Europe 1997).

Description of the Kit

Each 3-ml vial contains a lyophilized powder of sulesomab (IMMU-MN3 Fab′-SH antigranulocyte monoclonal antibody fragment, consisting mainly of Fab′, but also containing $F(ab')_2$ at 5% of total protein, with heavy- and light-chain fragments) buffered to pH 5.0–7.0. The powder is dissolved by agitation during 30 s; the labeling reaction is completed after 10 min at room temperature.

Isotopic dilution is observed with higher concentrations of ^{99}Tc in the first eluate; therefore, only eluates from generators eluted regularly within 24 h after the previous elution may be used for labeling (Ponto et al. 1987). The LeukoScan kit contains no antimicrobial agents (Immunomedics Europe 1997).

Clinical Applications

^{99m}Tc-LeukoScan has been evaluated as a marker of infection/inflammation in patients with suspected osteomyelitis, joint infection involving implants, inflammatory bowel disease, and diabetic patients with foot ulcers. These heterogeneous patients have been diagnosed with high sensitivity (>93%), comparable to bone scans with diphosphonates, but with considerably higher specificity (>91%), offering a reliable imaging method based on the specificity of sulesomab, a murine anti-granulocyte monoclonal antibody Fab′ fragment. In clinical trials with more than 350 patients, no induction of human anti-mouse antibody (HAMA) to antibody fragments has been observed.

When a bone scan is positive and imaging with ^{99m}Tc-LeukoScan is negative, infection is unlikely. Single-photon emission computed tomography (SPECT) imaging may aid in differentiating osteomyelitis from soft tissue infections (Becker et al. 1994). The advantages of rapid imaging in different conditions of infection have been investigated (Barron et al. 1999; Becker et al. 1996; Gratz et al. 2000, 2003; Hakki et al. 1997; Harwood et al. 1999; Kampen et al. 1999).

Time of Examination. Planar scintigraphy should be performed anytime between 1 and 8 h after injection of ^{99m}Tc-LeukoScan.

Recommended Activities for Indications

Planar and SPECT imaging:	750–1,100 MBq (20–30 mCi), injected intravenously
	0.25 mg of Fab′ antigranulocyte fragment

Since the recommended adult dose of Fab′ antigranulocyte fragment injected as a single dose is limited with 0.25 mg, 1.2 ml of the injection solution may be used for one patient.

Pediatric Dose. The amount of radioactivity for infants and children is based on either body weight (static imaging) or body surface area (dynamic examinations). Recommendations by the Pediatric Task Group of the European Association of Nuclear Medicine (EANM) based on body weight should be followed (see Appendix 1, Table A1.2).

Additional Information

Patients should drink sufficient water and be encouraged to empty the bladder before scintigraphy is started to reduce the radiation exposure to the bladder wall.

HAMA titers should be determined before repeated administration of ^{99m}Tc-LeukoScan.

Any remaining portion of the injection solution should be discarded.

Quality Control

Radiochemical Purity

Thin-layer chromatography

The manufacturer recommends TLC on instant (I)TLC-silica gel (SG) fiberglass sheets for the identification of free ^{99m}Tc-Na-pertechnetate, using acetone as solvent. Unbound ^{99m}Tc-pertechnetate moves with the solvent front (R_f = 1.0).

The radiochemical purity of ^{99m}Tc-LeukoScan should not be less than 90% (Immunomedics Europe 1997). The radiochemical purity should be analyzed prior to administration of ^{99m}Tc-LeukoScan.

Thin-layer chromatography		
Stationary phase:	Gelman ITLC-SG (fiberglass), 1×9.5 cm	
Solvent:	Acetone	
Developing time:	5 min	
R_f values:	^{99m}Tc-sulesomab	0.0–0.1
	^{99m}Tc reduced, hydrolized:	0.0–0.1
	^{99m}Tc-pertechnetate:	0.9–1.0 (<10%)

Procedure

- A 10-μl sample of the labeled antibody is spotted undiluted and immediataly developed in acetone by thin-layer chromatography.
- When the solvent front is within 1 cm of the top, the strip is removed, dried, and analyzed in a radiochromatogram scanner.
- Otherwise, the strip is cut into half and placed into two test tubes for measurement with a gamma scintillation counter or a dose calibrator.

$$^{99m}\text{Tc-pertechnetate (\%)} = \frac{\text{Activity in upper piece}}{\text{Activity in both pieces}} \times 100$$

Results of analysis *(12 samples)*

Results were obtained by thin-layer chromatography in acetone, at different times after labelling.

Labeling and stability	15 min (%)	3 h (%)
^{99m}Tc-Sulesomab complex	96.8±0.30	94.7±0.38
^{99m}Tc-Na-pertechnetate	3.12±0.15	5.3±0.34

Pharmacokinetic Data

Pharmacokinetic studies were performed after the intravenous administration of ^{99m}Tc-LeukoScan. The elimination from the blood is indicated by 34% of baseline activity at 1 h after infusion, 17% at 4 h, and 7% after 24 h. The distribution half-time was approximately 1.5 h; the route of excretion is essentially renal, with 41% of the radiolabel excreted in urine over the first 24 h after injection (Immunomedics Europe 1997).

Radiation Dose

^{99m}Tc-sulesomab is excreted by the kidneys. The most exposed organs are the kidneys and urinary bladder, and the spleen and liver.

The effective (whole body) dose equivalent for ^{99m}Tc-sulesomab was calculated as 0.0103 mSv/MBq (International Commission on Radiological Protection 1990). The effective dose in adults (70 kg) resulting from 750 MBq (20.3 mCi) of intravenously injected ^{99m}Tc-LeukoScan is 7.7 mSv. The values were calculated assuming a 2-h bladder voiding period.

The absorbed radiation dose to the kidneys resulting from an intravenous injection of 750 MBq (20 mCi) of ^{99m}Tc-LeukoScan for diagnostic immunoscintigraphy is calculated as 33.7 mGy, the absorbed radiation dose to the spleen as 11.8 mGy (based on data obtained by the standard medical internal radiation dose [MIRD] method) (Immunomedics Europe 1997).

Storage and Stability

Storage. Lyophilized kits should be stored at 2–8 °C and not frozen. ^{99m}Tc-LeukoScint injection solution is kept at 15–25 °C and not refrigerated or frozen.

Stability. ^{99m}Tc-LeukoScan injection solution is stable for 4 h.

References

Barron B, Hanna C, Passalaqua A, Lamki L, Wegener WA, Goldenberg DM (1999) Rapid diagnostic imaging of acute nonclassic appendicitis by leukoscintigraphy with sulesomab, a technetium 99m-labeled anti-granulocyte monoclonal antibody Fab′. Surgery 125(3):288–296

Becker W, Bair J, Behr TM, Repp R, Streckenbach H, Beck H, Gramatzki M, Winship MJ, Goldenberg DM, Wolf F (1994) Detection of soft-tissue and osteomyeltis using a technetium-99m labeled anti-granulocyte monoclonal antibody fragment. J Nucl Med 35:1436–1443

Becker W, Palestro PJ, Winship MJ, Feld T, Pinsky CM, Wolf F, Goldenberg DM (1996) Rapid imaging of infections with a monoclonal antibody fragment (LeukoScan). Clinical Orthopaedics 329:263–272

Gratz S, Raddatz D, Hagenah G, Behr TM, Behe M, Becker W (2000) ^{99m}Tc-labelled antigranulocyte monoclonal antibody FAB′ fragments versus echocardiography in the diagnosis of subacute infective endocarditis. Int J Cardiol 75:75–84

Gratz S, Schipper ML, Dorner J, Hoffken H, Becker W, Kaiser JW, Behe M, Behr TM (2003) LeukoScan for imaging infection in different clinical settings: A retrospective evaluation and extended review of the literature. Clin Nucl Med 28(4):267–276

Hakki S, Harwood SJ, Morrissey M, Camblin JG, Laven DL, Webster WB Jr (1997) Comparative study of monoclonal antibody scan in diagnostic orthopaedic infection. Clin Orthop Relat Res 335:275–285.

Harwood SJ, Valdivia S, Hung GL, Quenzer RW (1999) Use of sulesomab, a radiolabeled antibody fragment, to detect osteomyelitis in diabetic patients with foot ulcers by leukoscintigraphy. Clin. Infect. Dis. 28(6):1200–1205

Immunomedics Europe (1997) Product monograph for LeukoScan (sulesomab). Issued by Immunomedics Europe, Darmstadt, Germany.

International Commission on Radiological Protection (1990) ICRP publication 60: recommendations of the International Commission on Radiological Protection. In: Annals of the ICRP, vol 21, no 1–3. Pergamon, Oxford

Kampen W, Brenner W, Terheyden H, Bohuslavizki KH, Henze E (1999) Decisive diagnosis of infected mandibular osteoradionecrosis with a Tc-99m-labeled anti-granulocyte Fab'-fragment. Nuklearmedizin 38:309–311

Ponto JA, Swanson DP, Freitas JE (1987) Clinical manifestations of radiopharmaceutical formulation problems. Hladik WB III, Saha GB, Study KT (eds) In: Essentials of nuclear medicine science. Williams & Wilkins, Baltimore, pp 268–289

Appendix 1

Table A.1. Administration of Radioactive Substances Advisory Committee (ARSAC): radiation doses for children

Calculation of administered activity to children based on body weight	
Body weight (kg)	Fraction of adult administered radioactivity scaling factors (F)
3	(0.1) [a]
4	(0.1) [a]
6	(0.1) [a]
8	0.11
10	0.14
12	0.17
14	0.20
16	0.23
18	0.26
20	0.28
25	0.36
30	0.43
35	0.50
40	0.57
45	0.64
50	0.64
55	0.79
60	0.86
65	1.00

ARSAC – Administration of Radioactive Substances Advisory Committee; Notes for Guidance on the Administration of Radioactive Substances to Persons for Purposes of Diagnosis, Treatment or Research, January 1993

[a] Depending on local techniques and facilities the acceptable minimum administered activity will vary, regardless of the size of the child. The minimum value of the scaling factor is 0.1, since activities below 10% of the value of the adult radioactivity have resulted in poor image quality

Table A.2. European Association of Nuclear Medicine (EANM): radiation doses for children (issued by the Pediatric Task Group of the EANM)

Calculation of administered activity to children based on body weight	
Body weight (kg)	Fraction of adult administered radioactivity scaling factors (*F*)
3	0.1
4	0.14
6	0.19
8	0.23
10	0.27
12	0.32
14	0.36
16	0.40
18	0.44
20	0.46
22	0.50
24	0.53
26	0.56
28	0.58
30	0.62
32	0.65
34	0.68
36	0.71
38	0.73
40	0.76
42	0.78
44	0.80
46	0.83
48	0.85
50	0.88
52–54	0.90
56–58	0.92
60–62	0.96
64–66	0.98
68	0.99
>70	1.00

Example: To calculate the amount of radioactivity to be administered to a child with a body weight of 22 kg, the corresponding $F = 0.5$. Accordingly, the amount of radioactivity recommended for adults is multiplied by 0.5

Recommended Reading

Textbooks in Radiopharmacy/Radiopharmacology

Büll U, Schicha H, Biersack H-J, Knapp WH, Reiners Chr, Schober O (eds) (1996) Nuklearmedizin, 2nd edn. Georg Thieme, Stuttgart

Colombetti LG (ed) (1976–1986) CRC Series in radiotracers in biology and medicine. CRC Press, Boca Raton:
Colombetti LG (ed) (1979) Principles of radiopharmacology, vol. I–II
Colombetti LG (ed) (1982) Biological transport of radiotracers, vol. III
Anghileri LJ (ed) (1982) General processes of radiotracer localization, vol. I–II
Eckelman WC (ed) (1982) Receptor-binding radiotracers, vol. I–II
Robertson JS (ed) (1983) Compartmental distribution of radiotracers

Fritzberg AR (ed) (1986) Radiopharmaceuticals: progress and clinical perspectives, vol. I–II

Cox PH (ed) (1982) Progress in radiopharmacology 3: selected topics. Martinus Nijhoff, The Hague

Cox PH (ed) (1985) Radiopharmacy and radiopharmacology yearbook. Gordon and Breach, New York

Cox PH, Mather SJ, Sampson CB, Lazarus CR (eds) (1986) Progress in radiopharmacy. Martinus Nijhoff, Dordrecht, The Netherlands

Hladik WB III, Saha G, Study KT (eds) (1987) Essentials of nuclear medicine science. Williams & Wilkins, Baltimore

Owunwanne A, Patel M, Sadek S (eds) (1995) Handbook of radiopharmaceuticals. Chapman & Hall, London

Saha GB (1998) Fundamentals of nuclear pharmacy, 4th edn. Springer, Berlin Heidelberg New York

Sampson CB (ed) (1994) Textbook of radiopharmacy: theory and practice, 2nd edn. Gordon and Breach Science, UK

Wagner HN Jr, Szabo Z, Buchanan JW (eds) (1995) Principles of nuclear medicine, 2nd edn. Saunders, Philadelphia

Good Radiopharmacy Practice

Kristensen K, Norbygaard E (eds) (1984, 1987) Safety and efficacy of radiopharmaceuticals. Martinus Nijhoff, Boston (1984), Dordrecht (1987)

Nordic Council on Medicines (1989) Radiopharmacy: preparation and control of radiopharmaceuticals in hospitals. Nordic Guidelines. NLN publication no. 26. Nordic Council on Medicines, Uppsala

Quality Control of Radiopharmaceuticals

Brandau W, Hotze L-A, Meyer G-J (1996) Qualitätskontrolle In: Nuklearmedizin, Büll U, Schicha H, Biersack H-J, Knapp WH, Reiners Chr, Schober O (eds) 4. Radiochemie, 2nd edn. [in German]. Georg Thieme, Stuttgart, pp 97–104

Frier M, Hesslewood SR (1980) Quality assurance of radiopharmaceuticals: a guide to hospital practice. Nucl Med Comm Special Issue 1980, Chapman and Hall, London

Hammermaier A, Reich E, Bögl W (1986) Qualitätskontrolle von in-vivo Radiopharmaka: Ein Handbuch für das nuklearmedizinische Labor zur Analytik radiochemischer Verunreinigungen [in German]. Reimers, ISH Heft 94, Bundesgesundheitsamt, Berlin

Mallol J, Bonino C (1997) Comparison of radiochemical purity control methods for technetium radiopharmaceuticals used in hospital radiopharmacy. Nucl Med Commun 18:419–422

Millar AM (1989) Quality assurance of radiopharmaceuticals In: Theobald AE (ed) Radiopharmaceuticals: using radioactive compounds in pharmaceutics and medicine. Horwood, Chichester, pp 83–102

Johannsen B, Narasimhan DVS (eds) (1992) Preparation of kits for ^{99m}Tc-radiopharmaceuticals. International Atomic Energy Agency: IAEA-TECDOC-649. Vienna

Robbins PJ (1985) Chromatography of Tc-99m-radiopharmaceuticals – a practical guide, 2nd printing. Society Nuclear Medicine, New York

Theobald AE (1990) Quality control of radiopharmaceuticals In: Sampson CB (ed) Textbook of radiopharmacy: theory and practice. Gordon and Breach, New York, pp 115–148

Webber D, Zimmer AM, Spies SM (1983) Common errors associated with miniaturized chromatography. J Nucl Med Technol 11:66

Zimmer AM, Pavel DG (1977) Rapid miniaturized chromatographic quality control procedures for Tc-99m radiopharmaceuticals. J Nucl Med 18:1230

Physics/Radiological Protection

Administration of Radioactive Substances Advisory Committee (1993) Investigations – adult patients. In: Notes for guidance on the administration of radioactive substances to persons for purposes of diagnosis, treatment or research, Appendix 1, Part A. Administration of Radioactive Substances Advisory Committee – ARSAC, Oxon, UK, p 25

Ibidem (1993) Investigations – children and young persons. In: Notes for guidance on the administration of radioactive substances to persons for purposes of diagnosis, treatment or research, Appendix 1, Part D, pp 34–35

International Commission on Radiation Protection (1987) Annals of the ICRP, radiation dose to patients from radiopharmaceuticals, biokinetic models and data. ICRP publication 53, vol. 18, no. 1–4. Pergamon, Oxford

International Commission on Radiation Protection (1993) Annals of the ICRP, radiological protection in biomedical research, ICRP Publication 62, vol. 22, no. 3. Pergamon, Oxford

Saha GB (2000) Physics and radiobiology of nuclear medicine, 2nd edn. Springer, Berlin Heidelberg New York

Young MEJ (1983) Radiological Physics, 3rd edn. Lewis, London

Official Publications

Council of Europe (2005) European pharmacopoeia 5.0. Maisonneuve, Sainte-Ruffine

European Commission (1997) Pharmaceutical legislation – Eudralex – the rules governing medicinal products in the European Union. Good manufacturing practices, vol. 4. European Commission, Luxembourg, pp 70, 81–82

United States Pharmacopeial Convention (2005) United States pharmacopeia (USP) 28 – national formulary (NF) 23

Appendix 2

Scope of COST B3: WG-1

COST is a European *co*operation in the field of *s*cientific and *t*echnical research. COST Action B3 was devoted to the development of new radiotracers for nuclear medicine application and methods of quality assurance. The main objectives of the cooperation were defined in a preparatory meeting in Vienna on 12 October 1990. Fifteen participants from six member states worked out a draft of the Memorandum of Understanding (MOU), receiving expert advice from the Austrian Ministry of Science and Research.

COST Action B3 was enacted in December 1992 and ended in December 1997. Sixteen European states – Austria, Belgium, Denmark, Finland, France, Germany, Greece, Hungary, Italy, The Netherlands, Norway, Slovenia, Spain, Sweden, Switzerland, and the United Kingdom – signed the MOU; their national institutions participated in five Working Groups.

Chairpersons

1993–1994	Dr. Ilse Zolle, M. Pharm. M. Sc., Department of Nuclear Medicine, AKH, Vienna
1994–1997	Prof. Dr. P.A. Schubiger, Paul Scherrer Institute, Center for Radiopharmaceutical Science, Villigen, CH

Working Group 1 was concerned with the standardization of methods for labeling and quality control:

(a) Quality control of ^{99m}Tc-radiopharmaceuticals, Coordinator F. Rakias (Hungary)
(b) Standardization of radio-iodinated pharmaceuticals, Coordinator J. Mertens (Belgium)

The scientific goal of Working Group 1(a) has been the development and updating of quality control methods to assure safety of ^{99m}Tc radiopharmaceuticals for parenteral application in nuclear medicine. Scientific institutions in 12 European countries have contributed their experience and results for comparison of the available analytical methods. Some results are presented in the monographs.

Participating Institutions in Working Group 1(a)

Austria:	Department of Nuclear Medicine, Radiopharmacology, Ludwig-Boltzmann-Institute of Nuclear Medicine AKH Wien, University of Vienna	Ilse Zolle, Rudolf Höfer
	Department of Nuclear Medicine, University of Innsbruck	Clemens Decristoforo
Belgium:	VUB-Cyclotron Eenheid, Vrije Universiteit Brussel	John Mertens
	Department of Radiopharmacy, Katholieke Universiteit Leuven, Leuven	Alfons Verbruggen

France:	Service Medecine Nucleaire, Centre Hospitalier, Regional et Universitaire de Tours, Tours	Denis Guilloteau
Germany:	Klinikum Berlin-Buch, Klinik für Nuklearmedizin, und Endokrinologie, Berlin-Buch	A. G. Bratouss
	Radiologische Klinik, Abt. f. Nuklearmedizin, Christian Albrechts Universität Kiel, Kiel	Heike Wolf
Greece:	Radiodiagnostic Products, NCSR–"Democritos", Radiopharmaceuticals Section, Athens	Efstratios Chiotellis
Hungary:	Drug Quality Department, Radiopharmaceutical Section, National Institute of Pharmacy, Budapest	Ferenc Rakiás
	Frederic Joliot-Curie National Research Institute for Radiobiology and Radiohygiene, Department of Applied Radioisotopes, Budapest	Gyözö A. Jánoki
Italy:	Department of Nuclear Medicine, Universita di Ferrara, Ferrara	Adriano Piffanelli
	Department Scienze Farmaceutiche, Universita di Padova, Padua	Ulderico Mazzi
Norway:	Institut Energiteknikk, Kjeller	Per Bremer
Slovenia:	Department of Nuclear Medicine, Division of Radiochemistry, University Medical Centre Ljubljana, Ljubljana	Silvester Kladnik, Tanja Stopar
Spain:	Dpto. Technologia Pharmaceutica, Facultad Pharmacia, Universidad La Laguna, La Laguna, Tenerife	Jesus Mallol
Sweden:	Medical Products Agency, Radiopharmaceuticals, Uppsala	Trygve Bringhammar
United Kingdom:	Department of Nuclear Medicine, Addenbrooke's Hospital, Cambridge	Charles B. Sampson

Publications Working Group 1

Mallol J, Bonino C (1997) Comparison of radiochemical purity control methods for technetium radiopharmaceuticals used in hospital radiopharmacy. Nucl Med Commun 18:419–422

Noll B, Johannsen B, Spies H (1995) Sources of radiochemical impurities in the ^{99m}Tc/S-unprotected MAG_3-system. Presented at the second COST B3 Workshop in Villigen (Switzerland), 6–7 October 1994. Nucl Med Biol 22:1057–1062

Sampson CB (1996) Complications and difficulties in radio-labelling blood cells: a review. Nucl Med Commun 17:648–658

Rakias F, Zolle I (1997) Stannous ion determination: importance and relevance for radiopharmaceutical kits. In: Bergmann H, Kroiss A, Sinzinger H (eds) Radioactive isotopes in clinical medicine and research XXII. Birkhäuser, Basel, Switzerland, pp 401–407

Seifert S, Muth O, Jantsch K, Johannsen B (1995) Radiochemical purity of ^{99m}Tc-HM-PAO: critical parameters during kit preparation. Presented at the second COST B3 workshop in Villigen (Switzerland), 6–7 October 1994. Nucl Med Biol 22:1063–1066

Solanki C, Barber R, Sampson CB (1994) Stabilization and multi-dose usage of exametazime for cerebral perfusion studies. Nucl Med Commun 15:718–722

Working Group Meetings

25 June 1993	National Institute of Pharmacy, Radiopharmaceutical Section, Budapest, Hungary
22 March 1994	National Institute of Pharmacy, Radiopharmaceutical Section, Budapest, Hungary
12 May 1995	National Institute of Pharmacy, Radiopharmaceutical Section, Budapest, Hungary
22 March 1996	Dpto. Technologia Pharmaceutica, Facultad Pharmacia, Universidad La Laguna, Tenerife, Spain

Editors' Meetings

In preparation of the *Handbook of quality control of* ^{99m}*Tc-radiopharmaceuticals*

5 December 1994	Department of Nuclear Medicine, Radiopharmacology, AKH Wien, Vienna, Austria
14 July 1995	National Institute of Pharmacy, Radiopharmaceutical Section, Budapest, Hungary
8 January 1996	Hotel Elisabethpark, Badgastein, Austria

Short-Term Scientific Missions

Jesus Mallol, Ph.D, September 1995: *Comparison of radiochemical purity control methods for technetium radiopharmaceuticals used in hospital radiopharmacy.* Sorin Biomedica S.p.A., Radiopharmaceutical Development, Saluggia, Italy

Martina Netter, M.Sc., 3–21 June 1996: *Study of methods to enhance the stability of the* ^{99m}*Tc-HM-PAO complex.* Addenbrooke's Hospital, Cambridge, UK

FSC
www.fsc.org
MIX
Papier aus verantwortungsvollen Quellen
Paper from responsible sources
FSC® C105338

If you have any concerns about our products,
you can contact us on
ProductSafety@springernature.com

In case Publisher is established outside the EU,
the EU authorized representative is:
Springer Nature Customer Service Center GmbH
Europaplatz 3, 69115 Heidelberg, Germany

Printed by Libri Plureos GmbH
in Hamburg, Germany